INTERNATIONAL SERIES IN INTELLIGENT TECHNOLOGIES

Prof. Dr. Dr. h.c. Hans-Jürgen Zimmermann, Editor
European Laboratory for Intelligent
Techniques Engineering
Aachen, Germany

Other books in the series:

Applied Research in Fuzzy Technology
by Anca L. Ralescu

Analysis and Evaluation of Fuzzy Systems
by Akira Ishikawa and Terry L. Wilson

Fuzzy Logic and Intelligent Systems
edited by Hua Li and Madan Gupta

Fuzzy Set Theory and Advanced Mathematical Applications
edited by Da Ruan

Fuzzy Databases: Principles and Applications
by Frederick E. Petry with Patrick Bose

Distributed Fuzzy Control of Multivariable Systems
by Alexander Gegov

Fuzzy Modelling: Paradigms and Practices
by Witold Pedrycz

Fuzzy Logic Foundations and Industrial Applications
by Da Ruan

Fuzzy Sets in Engineering Design and Configuration
by Hans-Juergen Sebastian and Erik K. Antonsson

Consensus Under Fuzziness
by Mario Fedrizzi, Janusz Kacprzyk, and Hannu Nurmi

Uncertainty Analysis in Enginerring Sciences: Fuzzy Logic, Statistices, and Neural Network Approach
by Bilal M. Ayyub and Madan M. Gupta

Fuzzy Modeling for Control

FUZZY MODELING FOR CONTROL

ROBERT BABUŠKA
Control Engineering Laboratory
Faculty of Information Technology and Systems
Delft University of Technology, Delft, the Netherlands

Kluwer Academic Publishers
Boston/Dordrecht/London

Distributors for North, Central and South America:
Kluwer Academic Publishers
101 Philip Drive
Assinippi Park
Norwell, Massachusetts 02061 USA

Distributors for all other countries:
Kluwer Academic Publishers Group
Distribution Centre
Post Office Box 322
3300 AH Dordrecht, THE NETHERLANDS

Library of Congress Cataloging-in-Publication Data

A C.I.P. Catalogue record for this book is available from
the Library of Congress.

Printed on acid-free paper.

Printed in the United States of America

to Dana, Michaela, and Markéta

Contents

Preface

Since its introduction in 1965, fuzzy set theory has found applications in a wide variety of disciplines. Modeling and control of dynamic systems belong to the fields in which fuzzy set techniques have received considerable attention, not only from the scientific community but also from industry. Many systems are not amenable to conventional modeling approaches due to the lack of precise, formal knowledge about the system, due to strongly nonlinear behavior, due to the high degree of uncertainty, or due to the time varying characteristics. Fuzzy modeling along with other related techniques such as neural networks have been recognized as powerful tools which can facilitate the effective development of models. One of the reasons for this is the capability of fuzzy systems to integrate information from different sources, such as physical laws, empirical models, or measurements and heuristics.

Fuzzy models can be seen as logical models which use "if–then" rules to establish qualitative relationships among the variables in the model. Fuzzy sets serve as a smooth interface between the qualitative variables involved in the rules and the numerical data at the inputs and outputs of the model. The rule-based nature of fuzzy models allows the use of information expressed in the form of natural language statements and consequently makes the models transparent to interpretation and analysis. At the computational level, fuzzy models can be regarded as flexible mathematical structures, similar to neural networks, that can approximate a large class of complex nonlinear systems to a desired degree of accuracy.

Recently, a great deal of research activity has focused on the development of methods to build or update fuzzy models from numerical data. Most approaches are based on neuro-fuzzy systems, which exploit the functional similarity between fuzzy reasoning systems and neural networks. This "marriage" of fuzzy systems and neural networks enables a more effective use of optimization techniques for building fuzzy systems, especially with regard to their approximation accuracy. However, the aspects related to the transparency and interpretation tend to receive considerably less attention. Consequently, most neuro-fuzzy models can be regarded as black-box models which provide little insight to help understand the underlying process.

The approach adopted in this book aims at the development of transparent rule-based fuzzy models which can accurately predict the quantities of interest, and at the same time provide insight into the system that generated the data. Attention is paid to the selection of appropriate model structures in terms of the dynamic properties, as well as the internal structure of the fuzzy rules (linguistic, relational, or Takagi–Sugeno type). From the system identification point of view, a fuzzy model is regarded as a composition of local submodels. Fuzzy sets naturally provide smooth transitions between the submodels, and enable the integration of various types of knowledge within a common framework.

In order to automatically generate fuzzy models from measurements, a comprehensive methodology is developed. It employs fuzzy clustering techniques to partition the available data into subsets characterized by a linear behavior. The relationships between the presented identification method and linear regression are exploited, allowing for the combination of fuzzy logic techniques with standard system identification tools. Attention is paid to the aspects of accuracy and transparency of the obtained fuzzy models.

Using the concepts of model-based predictive control and internal model control with an inverted fuzzy model, the control design based on a fuzzy model of a nonlinear dynamic process is addressed. To this end, methods which exactly invert specific types of fuzzy models are presented. In the context of predictive control, branch-and-bound optimization is applied. Attention is paid to algorithmic solutions of the control problem, mainly with regard to real-time control aspects.

The orientation of the book is towards methodologies that in the author's experience proved to be practically useful. The presentation reflects theoretical and practical issues in a balanced way, aiming at readership from the academic world and also from industrial practice. Simulation examples are given throughout the text and three selected real-world applications are presented in detail. In addition, an implementation in a MATLAB toolbox of the techniques presented is described. This toolbox can be obtained from the author.

ROBERT BABUŠKA
DELFT, THE NETHERLANDS

Acknowledgments

I would like to express my sincere thanks to all my colleagues and students who in one way or another have contributed to the research presented in this book. Among them, a special word of thanks to Hubert te Braake of Heineken, Uzay Kaymak of Shell International, Henk Verbruggen, Piet Bruijn, João Miguel Sousa and Magne Setnes of the Control Engineering Laboratory, Faculty of Information Technology and Systems, Delft University of Technology. Further, René Jager of Frog Navigation Systems, Cesare Fantuzzi of the University of Ferrara, Petr Horáček of the Czech Technical University Prague, and Vilém Novák of the University of Ostrava contributed by their comments and suggestions for improving drafts of the manuscript. I also wish to thank Mrs. J.B. Zaat-Jones for revisions to my English text.

The applications reported in Chapter 7 were realized in cooperation with the Department of Mining and Petroleum Engineering and with the Kluyver Laboratory for Biotechnology, both at the Delft University of Technology. I also thank Tor Arne Johansen of SINTEF for providing data and simulation programs for the heat transfer system and the pH process.

Finally, I thank my wife, Dana, for her understanding, patience and continual support during the writing of this book.

Some of the figures appearing in this book were reproduced from my previous works with the kind permission of the respective copyright holders. Figure 2.1 and Figures 2.11 through 2.14 were reproduced from (Babuška and Verbruggen, 1996e), Elsevier Science. Figure 2.15 and Figures 2.17 through 2.20 were reproduced from (Babuška, et al., 1996), IEEE. Figures 2.21, 2.22, 2.24, 4.12, 5.35 and 5.36 were reproduced from (Babuška and Verbruggen, 1997f), Taylor & Francis. Figures 5.11 through 5.17 were reproduced from (Babuška, et al., 1996), IEEE. Figure 5.28 is reproduced from (Babuška and Verbruggen, 1995b), IEEE. Figure 6.9 is reproduced from (Sousa, et al., 1997), Elsevier Science. Further, Section 7.1 is a shortened version of the article (den Hartog, et al., 1997) published by Elsevier Science, and the figures and tables in this section were reproduced from this article. Section 7.3 is an extended version of the article (Babuška, et al., 1996) published by IFAC. Figures 7.18, 7.19, 7.21 and Table 7.5 were reproduced from this article.

1 INTRODUCTION

This book addresses the modeling of complex, nonlinear, or partially unknown systems by means of techniques based on fuzzy set theory and fuzzy logic. This approach, termed fuzzy modeling, is shown to be able to cope with systems that pose problems to conventional techniques, mainly due to nonlinearities and lack of precise knowledge about these systems. Methods are described for the development of fuzzy models from data, and for the design of control systems which make use of an available fuzzy model. The presented framework allows for an effective use of heterogeneous information in the form of numerical data, qualitative knowledge, heuristics and first-principle models for the building, validation and analysis of models, and for the design of controllers. The obtained model can be a part of a real-time control algorithm, or can serve for analysis of the process, in order to gain better understanding, and to improve the operation, monitoring and diagnosis.

1.1 Modeling and Identification of Complex Systems

Developing mathematical models of real systems is a central topic in many disciplines of engineering and science. Models can be used for simulations, analysis of the system's behavior, better understanding of the underlying mechanisms in the system, design of new processes, and for controlling systems. The development of a mathematical model which adequately represents the reality is an important task. If the model is not accurate enough, the subsequent steps of analysis, prediction, controller synthesis, etc., cannot be successful. However, there is an obvious tradeoff between the necessary accuracy of the model and its complexity. Models should provide information at the most relevant level of precision (abstraction), suppressing unnecessary details when appropriate. If the model is too simple, it cannot properly represent the studied characteristics of the system and does not serve its purpose. However, the model should not be too complex if it is to be practically useful.

In control engineering, modeling and identification are important steps in the design of control, supervision and fault-detection systems. Modern production and manufacturing methods in industry, combined with the growing demands concerning product lifetime, quality, flexibility in production, and safety, have increased the performance requirements imposed on the control systems. Production is often characterized by frequent changes in product throughput, product mix, operating points and operating conditions. To satisfy the tight quality requirements, control systems must guarantee high performance over a wide range of operating conditions. Under these conditions, process modeling often becomes a major bottleneck for the application of advanced model-based techniques. Many systems are not amenable to conventional modeling approaches due to the lack of precise, formal knowledge about the system, strongly nonlinear behavior, the high degree of uncertainty, time varying characteristics, etc. Example of such systems can be found in the process industry, flexible manufacturing, aerospace engineering, (bio)chemical engineering, but also in ecological, social or financial domains.

1.2 Different Modeling Paradigms

Traditionally, modeling is seen as a conjunction of a thorough understanding of the system's nature and behavior, and of a suitable mathematical treatment that leads to a usable model. This approach is usually termed "white-box" (physical, mechanistic, first-principle) modeling. However, the requirement for a good understanding of the physical background of the problem at hand proves to be a severe limiting factor in practice, when complex and poorly understood systems are considered. Difficulties encountered in conventional white-box modeling can arise, for instance, from poor understanding of the underlying phenomena, inaccurate values of various process parameters, or from the complexity of the resulting model. A complete understanding of the underlying mechanisms is virtually impossible for a majority of real systems. However, gathering an acceptable degree of knowledge needed for physical modeling may be a very difficult, time-consuming and expensive task. Even if the structure of the model is determined, a major problem of obtaining accurate values for the parameters remains. It is the task of system identification to estimate the parameters from data measured on the system. Identification methods are currently developed to a mature level for linear systems only. Most real processes are, however, nonlinear and can be approximated by linear models only locally.

A different approach assumes that the process under study can be approximated by using some sufficiently general "black-box" structure used as a general function approximator. The modeling problem then reduces to postulating an appropriate structure of the approximator, in order to correctly capture the dynamics and the nonlinearity of the system. In black-box modeling, the structure of the model is hardly related to the structure of the real system. The identification problem consists of estimating the parameters in the model. If representative process data is available, black-box models usually can be developed quite easily, without requiring process-specific knowledge. A severe drawback of this approach is that the structure and parameters of these models usually do not have any physical significance. Such models cannot be used for analyzing the system's behavior otherwise than by numerical simulation, cannot be scaled up or down when moving from one process scale to another, and therefore are less useful for industrial practice.

There is a range of modeling techniques that attempt to combine the advantages of the white-box and black-box approaches, such that the known parts of the system are modeled using physical knowledge, and the unknown or less certain parts are approximated in a black-box manner, using process data and black-box modeling structures with suitable approximation properties. These methods are often denoted as hybrid, semi-mechanistic or gray-box modeling.

A common drawback of most standard modeling approaches is that they cannot make effective use of extra information, such as the knowledge and experience of engineers and operators, which is often imprecise and qualitative in its nature. The fact that humans are often able to manage complex tasks under significant uncertainty has stimulated the search for alternative modeling and control paradigms. So-called "intelligent" methodologies, which employ techniques motivated by biological systems and human intelligence to develop models and controllers for dynamic systems, have been introduced. These techniques explore alternative representation schemes, using, for instance, natural language, rules, semantic networks or qualitative models, and possess formal methods to incorporate extra relevant information. Fuzzy modeling and control are typical examples of techniques that make use of human knowledge and deductive processes. Artificial neural networks, on the other hand, realize learning and adaptation capabilities by imitating the functioning of biological neural systems on a simplified level.

1.3 Fuzzy Modeling

Systems can be represented by mathematical models of many different forms, such as algebraic equations, differential equations, finite state machines, etc. The modeling framework considered in this book is based on rule-based fuzzy models, which describe relationships between variables by means of if–then rules, such as:

If *the heating power is high* **then** *the temperature will increase fast.*

These rules establish logical relations between the system's variables by relating qualitative values of one variable (power is *high*) to qualitative values of another variable (temperature will *increase fast*). The qualitative values typically have a clear linguistic interpretation, such as in the above example, and are called linguistic terms (labels, values). The concept of system modeling and analysis by means of linguistic variables was introduced by Zadeh (1973), and it has developed considerably in recent years. The meaning of the linguistic terms with regard to the input and output variables which may be numerical (heating power, temperature) is defined by suitably chosen fuzzy sets. In this sense, fuzzy sets, or more precisely, their membership functions, provide an interface between the input and output numerical variables and the linguistic qualitative values in the rules.

The logical structure of the rules facilitates the understanding and analysis of the model in a semi-qualitative manner, close to the way humans reason about the real world. In a given context, the characterization of the values by linguistic terms may be more appropriate than a precise numerical value. The deliberate overlap of the membership functions ensures generalization for situations not completely captured by the rules. In mathematical terms, the inference process in fuzzy models can be regarded as an interpolation between the outcomes of the individual rules.

Fuzzy set approaches have several advantages over other "intelligent" modeling techniques, such as neural networks (Haykin, 1994), CMAC (Albus, 1975), or radial basis function networks (Chen, et al., 1991):

- Fuzzy models integrate the logical processing of information with attractive mathematical properties of general function approximators. Fuzzy models can be seen as rule-based systems suitable for formalizing the knowledge of experts, and at the same time they are flexible mathematical structures, which can represent complex nonlinear mappings (Kosko, 1994; Wang, 1994; Zeng and Singh, 1995b). As fuzzy modeling integrates numerical and symbolic processing into one common framework, it is not restricted to areas requiring human expertise and knowledge. Fuzzy models can also make effective use of data-driven learning algorithms and can be combined with conventional regression techniques (Takagi and Sugeno, 1985; Wang, 1994; Lin, 1994).

- The rule-based structure of fuzzy systems is useful in the analysis of fuzzy models acquired from numerical data, since the obtained rules may reveal a useful qualitative description of the system that generated the data. Such a description can be confronted and possibly combined with the knowledge of experts, which helps in understanding the system and validating the model at the same time.

- The use of linguistic qualitative terms in the rules can be regarded as a kind of information quantization. Depending on the number of qualitative values considered (the granularity), models at different levels of abstraction and accuracy can be developed for a given system. Each of the models may serve a different purpose (prediction, analysis, controller design, monitoring, etc.).

1.4 Fuzzy Identification

The term *fuzzy identification* usually refers to techniques and algorithms for constructing fuzzy models from data. Two main approaches to the integration of knowledge and data in a fuzzy model can be distinguished:

1. The expert knowledge expressed in a verbal form is translated into a collection of if–then rules. In this way, a certain model structure is created. Parameters in this structure (membership functions, weights of the rules, etc.) can be fine-tuned using input-output data. The particular tuning algorithms exploit the fact that at the computational level, a fuzzy model can be seen as a layered structure (network),

similar to artificial neural networks, to which standard learning algorithms can be applied. This approach is usually called neuro-fuzzy modeling.

2. No prior knowledge about the system under study is initially used to formulate the rules, and a fuzzy model is constructed using numerical data only. It is expected that the extracted rules and membership functions can provide an a posteriori interpretation of the system's behavior. An expert can confront this information with his own knowledge, can modify the rules, or supply new ones, and can design additional experiments in order to obtain more informative data.

The two techniques, of course, can be combined, depending on the particular application. This book focuses mainly on the development of methods and algorithms for the second approach, i.e., for the automated acquisition of fuzzy models from data. It is believed that this technique is more useful in practice, as it can obviate the process of knowledge acquisition which is a well-known bottleneck for the practical applications of knowledge-based systems (Gaines and Boose, 1988; McGraw and Harbisson-Briggs, 1989). Instead, the expert is invited to assume a more active role of model analysis and validation, which may lead to revealing new pieces of information, and may result in a kind of "emergent" knowledge acquisition. This aspect has been verified in practice in a number of practical applications of the methods developed in this book.

To date, relatively little attention has been devoted to the identification of transparent fuzzy models from data. Most of the techniques reported in the literature aim at obtaining numerical models that simply fit the data with the best possible accuracy, without paying attention to the interpretation of the results (Takagi and Sugeno, 1985; Sugeno and Kang, 1988; Wang, 1994; Lin, 1994). In the context of this book, the lack of interpretability is seen as a major drawback, since many other techniques can be used for completely black-box modeling, such as standard nonlinear regression (Seber and Wild, 1989), spline techniques (de Boor, 1978; Brown and Harris, 1994), or neural networks (Hunt, et al., 1992). In many cases, a natural requirement is that a model not only accurately predicts the system's outputs but also provides some insights into the working of the system. Such a model can be used not only for the given situation, but can also be more easily adapted to changing design parameters and operating conditions.

In this book, fuzzy models are viewed as a class of local modeling approaches, which attempt to solve a complex modeling problem by decomposing it into a number of simpler subproblems. The theory of fuzzy sets offers an excellent tool for representing the uncertainty associated with the decomposition task, for providing smooth transitions between the individual local submodels, and for integrating various types of knowledge within one common framework. From this perspective, fuzzy identification can be regarded as a search for a decomposition of a nonlinear system, which gives a desired balance between the complexity and the accuracy of the model, effectively exploring the fact that the complexity of systems is usually not uniform. Since it cannot be expected that sufficient prior knowledge is available concerning this decomposition, methods for automated generation of the decomposition, primarily from system data, are developed. A suitable class of fuzzy clustering algorithms is used for this purpose.

In addition, techniques are proposed for the reduction and simplification of the initial fuzzy model acquired from data.

1.5 Control Design Based on Fuzzy Models

Several methods can be used to design control systems based on fuzzy models. By exploiting the mathematical properties of particular fuzzy model structures, such as local linearity, controllers can be developed which are closely related to traditional gain scheduling approaches in feedback control (Åström and Wittenmark, 1989) and to multiple model adaptive control (Narendra, et al., 1995). Considerable research has been devoted to the study of techniques for designing feedback fuzzy controllers of the Takagi–Sugeno structure, based on an available fuzzy model of the process (Tanaka and Sugeno, 1992; Tanaka, et al., 1996; Zhao, 1995). The approaches investigated in this book are more related to (multistage) decision making and to model-based predictive control. Inversion of a fuzzy model and search techniques are applied to find inputs which force the system to a desired state, or to a state which meets best the given control objectives. It is believed that these approaches, though computationally more involved than purely feedback control, lead to a more effective use of the information contained in the model, and allow us to deal with constraints on the system's variables in a consistent way.

1.6 Outline of the Book

This book is not intended to be a complete exposition of fuzzy modeling and control. Its aim is to elaborate on certain specific ideas and concepts and develop them into useful techniques. Readers interested in a detailed and fundamental treatment can consult research monographs by Pedrycz (1993), Driankov, et al. (1993), Yager and Filev (1994), or Jang, et al. (1997).

Chapter 2 of this book addresses selected issues of fuzzy modeling, which are considered essential to understanding the remainder of the book. This chapter contains both well-known material and new insights and extensions, mainly concerning the inference in the Takagi–Sugeno fuzzy model.

Chapter 3 presents definitions of different partitions in data sets, and gives an overview of selected fuzzy clustering algorithms which are relevant to the identification of fuzzy models described in the subsequent chapters. Illustrative examples are presented to demonstrate some of the essential properties and features of the algorithms.

Chapter 4 addresses the solution of a nonlinear identification problem by means of fuzzy clustering. Performance of the different clustering algorithms is analyzed and illustrated via examples. Methods are presented to find an appropriate number of clusters in the data, using a modified compatible cluster merging algorithm and cluster validity measures.

Chapter 5 presents methods and algorithms to derive fuzzy models of the Takagi–Sugeno type, and fuzzy relational and linguistic models from the partitions obtained by fuzzy clustering. Simplification and linguistic approximation of the obtained rule bases by means of similarity measures are discussed in the context of interpretation and analysis of the models.

Chapter 6 addresses control design based on an available fuzzy model. Two particular techniques are considered in more detail: model inversion and predictive control. A method for exact inversion of a singleton model is developed and combined with a predictive control scheme based on discrete optimization. Simulation examples of two nonlinear processes are presented.

Chapter 7 presents three selected applications of the developed modeling and control techniques to real systems. In the first one, a fuzzy model for prediction of the performance of a rock-excavation machine is described, as a typical example of knowledge-based fuzzy modeling combined with data-driven adaptation of the fuzzy model. The second example presents fuzzy modeling and predictive control of pressure in a laboratory fermenter. The last application is an example of hybrid (semi-mechanistic) modeling of a biochemical process. A fuzzy model acquired from process measurements serves as a good numerical predictor, and at the same time provides a qualitative description of the process.

Appendix A presents some basic concepts and definitions from fuzzy set theory, which are used throughout the book. In Appendix B, a software tool is described which has been developed to support the fuzzy modeling and identification techniques presented in this book. Appendix C contains a list of mathematical symbols and abbreviations.

2 FUZZY MODELING

The concepts of fuzzy-set theory and fuzzy logic can be employed in the modeling of systems in a number of ways. Examples of fuzzy systems are rule-based fuzzy systems (Zadeh, 1973; Driankov, et al., 1993), fuzzy linear regression models (Tanaka, et al., 1982), or fuzzy models using cell structures (Smith, et al., 1994). This book focuses only rule-based fuzzy systems, i.e., systems where the relationships between variables are represented by a means of fuzzy if–then rules of the form:

If *antecedent proposition* **then** *consequent proposition*.

Depending on the particular structure of the consequent proposition, three types of models are distinguished:

- *Linguistic fuzzy model* (Zadeh, 1973; Mamdani, 1977), where both the antecedent and consequent are fuzzy propositions.
- *Fuzzy relational model* (Pedrycz, 1984; Yi and Chung, 1993), which can be regarded as a generalization of the linguistic model, allowing one particular antecedent proposition to be associated with several different consequent propositions via a fuzzy relation.
- *Takagi–Sugeno (TS) fuzzy model* (Takagi and Sugeno, 1985), where the consequent is a crisp function of the antecedent variables rather than a fuzzy proposition.

These three types of fuzzy models are reviewed in Sections 2.1, 2.2 and 2.3, respectively. The latter section also deals with the analysis of the TS approximation properties and presents a new inference mechanism for the TS model. Section 2.4 gives an overview of fuzzy identification methods to construct or adapt fuzzy models by using numerical data.

2.1 Linguistic Fuzzy Models

In the linguistic fuzzy model (also called the Mamdani model), both the antecedent and the consequent are *fuzzy propositions*. A general form of linguistic fuzzy if–then rules is:

$$R_i\text{:}\quad \textbf{If } \mathbf{x} \text{ is } A_i \textbf{ then } \mathbf{y} \text{ is } B_i, \quad i = 1, 2, \ldots, K, \tag{2.1}$$

where $\mathbf{x}$ is the *antecedent variable*, which represents the input to the fuzzy system, and $\mathbf{y}$ is the *consequent variable* representing the output of the fuzzy system. In most cases, real-valued vector inputs and outputs are used: $\mathbf{x} \in X \subset \mathbb{R}^p$ and $\mathbf{y} \in Y \subset \mathbb{R}^q$. Generally, however, the antecedent and consequent variables are *linguistic variables* (see Section 2.1.1), which have fuzzy sets as their values. In such a case, $\mathbf{x} \in \mathcal{F}(X)$ and $\mathbf{y} \in \mathcal{F}(Y)$, where $\mathcal{F}(D)$ is the set of all fuzzy sets on D. A_i and B_i are *linguistic terms* (fuzzy sets) defined by multivariate membership functions $\mu_{A_i}(\mathbf{x})\colon X \to [0, 1]$ and $\mu_{B_i}(\mathbf{y})\colon Y \to [0, 1]$, respectively. Finally, K denotes the number of rules in the model.

Fuzzy sets A_i define fuzzy regions in the antecedent space, for which the respective consequent propositions hold. Instead of using the multidimensional fuzzy sets, it is often more convenient to consider fuzzy propositions with scalar variables and create compound conditions by means of logical operations, as discussed in Section 2.1.2. Fuzzy propositions based on scalars usually convey certain meanings for the given variables, such as "low temperature", "high pressure", etc. This allows the interpretation of the fuzzy model by using natural language.

The membership functions for the linguistic terms are defined in a *data base* which is a part of the fuzzy system. The rule base together with the data base form the *knowledge base* of the fuzzy system. The input–output mapping is realized by the *fuzzy inference mechanism* that, given the knowledge stored in the knowledge base and an input value, derives the corresponding output value. In most engineering applications, the inputs and outputs are numerical values, rather that fuzzy sets. To deal with this, the fuzzy system must be equipped with conversion interfaces, the *fuzzification* and *defuzzification* units. This system represents a static multiple-input, multiple-output (MIMO) mapping $F\colon X \to Y$. Dynamic behavior must be provided by means of external dynamic filters, see Figure 2.1. One example is the NARX model:

$$\begin{aligned} R_i\text{:}\quad & \textbf{If } y(k) \text{ is } A_{i1} \textbf{ and } \ldots \textbf{ and } y(k-n_y+1) \text{ is } A_{in_y} \textbf{ and } u(k) \text{ is } B_{i1} \\ & \textbf{and } \ldots \textbf{ and } u(k-n_u+1) \text{ is } B_{in_u} \textbf{ then } y(k+1) \text{ is } C_i, \end{aligned} \tag{2.2}$$

where the input dynamic filter is a simple generator of the regressors $y(k), y(k-1), \ldots$ and $u(k), u(k-1), \ldots$ with the inputs $y(k)$ and $u(k)$, and no output filter is used. More details on the representation of dynamic systems are given in Section 4.2.

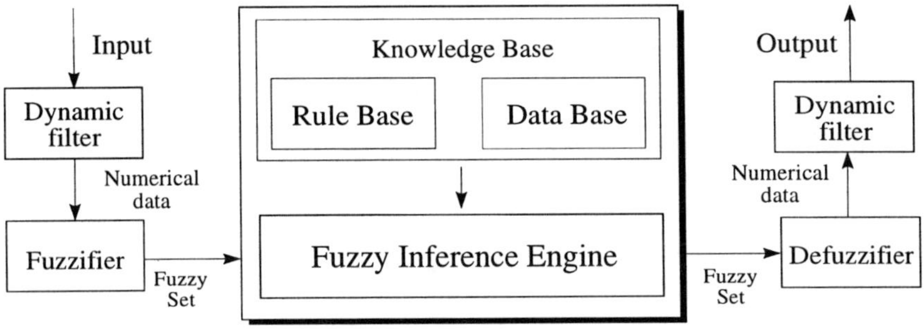

Figure 2.1. A generic fuzzy system with fuzzification and defuzzification units and external dynamic filters.

2.1.1 Linguistic Terms and Variables

Linguistic terms can be seen as qualitative values (information granulae) used to describe a particular relationship by linguistic rules. Typically, a set of N linguistic terms $\mathcal{A} = \{A_1, A_2, \ldots, A_N\}$ is defined in the domain of a given scalar variable x. Because this variable assumes linguistic values, it is called a linguistic variable. A linguistic variable L is defined as a quintuple (Klir and Yuan, 1995):

$$L = (x, \mathcal{A}, X, g, m), \tag{2.3}$$

where x is the *base variable*, $\mathcal{A} = \{A_1, A_2, \ldots, A_N\}$ is the set of *linguistic terms* of x, X is the *domain* (universe of discourse) of the base variable, g is a *syntactic rule* for generating linguistic terms and m is a *semantic rule* that assigns to each linguistic term its meaning (a fuzzy set in X). Figure 2.2 shows an example of a linguistic variable "temperature" with three linguistic terms "low", "medium" and "high". The base variable is the temperature given in appropriate physical units.

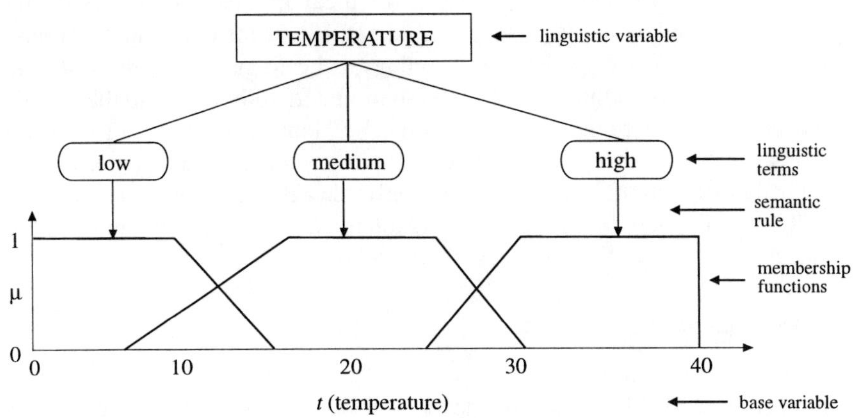

Figure 2.2. Example of a linguistic variable "temperature" with three linguistic terms.

It is usually required that the linguistic terms satisfy the properties of *coverage* and *semantic soundness* (Pedrycz, 1995). Coverage means that each element is assigned to at least one fuzzy set with a nonzero membership degree, i.e.,

$$\forall x,\ \exists i,\ \mu_{A_i}(x) > 0\,. \tag{2.4}$$

Alternatively, a stronger condition called ϵ-coverage may be imposed:

$$\forall x,\ \exists i,\ \mu_{A_i}(x) > \epsilon, \quad \epsilon \in (0,1)\,. \tag{2.5}$$

For instance, the membership functions in Figure 2.2 satisfy ϵ-coverage for $\epsilon = 0.5$. Most of the clustering algorithms used for the automatic generation of fuzzy models from data, presented in Chapters 3 and 4, impose yet a stronger condition:

$$\sum_{i=1}^{N} \mu_{A_i}(x) = 1, \quad \forall x \in X, \tag{2.6}$$

meaning that for each x, the sum of membership degrees equals one (*fuzzy partition*).

Semantic soundness is related to the linguistic meaning of the fuzzy sets. Usually, A_i are convex (unimodal) and normal fuzzy sets, which are sufficiently disjoint, and the number N of subsets per variable is small (say nine at most). The number of linguistic terms and the particular shape and overlap of the membership functions are related to the *granularity* of the information processing within the fuzzy system, and hence also to the level of precision with which a given system can be represented by a fuzzy model. For instance, trapezoidal membership functions, such as those given in Figure 2.2, provide some kind of "information hiding" for data within the cores of the membership functions (e.g., temperatures between 0 and 5 degrees cannot be distinguished, since all are classified as "low" with degree 1). Well-behaved mappings can be accurately represented with a very low granularity. Triangular membership functions, for instance, can exactly reproduce multi-linear functions.

Membership functions can be defined by the model developer (expert), using prior knowledge, or by experimentation, which is a typical approach in knowledge-based fuzzy control (Driankov, et al., 1993). In this case, the membership functions are designed such that they represent the meaning of the linguistic terms in the given context. When input–output data of the system under study are available, methods for constructing or adapting the membership functions from data can be applied, see Section 2.4 for an overview. Here the task is somewhat the reverse, since first the membership functions are created and only then is some meaning assigned to them. This meaning can be simply given by the model developer or can be obtained automatically by means of *linguistic approximation*.

2.1.2 *Antecedent Propositions*

The antecedent proposition can contain fuzzy sets defined directly in the vector domain X, as in the rule (2.1). Most often, however, the rules are represented in a decomposed form, with the antecedent defined as a combination of simple fuzzy propositions on the individual components x_i of the vector $\mathbf{x}$. The logical operators of conjunction, disjunction and negation (complement) can be used in order to construct a desired compound proposition, for instance:

$$R_i\text{: } \textbf{If } x_1 \text{ is } A_{i1} \textbf{ or } x_2 \text{ is } A_{i2} \textbf{ and } x_3 \text{ is } \textbf{not } A_{i3} \textbf{ then } y \text{ is } B_i\,. \tag{2.7}$$

Here, one-dimensional fuzzy sets are defined for each component of the antecedent vector. The *degree of fulfillment* of the rule is computed using the appropriate t-norm, t-conorm and the complement operator, e.g.:

$$\beta_i = \mu_{A_{i1}}(x_1) \vee \mu_{A_{i2}}(x_2) \wedge \big(1 - \mu_{A_{i3}}(x_3)\big)\,. \tag{2.8}$$

Here the maximum operator ($\vee$) represents the disjunction (**or**), the minimum operator $\wedge$ represent the conjunction (**and**), and $1-\mu$ is the complement (negation, **not**). The most common is the *conjunctive form* given by:

$$R_i\text{: } \textbf{If } x_1 \text{ is } A_{i1} \textbf{ and } x_2 \text{ is } A_{i2} \textbf{ and } \ldots \textbf{ and } x_p \text{ is } A_{ip} \textbf{ then } y \text{ is } B_i, \tag{2.9}$$

with the degree of fulfillment β_i given by a conjunction (t-norm) operator, such as the minimum:

$$\beta_i = \mu_{A_{i1}}(x_1) \wedge \mu_{A_{i2}}(x_2) \wedge \ldots \wedge \mu_{A_{ip}}(x_p)\,. \tag{2.10}$$

This expression is, in fact, a membership function of a multivariate fuzzy set created by intersecting the univariate sets on the Cartesian product space of the components of $\mathbf{x}$. The shape of the resulting multidimensional fuzzy set depends on the particular intersection operator used. This approach can be compared to generating multivariate spline basis functions as a tensor product of univariate basis functions (Brown and Harris, 1994).

With respect to the generality of partitioning the antecedent space into fuzzy regions, the antecedent with multivariate membership functions is the most general one, as there is no restriction on the shape of the fuzzy sets. The boundaries between the fuzzy sets can be arbitrarily curved and opaque to the axes, as depicted in Figure 2.3a. By combining conjunctions, disjunctions and negations, various partitions of the antecedent space can be obtained. The boundaries are, however, restricted to the rectangular grid defined by the fuzzy sets of the individual variables, as shown in Figure 2.3b. The conjunctive antecedent divides the antecedent space into a lattice of axis-orthogonal hyperboxes, as depicted in Figure 2.3c. In this case, the number of rules needed to cover the entire domain is a product of the input space dimension and of the number of fuzzy sets used for each variable. For multivariate systems, the partition shown in Figure 2.3a may provide the most effective representation. The identification technique based on clustering presented in Chapter 4 searches in the first place for partitions like the one shown in Figure 2.3a. Linguistic interpretation is then obtained, which leads to partitions of the type (b) and (c) in Figure 2.3.

2.1.3 Linguistic Hedges

In addition to logical connectives and operators, linguistic hedges (linguistic modifiers) can also be used to modify the meaning of the linguistic terms. Examples of hedges are: *very*, *sightly*, *more or less*, etc. Two basic approaches to the implementation of linguistic hedges can be distinguished: *powered* hedges and *shifted* hedges.

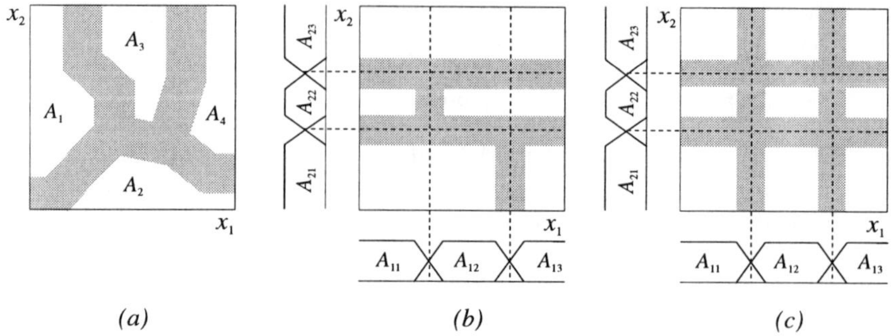

Figure 2.3. Different partitions of the antecedent space. Gray areas denote the overlapping regions of the fuzzy sets.

Powered hedges are implemented by functions operating on the membership degrees of the linguistic terms (Zimmermann, 1996). For instance, the hedge *very* squares the membership degrees of the term which meaning it modifies, i.e., $\mu_{\text{very } A}(x) = \mu_A^2(x)$. Shifted hedges (Lakoff, 1973), on the other hand, shift the membership functions along their domains. Combinations of the two approaches have been proposed as well (Novák, 1989; Novák, 1996).

2.1.4 *Inference in the Linguistic Model*

Inference in fuzzy rule-based systems is the process of deriving an output fuzzy set given the rules and the inputs. The inference mechanism in the linguistic model is based on the *compositional rule of inference* (Zadeh, 1973). For the sake of simplicity, consider a single rule with scalar antecedent and consequent variables:

$$\textbf{If } x \text{ is } A \textbf{ then } y \text{ is } B\,. \tag{2.11}$$

This rule can be regarded as a fuzzy relation $R\colon (X \times Y) \to [0,1]$, computed by

$$\mu_R(x,y) = \mathrm{I}(\mu_A(x), \mu_B(y)), \tag{2.12}$$

where the operator I can be either a fuzzy implication (Klir and Yuan, 1995), or a conjunction operator (a t-norm). Examples of fuzzy implications are the Łukasiewicz implication given by:

$$\mathrm{I}(\mu_A(x), \mu_B(y)) = \min(1, 1 - \mu_A(x) + \mu_B(y)), \tag{2.13}$$

or the Kleene–Diene implication:

$$\mathrm{I}(\mu_A(x), \mu_B(y)) = \max(1 - \mu_A(x), \mu_B(y))\,. \tag{2.14}$$

Examples of t-norms are the minimum, often, not quite correctly, called the Mamdani "implication",

$$\mathrm{I}(\mu_A(x), \mu_B(y)) = \min(\mu_A(x), \mu_B(y)), \tag{2.15}$$

or the product, also called the Larsen "implication",

$$\mathrm{I}(\mu_A(x), \mu_B(y)) = \mu_A(x) \cdot \mu_B(y) \,. \tag{2.16}$$

More details about fuzzy implications and the related operators can be found, for instance, in (Klir and Yuan, 1995; Lee, 1990a; Lee, 1990b; Jager, 1995). While fuzzy implications represent the unidirectional relationship "A implies B", rules encoded by t-norms should, rather, be interpreted as nondirectional relationships "it is true that A holds **and** B holds".

The inference mechanism is based on the generalized *modus ponens* rule:

$$\begin{array}{l} \textbf{If } x \text{ is } A \textbf{ then } y \text{ is } B \\ x \text{ is } A' \\ \hline y \text{ is } B' \end{array}$$

Given the rule " **If** x is A **then** y is B" and the fact the "x is A'", the output fuzzy set B' is derived by the relational max-t composition (Klir and Yuan, 1995):

$$B' = A' \circ R \,. \tag{2.17}$$

For the minimum t-norm, the max-min composition is obtained:

$$\mu_{B'}(y) = \max_{x} \min_{x,y} \left(\mu_{A'}(x), \mu_R(x,y) \right) . \tag{2.18}$$

Figure 2.4a shows an example of fuzzy relation R computed by (2.15). Figure 2.4b illustrates the inference of B', given the relation R and the input A', by means of the max-min composition (2.18).

Example 2.1 Consider a fuzzy rule (2.11) with the fuzzy sets:[1]

$$\begin{aligned} A &= \{0/1,\ 0.1/2,\ 0.4/3,\ 0.8/4,\ 1/5\}, \\ B &= \{0/-2,\ 0.6/-1,\ 1/0,\ 0.6/1,\ 0/2\} \,. \end{aligned}$$

Using the minimum t-norm (Mamdani "implication"), the relation R_M representing the fuzzy rule is computed by eq. (2.15):

$$R_M = \begin{bmatrix} 0 & 0 & 0 & 0 & 0 \\ 0 & 0.1 & 0.1 & 0.1 & 0 \\ 0 & 0.4 & 0.4 & 0.4 & 0 \\ 0 & 0.6 & 0.8 & 0.6 & 0 \\ 0 & 0.6 & 1 & 0.6 & 0 \end{bmatrix} . \tag{2.19}$$

[1] In this notation, fuzzy sets are defined by a list of ordered pairs: membership degree/set element, see also Appendix A.

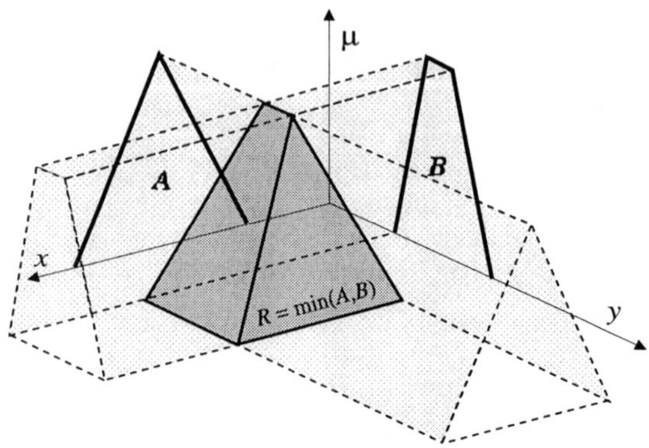

(a) Fuzzy relation (intersection).

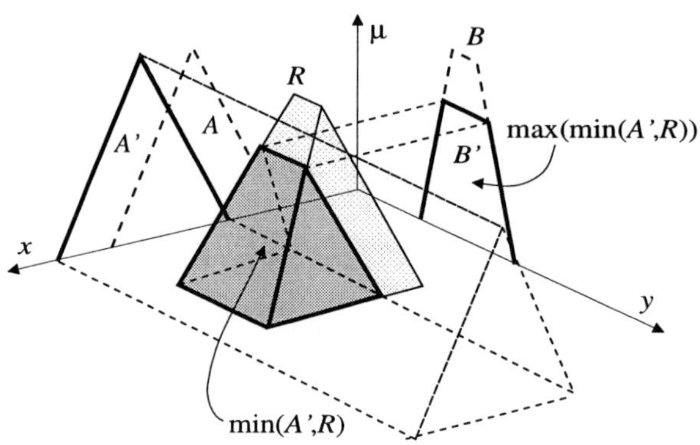

(b) Fuzzy inference.

Figure 2.4. (a) Fuzzy relation representing the rule "**If** x **is** A **then** y **is** B", (b) the compositional rule of inference.

The rows of this relational matrix correspond to the domain elements of A and the columns to the domain elements of B. Now consider an input fuzzy set to the rule:

$$A' = \{0/1,\ 0.2/2,\ 0.8/3,\ 1/4,\ 0.1/5\}\,. \tag{2.20}$$

The application of the max-min composition (2.18), $B'_M = A' \circ R_M$, yields the following output fuzzy set:

$$B'_M = \{0/-2,\ 0.6/-1,\ 0.8/0,\ 0.6/1,\ 0/2\}\,. \tag{2.21}$$

By applying the Łukasiewicz fuzzy implication (2.13), the following relation is obtained:

$$R_L = \begin{bmatrix} 1 & 1 & 1 & 1 & 1 \\ 0.9 & 1 & 1 & 1 & 0.9 \\ 0.6 & 1 & 1 & 1 & 0.6 \\ 0.2 & 0.8 & 1 & 0.8 & 0.2 \\ 0 & 0.6 & 1 & 0.6 & 0 \end{bmatrix}. \tag{2.22}$$

Using the max-t composition, where the t-norm is the Łukasiewicz t-norm (see Appendix A), the inferred fuzzy set $B'_L = A' \circ R_L$ equals:

$$B'_L = \{0.4/-2,\ 0.8/-1,\ 1/0,\ 0.8/1,\ 0.4/2\}. \tag{2.23}$$

Note the difference between the relations R_M and R_L, which are also depicted in Figure 2.5. The implication is false (zero entries in the relation) only when A holds and B does not. When A does not hold, the truth value of the implication is 1 regardless of B. The t-norm, however, is false whenever either A or B or both do not hold, and thus represents a bi-directional relation (correlation).

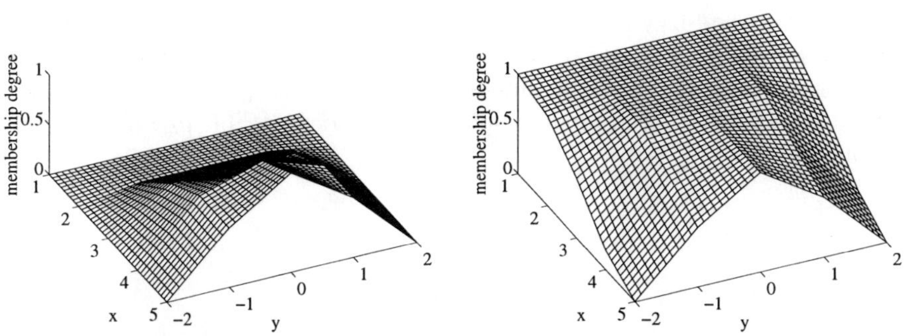

(a) Minimum t-norm. *(b) Łukasiewicz implication.*

Figure 2.5. Fuzzy relations obtained by applying a t-norm operator (minimum) and a fuzzy implication (Lukasiewicz).

This difference naturally influences the result of the inference process. Since the input fuzzy set A' is different from the antecedent set A, the derived conclusion B' is in both cases "less certain" than B. The difference is that, with the fuzzy implication, this uncertainty is reflected in the increased membership values for the domain elements that have low or zero membership in B, which means that these output values are possible to a greater degree. However, the t-norm results in decreasing the membership degree of the elements that have high membership in B, which means that these outcomes are less possible. This influences the properties of the two inference mechanisms and the choice of suitable defuzzification methods, as discussed later on. Section 2.1.6

gives an example to illustrate some advantages of an inference scheme based on fuzzy implications over the Mamdani inference using a t-norm. □

The entire rule base is represented by aggregating the relations of the individual rules into a single fuzzy relation. Let R_i denote the fuzzy relation corresponding to the ith rule. When R_i's are computed by implications, the relation R for the entire rule base is obtained by aggregating the individual relations R_i by means of an intersection:

$$R = \bigcap_{i=1}^{K} R_i \,. \tag{2.24}$$

For I being a t-norm, the aggregated relation R is computed as a union of the individual relations R_i:

$$R = \bigcup_{i=1}^{K} R_i \,. \tag{2.25}$$

The output fuzzy set B' is inferred in the same way as in the case of one rule, by using the compositional rule of inference (2.17). The representation of a system by the fuzzy relation can be regarded as the so called fuzzy graph, i.e., a union of fuzzy sets in the Cartesian product space of the antecedent and consequent variables, see Figure 2.6. The compositional rule of inference can be regarded as a generalized function evaluation using a graph. The fuzzy relation R, defined on the Cartesian product space of the system's variables $X_1 \times X_2 \times \cdots X_p \times Y$, can be regarded as a possibility distribution (restriction) of the different input–output tuples $(x_1, x_2, \ldots, x_p, y)$. An α-cut of R can be interpreted as a set of input–output combinations possible to a degree greater or equal to α.

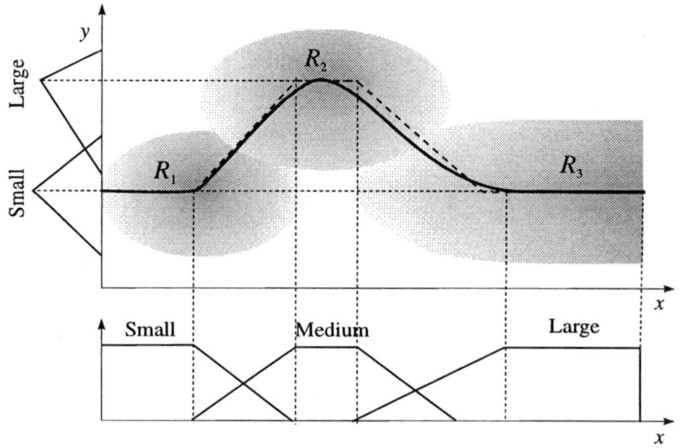

Figure 2.6. Mamdani fuzzy model as a fuzzy graph, i.e., a union of multidimensional fuzzy sets in the Cartesian product space of the rule antecedents and consequents.

Only when fuzzy implications are used in combination with fuzzy inputs (A' is a fuzzy set), does the reasoning scheme have to be realized by the relational composition described above. For computer implementation, it means that the domains must be discretized. However, in other cases, i.e., for fuzzy implications with crisp (numerical) inputs, and for t-norms with both crisp and fuzzy inputs, the reasoning scheme can be simplified, bypassing the relational calculus (Jager, 1995). For the t-norm, the simplification results in the well-known scheme, in the literature called the max-min or Mamdani inference, which is summarized in Algorithm 2.1.

Algorithm 2.1 (Mamdani max-min inference)

Step 1: For each rule i, the degree of fulfillment β_i of the antecedent is computed:

$$\beta_i = \mu_{A_{i1}}(x_1) \wedge \mu_{A_{i2}}(x_2) \wedge \ldots \wedge \mu_{A_{ip}}(x_p)\,. \tag{2.26}$$

Step 2: The output fuzzy set B_i' is derived for each rule, using the minimum t-norm:

$$\mu_{B_i'}(y) = \beta_i \wedge \mu_{B_i}(y), \quad \forall y \in Y\,. \tag{2.27}$$

Step 3: The aggregated output fuzzy set is computed by taking the maximum (union) of the individual conclusions B_i':

$$\mu_{B'}(y) = \max_{i=1,2,\ldots,K} \mu_{B_i'}(y), \quad \forall y \in Y\,. \tag{2.28}$$

Remark: If the inputs A_j' are fuzzy sets (or a singletons), the degree of fulfillment is given by:

$$\begin{aligned}\beta_i \quad = \quad & \max_{x_1 \in X_1}[\mu_{A_1'}(x_1) \wedge \mu_{A_{i1}}(x_1)] \wedge \max_{x_2 \in X_2}[\mu_{A_2'}(x_2) \wedge \mu_{A_{i2}}(x_2)] \wedge \ldots \\ & \wedge \max_{x_p \in X_p}[\mu_{A_p'}(x_p) \wedge \mu_{A_{ip}}(x_p)]\,. \end{aligned} \tag{2.29}$$

Example 2.2 Consider the rule and the fuzzy sets from Example 2.1. By applying Algorithm 2.1, first the degree of fulfillment is computed as

$$\max_{x \in X}(\mu_{A'}(x) \wedge \mu_A(x)) = 0.8\,. \tag{2.30}$$

Applying eq. (2.27) yields

$$B' = \{0/-2,\ 0.6/-1,\ 0.8/0,\ 0.6/1,\ 0/2\}, \tag{2.31}$$

which is the same result as in Example 2.1. □

2.1.5 Defuzzification

The result of fuzzy inference is the fuzzy set B'. If a crisp (numerical) output value is required, the output fuzzy set must be *defuzzified.* Defuzzification is a transformation that replaces a fuzzy set by a single numerical value representative of that set. Figure 2.7 shows two most commonly used defuzzification methods: the center of gravity (COG) and the mean of maxima (MOM).

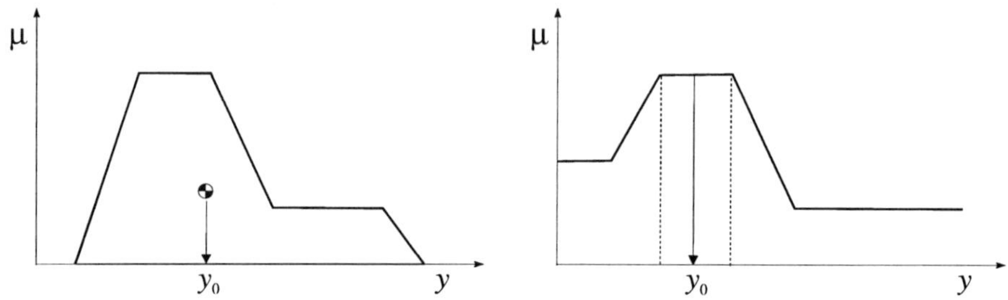

(a) Center of gravity. (b) Mean of maxima.

Figure 2.7. The center-of-gravity and the mean-of-maxima defuzzification methods.

The COG method calculates numerically the y coordinate of the center of gravity of the fuzzy set B':

$$\operatorname{cog}(B') = \frac{\sum_{q=1}^{N_q} \mu_{B'}(y_q)\, y_q}{\sum_{q=1}^{N_q} \mu_{B'}(y_q)}, \tag{2.32}$$

where N_q is the number of discretized values y_q in Y. The MOM method computes the mean value of the interval with the largest membership degree:

$$\operatorname{mom}(B') = \operatorname{mean}\{y | \mu_{B'}(y) = \max_{y \in Y} \mu_{B'}(y)\}\,. \tag{2.33}$$

The COG method is used with the Mamdani max-min inference, as it provides interpolation between the consequents, in proportion to the height of the individual consequent sets. This is necessary, as the Mamdani inference method itself does not interpolate, and the use of the MOM method in this case results in a step-wise output. The MOM method is used with the inference based on fuzzy implications, to select the "most possible" output. The inference with implications interpolates, provided that the consequent sets sufficiently overlap (Jager, 1995). The COG method cannot be directly used in this case, because the uncertainty in the output results in an increase of the membership degrees, as shown in Example 2.1. The COG method would give an inappropriate result.

To avoid the numerical integration in the COG method, a modification of this approach called the *fuzzy-mean* defuzzification is often used. The consequent fuzzy sets are first defuzzified, in order to obtain crisp values representative of the fuzzy sets, using for instance MOM: $b_j = \operatorname{mom}(B_j)$. A crisp output value is then computed by taking a weighted mean of b_j's:

$$y_0 = \frac{\sum_{j=1}^{M} \gamma_j \, b_j}{\sum_{j=1}^{M} \gamma_j}, \tag{2.34}$$

where M is the number of fuzzy sets B_j and γ_j is the maximum of the degrees of fulfillment β_i over all the rules with the consequent B_j. In terms of the aggregated fuzzy set B', given by eq. (2.28), γ_j can be expressed by $\gamma_j = \mu_{B'}(b_j)$. This method ensures linear interpolation between the b_j's, provided that the antecedent membership functions are piece-wise linear. This is not the case with the COG method, which introduces a nonlinearity, depending on the shape of the consequent functions (Jager, et al., 1992). Because the individual defuzzification is done off line, the shape and overlap of the consequent fuzzy sets have no influence, and these sets can be directly replaced by the defuzzified values (singletons), see also Section 2.1.8. In order to at least partially account for the differences between the consequent fuzzy sets, the weighted fuzzy-mean defuzzification can be applied:

$$y_0 = \frac{\sum_{j=1}^{M} \gamma_j \, S_j \, b_j}{\sum_{j=1}^{M} \gamma_j \, S_j}, \tag{2.35}$$

where S_j is the area under the membership function of B_j. An advantage of the fuzzy-mean methods (2.34) and (2.35) is that the parameters b_j can be estimated by linear estimation techniques.

2.1.6 Fuzzy Implication versus Mamdani Inference

A natural question arises: Which inference method is better, or in which situations should one method be preferred to the other? To find an answer, a detailed analysis of the presented methods must be carried out, which is outside the scope of this presentation. One of the distinguishing aspects, however, can be demonstrated by using an example.

Example 2.3 Consider a rule base given in Figure 2.8. Rules R_1 and R_2 represent a simple monotonic (approximately linear) relation between two variables. This may be, for example, a rule-based implementation of a proportional control law. Rule R_3, "If x is small then y is **not** small", represents a kind of "exception" from the simple relationship defined by interpolation of the previous two rules. In terms of control, such a rule may deal with undesired phenomena, such as static friction. For instance, when controlling an electrical motor with large Coulomb friction, it does not make sense to apply low current if it is not sufficient to overcome the friction, since in that case the motor only consumes energy. These three rules can be seen as a simple example of combining general background knowledge with more specific information in terms of exceptions.

A moment of reflection would already suggest that the Mamdani inference scheme will not work for this example. The reason is that the interpolation is provided by the defuzzification method and not by the inference mechanism itself. In Figure 2.9, the

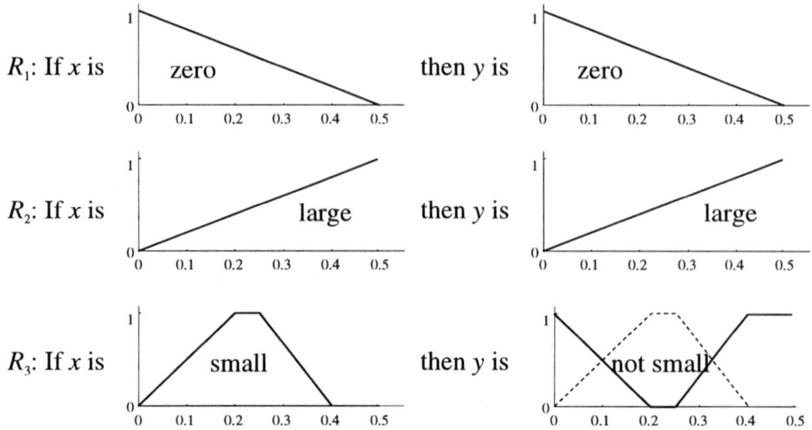

Figure 2.8. The considered rule base.

defuzzified values are plotted against the corresponding input values for two difference inference/defuzzification combinations. Figure 2.9a shows the result for the Mamdani inference method with the COG defuzzification. One can see that the presence of the third rule significantly distorts the original, almost linear characteristic, also in the region of x where R_1 has the greatest membership degree. At the same time, the purpose of avoiding small values of y is not achieved.

Figure 2.9b shows the result obtained with the logical inference method based on the Łukasiewicz implication and MOM defuzzification. One can see that the third rule fulfills its purpose, i.e., forces the fuzzy system to avoid the region of small outputs (around 0.25) for small input values (around 0.25). The exact form of the input–output mapping depends on the choice of the particular inference operators (implication, composition), but the overall behavior remains unchanged. □

The above example shows that representing the rules by a fuzzy implication may be more suitable in the case where knowledge of varying specificity is included in the rule base. Methods dealing with this situation within the Mamdani inference scheme also have been introduced in the literature. These methods, however, require special structures, such as splitting of the rule base into "positive" and "negative" rules (Krone and Kiendl, 1994), or organization of knowledge in a hierarchical way (Yager, 1994).

The implication-based reasoning scheme imposes certain requirements on the overlap of the consequent membership functions, which may be hard to fulfil in the case of multi-input rule bases (Jager, 1995). In addition, this method must generally be implemented using fuzzy relations and the compositional rule of inference, which increases the computational demands. In the modeling of systems from data, modifications of the Mamdani inference which involve differentiable algebraic operators (e.g., product and sum), and the fuzzy-mean defuzzification are preferred, as the parameters in the model can be adjusted by gradient optimization techniques, see Section 2.4.

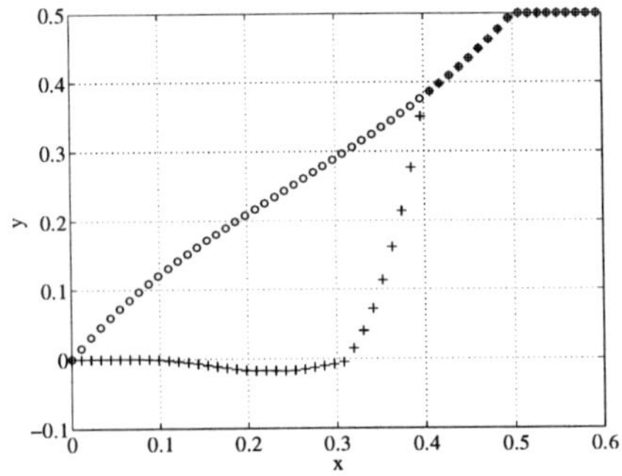

(a) Mamdani inference.

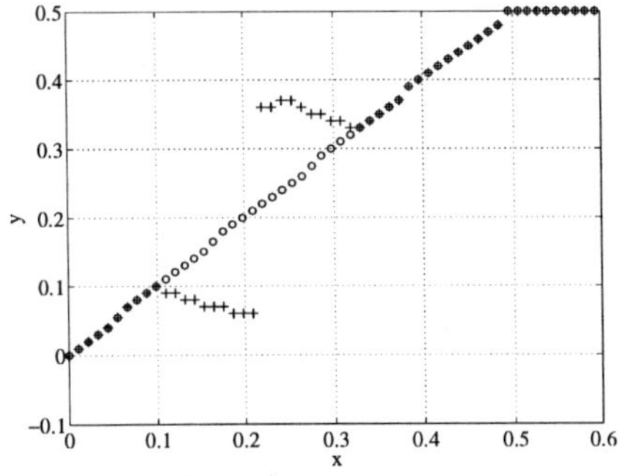

(b) Inference with Łukasiewicz implication.

Figure 2.9. Input-output mapping of the rule base of Figure 2.8 for two different inference methods. Markers 'o' denote the defuzzified output of rules R_1 and R_2 only, markers '+' denote the defuzzified output of the entire rule base.

2.1.7 Rule Chaining

So far, only a one-layer structure of a fuzzy model has been considered. In practice, however, an output of one rule base may serve as an input to another rule base. This results in a structure with several layers and chained rules. This situation occurs, for instance, in hierarchial models which include several rule bases. Hierarchical organization of knowledge is often used as a natural approach to reducing the complexity of the model. A large rule base with many input variables may be split into several interconnected rule bases with fewer inputs. As an example, suppose a rule base with three inputs, each with five linguistic terms. Using the conjunctive form (2.9), 125 rules have to be defined to cover all the input situations. Splitting the rule base in two smaller rule bases, as depicted in Figure 2.10, results in a total of 50 rules.

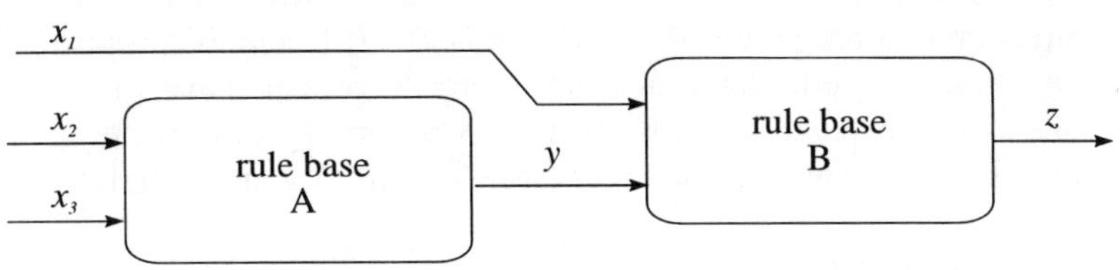

Figure 2.10. Cascade connection of two rule bases.

Another example of rule chaining is the simulation of dynamic fuzzy systems, where a cascade connection of rule bases results from the fact that a value predicted by the model at time k is used as an input at time $k+1$. As an example, consider a nonlinear discrete-time model

$$\hat{x}(k+1) = f\big(\hat{x}(k), u(k)\big), \tag{2.36}$$

where f is a mapping realized by the rule base, $\hat{x}(k)$ is a predicted state of the process at time k (at the same time it is the state of the model), and $u(k)$ is an input. At the next time step we have:

$$\hat{x}(k+2) = f\big(\hat{x}(k+1), u(k+1)\big) = f\big(f\big(\hat{x}(k), u(k)\big), u(k+1)\big), \tag{2.37}$$

which gives a cascade chain of rules.

The hierarchical structure of the rule bases shown in Figure 2.10 requires that the information inferred in Rule base A is passed to Rule base B. This can be accomplished by *defuzzification* at the output of the first rule base and subsequent *fuzzification* at the input of the second rule base. A drawback of this approach is that membership functions have to be defined for the intermediate variable and that a suitable defuzzification method must be chosen. If the values of the intermediate variable cannot be verified by using data, there is no direct way of checking whether the choice is appropriate or not. Also, the fuzziness of the output of the first stage is removed by defuzzification and subsequent fuzzification. This method is used mainly for the simulation of dynamic systems, such as (2.37), when the intermediate variable serves at the same time as a crisp output of the system.

Another possibility is to feed the *fuzzy set* at the output of the first rule base directly (without defuzzification) to the second rule base. An advantage of this approach is that it does not require any additional information from the user. However, in general, the relational composition must be carried out, which requires discretization of the domains and a more complicated implementation. In the case of the Mamdani max-min inference method, the reasoning can be simplified, since the membership degrees of the output fuzzy set directly become the membership degrees of the antecedent propositions where the particular linguistic terms occur. Assume, for instance, that inference in Rule base A results in the following degrees of fulfillment of the consequent linguistic terms B_1 to B_5 :

$$\boldsymbol{\beta} = [0/B_1,\ 0.7/B_2,\ 0.1/B_3,\ 0/B_4,\ 0/B_5].$$

The membership degree of the propositions "If y is B_2" in Rule base B is thus 0.7, the membership degree of the propositions "If y is B_3" is 0.1, and the propositions with the remaining linguistic terms have the membership degree equal to zero. This method is mainly used in knowledge-based fuzzy models, where the information is stored in several hierarchically organized rule bases (see Chapter 7 for an example).

2.1.8 Singleton Model

As mentioned in Section 2.1.5, the consequent fuzzy sets B_i of a linguistic fuzzy model can be reduced to fuzzy singletons and represented as real numbers b_i:

$$R_i\colon\ \textbf{If } \mathbf{x} \text{ is } A_i \textbf{ then } y = b_i\,. \tag{2.38}$$

This model is called the singleton model. Contrary to consequent terms in the linguistic model, the number of distinct singletons in the rule base is usually not limited, i.e., each rule may have its own singleton consequent. For the singleton model, the COG defuzzification reduces to the fuzzy-mean method:

$$y_0 = \frac{\sum_{i=1}^{K} \beta_i\, b_i}{\sum_{i=1}^{K} \beta_i}\,. \tag{2.39}$$

Note that here all the K rules contribute to the defuzzification, as opposed to the method given by eq. (2.34). This means that if two rules which have the same consequent singleton are active, this singleton counts twice in the weighted mean (2.39). When using (2.34), each consequent would count only once with a weight equal to the larger of the two degrees of fulfillment. Note that the singleton model can also be seen as a special case of the Takagi–Sugeno model, presented in Section 2.3.

An advantage of the singleton model over the linguistic model is that the consequent parameters b_i can easily be estimated from data, using least-squares techniques. The singleton fuzzy model belongs to a general class of general function approximators, called the *basis functions expansion*, (Friedman, 1991) taking the form:

$$y = \sum_{i=1}^{K} \phi_i(\mathbf{x}) b_i\,. \tag{2.40}$$

Radial basis function networks, or spline models also belong to this class of systems. The connections between these types of nonlinear models have been investigated by many researchers, see (Jang and Sun, 1993; Brown and Harris, 1994) among others. In the singleton model, the basis functions $\phi_i(\mathbf{x})$ are given by the (normalized) degrees of fulfillment of the rule antecedents, see, for instance, eq. (2.10), and the constants b_i are the consequents. This model has the general function approximation property, i.e., it can represent a large class of nonlinear systems (Kosko, 1994; Zeng and Singh, 1994a; Zeng and Singh, 1995b). Multilinear interpolation between the rule consequents is obtained if the following conditions hold (Jager, 1995):

1. The antecedent membership functions are triangular and form the partition.

2. The product t-norm is used to represent the logical **and** connective.

3. The fuzzy-mean defuzzification (2.39) is used.

Such a singleton model can be exactly inverted to provide a control law based on the inverse of the process model. The inversion method is presented in Section 6.1.2.

2.2 Fuzzy Relational Models

Fuzzy relational models (Pedrycz, 1985; Pedrycz, 1993) encode associations between linguistic terms defined in the system's input and output domains by using fuzzy relations. The individual elements of the relation represent the strength of association between the fuzzy sets. As a simple example, assume a static model with one input $x \in X$ and one output $y \in Y$. Denote $\mathcal{A}$ a collection of M linguistic terms (fuzzy sets) defined on domain X, and $\mathcal{B}$ a collection of N fuzzy sets defined on Y:

$$\begin{aligned} \mathcal{A} &= \{A_1, A_2, \ldots, A_M\}, \\ \mathcal{B} &= \{B_1, B_2, \ldots, B_N\}\,. \end{aligned}$$

As depicted in Figure 2.11, a fuzzy relation $R = [r_{ij}] \in [0,1]^{M\times N}$ defines a mapping: $R\colon \mathcal{A} \to \mathcal{B}$, where each A_i is related to each B_j, with a strength given by the element r_{ij} of the relation.

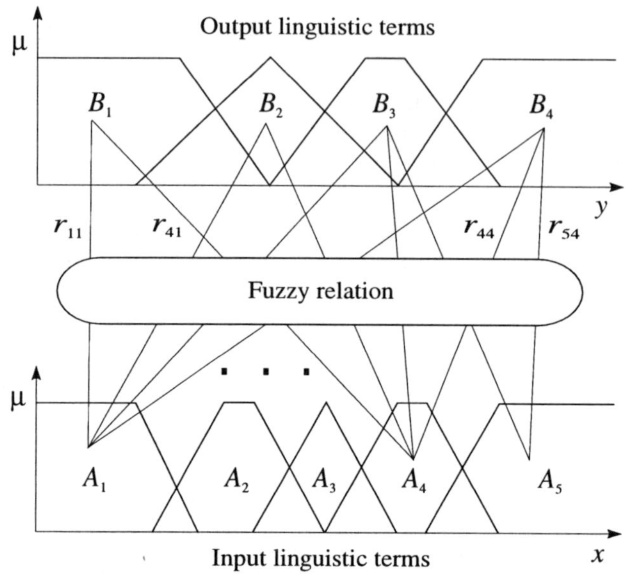

Figure 2.11. Fuzzy relation as a mapping from input to output linguistic terms.

It should be stressed that the relation R in fuzzy relational models is different from the relation (2.12) encoding fuzzy if–then rules. The latter relation is a multidimensional membership function defined in the product space of the input and output domains. Each element of this relation represents the degree of association between the individual *crisp* elements in the antecedent and consequent domains. In fuzzy relational models, however, the fuzzy relation represents associations between the individual *fuzzy sets* defined in the input and output domains of the model. It is, in fact, a table storing the rule base in which all the antecedents are related to all the consequents with different weights.

The inference in fuzzy relational models proceeds as follows. For a crisp input x, a fuzzy set X, given by

$$X = [\mu_{A_1}(x), \mu_{A_2}(x), \dots \mu_{A_M}(x)], \tag{2.41}$$

represents the degree to which x is compatible with the input terms. The corresponding output fuzzy set $Y = [\mu_1, \mu_2, \dots \mu_N]$ is derived by the max-t composition:

$$Y = X \circ R\,. \tag{2.42}$$

The crisp output of the fuzzy relational model y_0 is calculated using the weighted mean:

$$y_0 = \frac{\sum_{j=1}^{N} \mu_j\, b_j}{\sum_{j=1}^{N} \mu_j}, \tag{2.43}$$

where $b_j = \text{cog}(B_j)$ are the centroids (means) of fuzzy sets B_j. In the MIMO case, the sets X and Y are multidimensional fuzzy sets. The main advantage of the relational model is that the input–output mapping can be fine-tuned without changing the consequent fuzzy sets (linguistic terms). In the linguistic model, the outcomes of the individual rules are restricted to the grid given by the centroids of the output fuzzy sets, which is not the case in the relational model, see Figure 2.12.

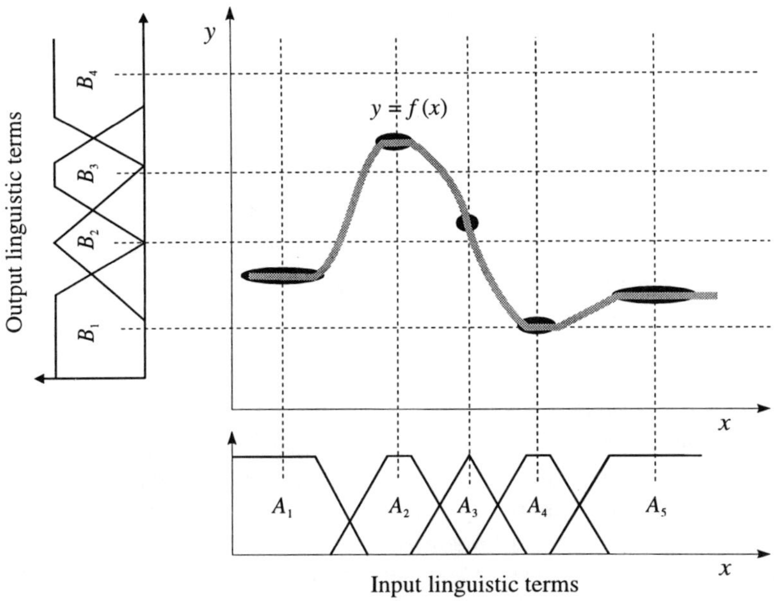

Figure 2.12. An input–output mapping of a fuzzy relational model.

For this additional degree of freedom, one pays by having more free parameters (elements in the relation), which poses problems in identification. Moreover, if no constraints are imposed on these parameters, several elements in a row of R can be nonzero, which may hamper the interpretation of the model. Furthermore, the shape of the output fuzzy sets has no influence on the resulting defuzzified value, since only centroids of these sets are considered in defuzzification.

It is easy to verify that if the antecedent fuzzy sets form a partition and the bounded-sum–product composition is used, a relational model can be computationally replaced by an equivalent model with singleton consequents (Voisin, et al., 1995). If also the consequent membership functions form a partition, a singleton model can be expressed as an equivalent relational model by computing the membership degrees of the singletons in the consequent fuzzy sets B_j. These membership degrees then become elements of the fuzzy relation:

$$R = \begin{bmatrix} \mu_{B_1}(b_1) & \mu_{B_2}(b_1) & \dots & \mu_{B_N}(b_1) \\ \mu_{B_1}(b_2) & \mu_{B_2}(b_2) & \dots & \mu_{B_N}(b_2) \\ \vdots & \vdots & \vdots & \vdots \\ \mu_{B_1}(b_K) & \mu_{B_2}(b_K) & \dots & \mu_{B_N}(b_K) \end{bmatrix}, \tag{2.44}$$

Note that a linguistic model can be regarded as a special case of the fuzzy relational model, with R being a crisp relation constrained such that only one nonzero element is allowed in each row of R (each rule has only one consequent). These relationships are exploited in the identification of relational and singleton models, presented in Section 5.2.

Example 2.4 Consider a univariate static mapping $y = f(x)$ described by the following fuzzy relation:

$$R = \begin{bmatrix} 0.8 & 0.1 & 0.0 \\ 0.6 & 0.2 & 0.0 \\ 0.5 & 0.7 & 0.0 \\ 0.0 & 0.1 & 0.9 \end{bmatrix} . \tag{2.45}$$

The elements r_{ij} describe the associations between the fuzzy sets A_i, $i = 1, 2, 3, 4$, defined on X and the fuzzy sets B_j, $j = 1, 2, 3$, defined on Y. In terms of rules, this relation can be interpreted as:

If x is A_1 **then** y is B_1 (0.8), y is B_2 (0.1), y is B_3 (0.0) .
If x is A_2 **then** y is B_1 (0.6), y is B_2 (0.2), y is B_3 (0.0) .
If x is A_3 **then** y is B_1 (0.5), y is B_2 (0.7), y is B_3 (0.0) .
If x is A_4 **then** y is B_1 (0.0), y is B_2 (0.1), y is B_3 (0.9) .

The numbers in parentheses are the respective elements of R, which can be regarded as weights of the different consequents. An approximate linguistic model can be obtained from this relational model simply by considering only the consequents with the largest weight factor (maxima over the rows of R).

Assuming that the antecedent fuzzy sets form a partition and the bounded-sum–product composition is used, the above relational model can be replaced by the following singleton model:

If x is A_1 **then** $y = (0.8b_1 + 0.1b_2)/(0.8 + 0.1)$,
If x is A_2 **then** $y = (0.6b_1 + 0.2b_2)/(0.6 + 0.2)$,
If x is A_3 **then** $y = (0.5b_1 + 0.7b_2)/(0.5 + 0.7)$,
If x is A_4 **then** $y = (0.1b_2 + 0.9b_3)/(0.1 + 0.9)$,

where b_j are defuzzified values of the fuzzy sets B_j, $b_j = \text{cog}(B_j)$. □

The mapping represented by a fuzzy relation is static. In order to model dynamic systems, the usual way of input–output or state-space modeling is applied, see Section 4.2. For instance, a relational system representing a NARX model is given by:

$$Y(k+1) = [U(k) \times \cdots \circ U(k - n_u) \times Y(k) \times \cdots \times Y(k - n_y)] \circ R, \tag{2.46}$$

where $U(k)$ is a fuzzy representation of the crisp or fuzzy input $u(k)$ at time instant k with respect to the reference fuzzy sets defined in the input universe U, similar to eq. (2.41). $Y(k)$ is the fuzzy representation of the output $y(k)$ with respect to reference fuzzy sets defined in the output universe Y, and R is a multidimensional fuzzy relation:

$$R\colon \underbrace{U \times \cdots \times U}_{n_u \text{ times}} \times \underbrace{Y \times \cdots \times Y}_{n_y \text{ times}} \to [0, 1] . \tag{2.47}$$

2.3 Takagi–Sugeno Models

A fuzzy rule-based model suitable for the approximation of a large class of nonlinear systems was introduced by Takagi and Sugeno (1985). In the Takagi–Sugeno (TS) fuzzy model, the rule consequents are crisp functions of the model inputs:

$$R_i\text{: } \textbf{If } \mathbf{x} \text{ is } A_i \textbf{ then } y_i = f_i(x), \quad i = 1, 2, \ldots, K, \tag{2.48}$$

where $\mathbf{x} \in \mathbb{R}^p$ is the input (antecedent) variable and $y_i \in \mathbb{R}$ is the output (consequent) variable. R_i denotes the ith rule, and K is the number of rules in the rule base. A_i is the antecedent fuzzy set of the ith rule, defined by a (multivariate) membership function:

$$\mu_{A_i}(\mathbf{x})\colon \mathbb{R}^p \rightarrow [0, 1]\,. \tag{2.49}$$

As in the linguistic model, the antecedent proposition "$\mathbf{x}$ is A_i"is usually expressed as a logical combination of simple propositions with univariate fuzzy sets defined for the individual components of $\mathbf{x}$, often in the *conjunctive form*:

$$\begin{aligned} R_i\text{:} \qquad & \textbf{If } x_1 \text{ is } A_{i1} \textbf{ and } x_2 \text{ is } A_{i2} \textbf{ and } \ldots \textbf{ and } x_p \text{ is } A_{ip} \\ & \textbf{then } y_i = f_i(x), \quad i = 1, 2, \ldots, K\,. \end{aligned} \tag{2.50}$$

The consequent functions f_i are typically chosen as instances of a suitable parameterized function, whose structure remains equal in all the rules and only the parameters vary. A simple and practically useful parameterization is the affine linear form:

$$y_i = \mathbf{a}_i^T \mathbf{x} + b_i, \tag{2.51}$$

where $\mathbf{a}_i$ is a parameter vector and b_i is a scalar offset. This model we here call an *affine TS model.* The consequents of the affine TS model are hyperplanes (p-dimensional linear subspaces) in $\mathbb{R}^{p+1}$. Figure 2.13 shows an example of a univariate function $y = f(x)$ represented by three affine TS rules. The antecedent of each rule defines a (fuzzy) validity region for the corresponding affine (linear) consequent model. The global model is composed as a concatenation of the local models, and can be seen as a smoothed piece-wise approximation of a nonlinear surface. Approximation properties of the affine TS model were investigated by Rovatti (1996).

A special case of the consequent function occurs when $b_i = 0,\ i = 1, \ldots, K$. Then the model is called a *homogeneous* TS model:

$$\textbf{If } \mathbf{x} \text{ is } A_i \textbf{ then } y_i = \mathbf{a}_i^T \mathbf{x}, \quad i = 1, 2, \ldots, K\,. \tag{2.52}$$

This model has more limited approximation capabilities than the affine TS model (Fantuzzi and Rovatti, 1996). However, the absence of the offset term facilitates controller design and stability analysis based on the homogeneous TS systems (Tanaka and Sugeno, 1992; Wang, et al., 1995; Zhao, 1995), because the model can be analyzed in a quasi-linear fashion.

When $\mathbf{a}_i = \mathbf{0},\ i = 1, \ldots, K$, the consequents in model (2.51) are constant functions, and the *singleton model*, is obtained:

$$\textbf{If } \mathbf{x} \text{ is } A_i \textbf{ then } y_i = b_i, \quad i = 1, 2, \ldots, K\,. \tag{2.53}$$

This model can also be seen as a special case of the linguistic fuzzy model, where the consequent fuzzy sets reduce to singletons, see Section 2.1.8.

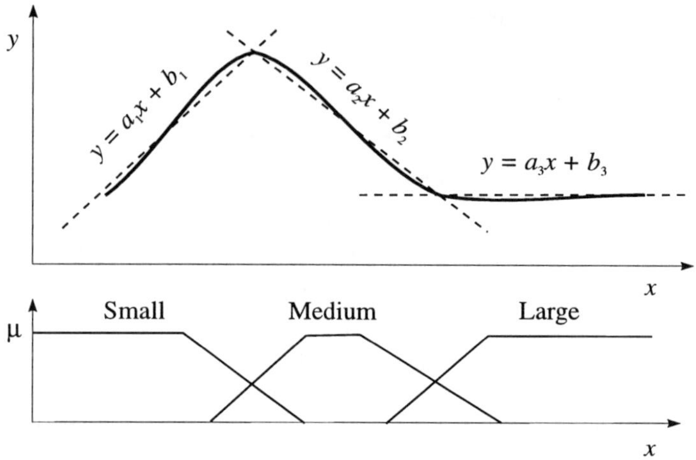

Figure 2.13. Smoothed piece-wise linear approximation with three affine TS fuzzy rules.

2.3.1 *Inference in the TS Model*

Before the output can be inferred, the *degree of fulfillment* of the antecedent denoted $\beta_i(x)$ must be computed. For rules with multivariate antecedent fuzzy sets given by (2.48) and (2.49), the degree of fulfillment is simply equal to the membership degree of the given input $\mathbf{x}$, i.e., $\beta_i = \mu_{A_i}(\mathbf{x})$. When logical connectives are used, the degree of fulfillment of the antecedent is computed as a combination of the membership degrees of the individual propositions using the fuzzy logic operators, see Section 2.1.2.

In the Takagi–Sugeno model, the inference is reduced to a simple algebraic expression, similar to the fuzzy-mean defuzzification formula (Takagi and Sugeno, 1985):

$$y = \frac{\sum_{i=1}^{K} \beta_i(\mathbf{x}) y_i}{\sum_{i=1}^{K} \beta_i(\mathbf{x})} . \tag{2.54}$$

Regarding the consequents in the TS model as input-dependent singletons, expression (2.54) can be seen as a straightforward extension of the fuzzy-mean defuzzification, see Section 2.1.5. By denoting the normalized degree of fulfillment

$$\lambda_i(\mathbf{x}) = \frac{\beta_i(\mathbf{x})}{\sum_{j=1}^{K} \beta_j(\mathbf{x})}, \tag{2.55}$$

the affine TS model with a common consequent structure can be expressed as a pseudo-linear model with input-dependent parameters:

$$y = (\sum_{i=1}^{K} \lambda_i(\mathbf{x}) \mathbf{a}_i^T)\mathbf{x} + \sum_{i=1}^{K} \lambda_i(\mathbf{x}) b_i = \mathbf{a}^T(\mathbf{x})\mathbf{x} + b(\mathbf{x}) . \tag{2.56}$$

The parameters $\mathbf{a}(\mathbf{x})$, $b(\mathbf{x})$ are convex linear combinations of the consequent parameters $\mathbf{a}_i$ and b_i, i.e.:

$$\mathbf{a}(\mathbf{x}) = \sum_{i=1}^{K} \lambda_i(\mathbf{x})\mathbf{a}_i, \quad b(\mathbf{x}) = \sum_{i=1}^{K} \lambda_i(\mathbf{x})b_i. \tag{2.57}$$

This property facilitates the analysis of TS models in the framework of *polytopic systems* (Boyd, et al., 1994). A TS model can be regarded as a mapping from the antecedent (input) space to a convex region (polytope) in the space of the parameters of a quasi-linear system (2.56), as schematically depicted in Figure 2.14.

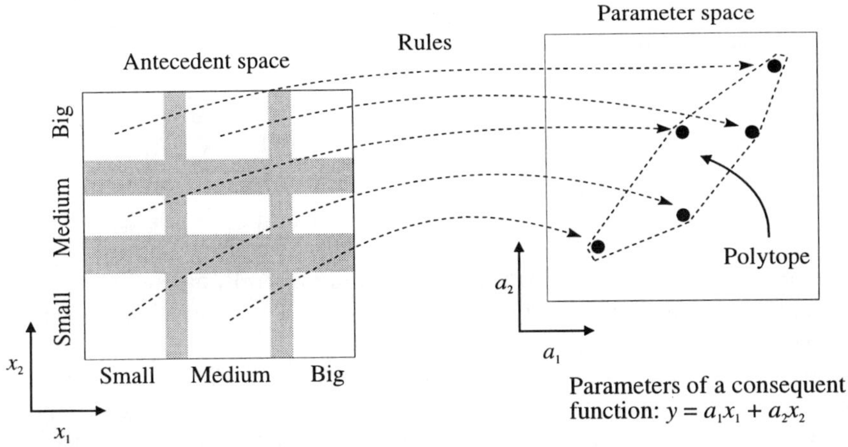

Figure 2.14. The TS model can be regarded as a mapping from the antecedent space to the space of the consequent parameters.

Consider, for instance, a dynamic system described by the following TS rules:

$$\begin{aligned} R_i: \quad & \textbf{If } y(k) \text{ is } A_{i1} \textbf{ and } y(k-1) \text{ is } A_{i2} \textbf{ and } \ldots\ y(k-n_y+1) \text{ is } A_{in_y} \\ & \textbf{and } u(k) \text{ is } B_{i1} \textbf{ and } u(k-1) \text{ is } B_{i2} \textbf{ and } \ldots\ u(k-n_u+1) \text{ is } B_{in_u} \\ & \textbf{then } y(k+1) = \sum_{j=1}^{n_y} a_{ij} y(k-j+1) + \sum_{j=1}^{n_u} b_{ij} u(k-j+1), \end{aligned} \tag{2.58}$$

where the consequents are linear ARX models (n_u and n_y are integers related to the order of the system). An alternative representation can be obtained by applying z-transform to these ARX models (for zero initial conditions) and by using eq. (2.56):

$$P(z, \boldsymbol{\lambda}) = \frac{\mathbf{a}(z, \boldsymbol{\lambda})}{\mathbf{b}(z, \boldsymbol{\lambda})} = \frac{\sum_{i=1}^{K} \lambda_i \mathbf{b}_i(z)}{\sum_{i=1}^{K} \lambda_i \mathbf{a}_i(z)}, \tag{2.59}$$

where $\boldsymbol{\lambda} = [\lambda_1, \lambda_2, \ldots, \lambda_K]^T$ is the vector of normalized membership degrees given by eq. (2.55). The numerator and denominator of $P(z, \boldsymbol{\lambda})$ are convex sums of polynomials (polytopes). For some specific forms of these models, methods have been developed to design controllers with desired closed loop characteristics (Filev, 1996) and to analyze their stability (Tanaka and Sugeno, 1992; Zhao, 1995; Tanaka, et al., 1996). Even though the linear convex combination of the consequents (2.56) has certain advantages for the analysis of TS fuzzy systems in the framework of polytopic systems, it has also some undesirable properties for function approximation, as shown in the following section.

2.3.2 *Analysis of the TS Inference*

Takagi–Sugeno fuzzy rules in the conjunctive form (2.50) are considered. Let $C_{kj} = \text{core}(A_{kj})$ denote the core of the fuzzy set A_{kj}. In order to analyze the interpolation mechanisms on the basis of pairs of adjacent rules, the following conditions on the antecedent membership functions are assumed:

$$
\begin{aligned}
&A_{kj} \text{ are normal and convex,} && 1 \le k \le N_j, && \text{(2.60a)}\\
&\max(C_{kj}) < \min(C_{k+1,j}), && 1 \le k \le N_j-1, && \text{(2.60b)}\\
&\mu_{A_{kj}}(x_j) \wedge \mu_{A_{k+1,j}}(x_j) > 0, && 1 \le k \le N_j-1, && \text{(2.60c)}\\
&\mu_{A_{kj}}(x_j) \wedge \mu_{A_{k+2,j}}(x_j) = 0, && 1 \le k \le N_j-2\,. && \text{(2.60d)}
\end{aligned}
$$

Condition (2.60b) ensures that the cores are disjoint intervals, ordered in the domain of x_j, and conditions (2.60c) and (2.60d) state that only pairs of adjacent fuzzy sets have nonempty intersections.

In this section, only affine TS models are considered. The reason is that the consequents of homogeneous TS rules (i.e., rules with $b_i = 0, \forall i$) intersect only at the origin and in linear subspaces containing the origin. As such, homogeneous TS models are generally less suitable for function approximation, since they cannot approximate both the value of a function and its derivative for $\mathbf{x} \in \text{core}(A_j)$. Setting $b_i = 0$ in eq. (2.56) yields:

$$
y = a_j^T \mathbf{x}, \tag{2.61}
$$

$$
\frac{\mathrm{d}y}{\mathrm{d}\mathbf{x}} = [a_{j1}, \ldots, a_{jd}]^T. \tag{2.62}
$$

If a_j in (2.61) is fixed in order to approximate y, no degrees of freedom are left to approximate the derivative (2.62). Hence, affine TS models are better suited for function approximation.

To demonstrate the interpolation properties of the TS inference, consider a simple affine TS model with a scalar input x and two rules:

$$
\begin{aligned}
R_1:&\quad \textbf{If } x \text{ is } A_1 \textbf{ then } y_1 = a_1 x + b_1,\\
R_2:&\quad \textbf{If } x \text{ is } A_2 \textbf{ then } y_2 = a_2 x + b_2.
\end{aligned} \tag{2.63}
$$

The antecedent fuzzy sets A_1, A_2 are defined by trapezoidal membership functions shown in the bottom part of Figures 2.14a and 2.15b. In this example, the membership degrees sum up to one for all the domain elements, which is not required; however, it makes the following analysis more transparent. Denote by $S_{i,j} = \text{supp}(A_i \cap A_j)$ the support of the intersection of A_i and A_j. In Figure 2.15, $S_{1,2}$ is deliberately chosen much larger than the cores C_1 and C_2 to emphasize the interpolation properties. The linear consequent models y_1 and y_2 are assumed to be a good approximation of f for $x \in C_1$ and $x \in C_2$ respectively.

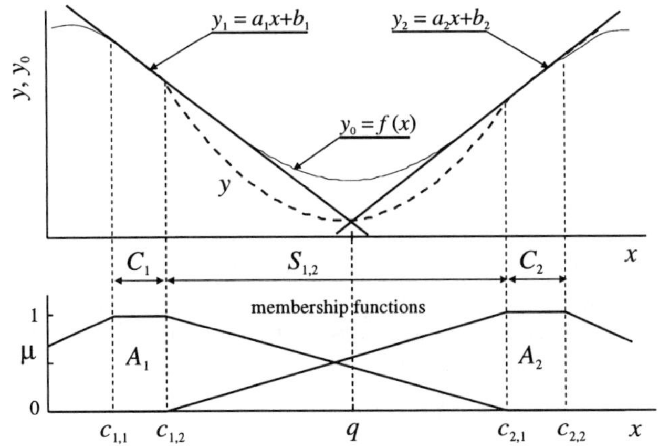

(a) V–type interpolation.

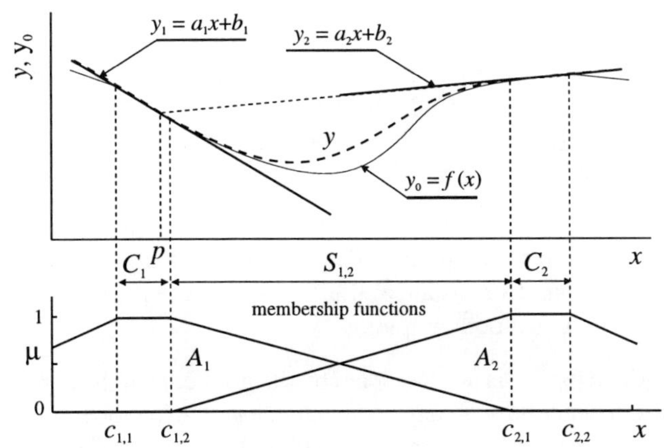

(b) S–type interpolation.

Figure 2.15. Different types of interpolation between two TS rules. The graphs present the approximated function $y_0 = f(x)$ (solid line), the affine consequents y_1 and y_2 (solid bold line) and the global output of the TS model y (dashed line) obtained from eq. (2.54). Reproduced from (Babuška, et al., 1996), ©1996 IEEE.

From the comparison of Figure 2.14a and Figure 2.15b, one can see that two different situations arise, depending on the way the TS model approximates functions:

1. Figure 2.14a shows what we denote the *"V–type"* rules, where the consequent models are facets of the function being approximated, and their intersection q projects into $S_{1,2}$. The nonlinear function f is assumed to be convex or concave on $S_{1,2}$.

2. Figure 2.15b illustrates what we denote the *"S–type"* rules. Here the local models intersect outside $S_{1,2}$ and the function is neither convex nor concave in this region.

For the S–type rules, the weighted mean (2.54) provides intuitively expected results, since the interpolant remains between the consequent models. In the limiting crisp case, y is a step-like function, and the a fuzzier the transition between A_1 and A_2 is, the more gradual is the transition from y_1 to y_2. For the V–type rules, however, the weighted mean (2.54) introduces some undesirable properties:

- The fuzziness of the antecedent introduces deviations of the model output from the function being approximated (we refer to this deviation as an *approximation error*). This can result in a decrease of accuracy, or in a biased estimate of the consequent parameters, if the model is tuned to the data using least-squares techniques, see Figure 2.16. In the limiting crisp case,

$$\mu_{A_1}(x) = \begin{cases} 1 & x < q, \\ 0 & \text{otherwise} \end{cases} \tag{2.64a}$$

$$\mu_{A_2}(x) = 1 - \mu_{A_1}(x), \quad \forall x\,. \tag{2.64b}$$

  y is the best approximation of y_0 and with increasing fuzziness of the transition between A_1 and A_2, the deviation between y and y_0 increases, which somewhat contradicts the benefits of using fuzzy sets in the model. Note, also, that the approximation error depends on the slope of the consequents and tends to large values as the slope approaches $\pi/2$.

- The smoothness of the model output depends directly on the smoothness of the antecedent membership functions. This restricts the choice of the type of the membership functions (for instance, the frequently used trapezoidal membership functions result in nonsmooth outputs).

- The local gradient $\mathrm{d}y/\mathrm{d}x$ is not restricted by the gradients of the consequents. In the context of nonlinear control, it may mean that the gain of a global controller becomes *locally* negative even though the gains of all the local controllers (consequents) are positive. This may lead to an unstable behavior of the control loop in that region.

Each type of TS rule is suitable for approximating different kinds of functions. While the S–type rules are suitable for step-wise and possibly discontinuous functions, the V–type rules are suited for the approximation of continuous, smooth functions. The S–type rules can always be converted to the V–type rules by providing an additional rule between the two fuzzy sets. In the sequel, we will focus on the V–type rules, as a generalization of the conventional approach based on local linearizations of a nonlinear function.

The previous example illustrated the approximation properties when the rule consequents are fixed. In fuzzy modeling, however, the consequent parameters are often

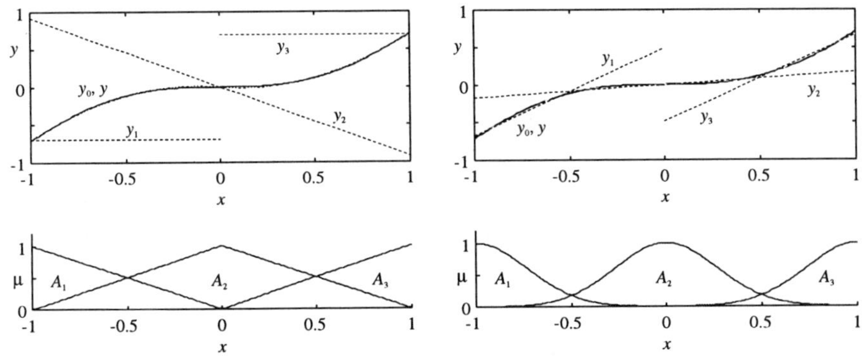

(a) Overlapping membership functions. (b) Membership functions with less overlap.

Figure 2.16. Consequents obtained by parameter estimation. The solid line is the function y_0 which is being approximated, dashed lines are the consequents y_i, and the dashed-dotted line is the output of the TS model y (almost indistinguishable from y_0).

estimated from available input–output data by minimizing the modeling error by least-squares methods, see Section 5.1.2 for details. Due to the large approximation error, the least-squares optimization may favor rule consequents that are *not* local linear approximations of the global nonlinear function. Figure 2.16a shows an example of when a good global fit is obtained (the difference between y_0 and y is barely visible), but the consequents, denoted y_1, y_2 and y_3, absolutely do not represent the local behavior of the function. Such a TS model can then hardly be interpreted in terms of the individual rules.

In general, the use of membership functions with less overlap and larger core regions is preferable to the partition with triangular functions, because the consequents are forced to represent the local behavior of the system, as depicted in Figure 2.16b. The membership functions should be carefully chosen, or a suitable identification technique must be used, to extract them from data, as described in Chapter 4.

Remark: A practically important issue is how to distinguish between S–type and V–type rules. First denote

$$H_i = \{(\mathbf{x}, y) | y = \mathbf{a}_i \mathbf{x} + b_i\} \tag{2.65}$$

the subset of all pairs $(x, y) \in \mathbb{R}^{(p+1)}$ satisfying the consequent equation of the ith rule. Further, denote

$$\Omega_{ij} = \text{proj}_{\mathbf{x}}(H_i \cap H_j) \tag{2.66}$$

the intersection of H_i and H_j projected on $\mathbf{x}$. It can be verified that a pair of affine TS rules (R_i, R_j) is of the V–type if and only if

$$\Omega_{ij} \cap S_{ij} \neq \emptyset \text{ and } \Omega_{ij} \cap (C_i \cup C_j) = \emptyset\,. \tag{2.67}$$

The meaning of this condition is that the projection of the intersection of the two hyperplanes (consequents) intersects with S_{ij} and does not intersect with either of the cores. For instance, in Figure 2.14a, $\Omega_{1,2} = q$, conditions (2.67) are satisfied, and the rules are of the V–type. However, in Figure 2.15b, $q \in C_1$ and the rules are of the S type.

2.3.3 *Alternative Interpolation Scheme for the TS Model*

The interpolation properties of the TS inference can be improved by replacing the weighted mean (2.54) by another functional (Babuška, et al., 1996). With respect to the applications of TS models to control, modeling and identification, it is useful if the inference/interpolation mechanism results in a piece-wise convex or concave interpolation surface such that:

- The gradients $\mathrm{d}y/\mathrm{d}x$ are bounded by the gradients of the rule consequents $\mathbf{a}_i$.
- The surface is smooth with continuous derivatives of a sufficient order.
- The model approximates the function more accurately than the surface generated by (2.54).

One possibility is to use the *smoothing maximum* function (Zhang, 1980), which connects two consequent hyperplanes by a smooth convex or concave surface defined by a polynomial. Recalling that the V–type rules give the best interpolation in the crisp case, it is natural to start from that situation. A crisp transition between the two TS rules as in Figure 2.14a can be formulated using the maximum function:

$$y = \max(y_1, y_2) = \max(\mathbf{a}_1^T\mathbf{x} + b_1, \mathbf{a}_2^T\mathbf{x} + b_2)\,. \tag{2.68}$$

An approximation of a concave function is achieved by taking the minimum instead of the maximum, or by $y = -\max(-y_1, -y_2)$. The type of approximation is determined from the sign of the difference $y_1 - y_2$ for some value in the core of one of the membership functions involved. In Figure 2.14a, for instance, the difference $(a_1c_{1,2} + b_1) - (a_2c_{1,2} + b_2)$ is positive, indicating a convex function, and thus the maximum function is used. The expressions derived below for the maximum formulation apply analogically to the minimum formulation. Note that the gradient of (2.68) is bounded by (equal to) the gradients of the consequent models, except for the intersection point where it is not defined. This drawback can be eliminated by smoothing the maximum function, as shown in the next section.

Smoothing Maximum. The main idea is to replace the maximum function in eq. (2.68) by some smooth approximation of it. First, the following function is introduced

$$s(z) = \max(0, z), \quad z \in \mathbb{R}\,. \tag{2.69}$$

Using this function, eq. (2.68) now can be rewritten as follows:

$$y = y_1 + s(y_2 - y_1)\,. \tag{2.70}$$

The maximum in (2.69) can be replaced by a piece-wise polynomial (Zhang, 1980; Seber and Wild, 1989):

$$s_\gamma(z) = \begin{cases} 0, & z \le -\gamma, \\ (z+\gamma)^2/4\gamma, & -\gamma < z < \gamma, \\ z, & z \ge \gamma, \end{cases} \tag{2.71}$$

where γ is positive real. The function $s_\gamma(z)$ is plotted in Figure 2.17. The interpolation formula for two TS rules now becomes:

$$y = y_1 + s_\gamma(y_2 - y_1)\,. \tag{2.72}$$

Note that the derivative of $s_\gamma(z)$ in (2.71) is continuous. Higher orders of continuity can be obtained by using higher-order polynomials. For example,

$$s_\gamma(z) = \frac{-z^4}{16\gamma^3} + \frac{3z^2}{8\gamma} + \frac{z}{2} + \frac{3\gamma}{16}, \quad -\gamma < z < \gamma \tag{2.73}$$

has a continuous second derivative. In the general case of K overlapping rules, eq. (2.72) can be applied recursively:

$$y = y_1 + s_{\gamma_1}(y_2 + s_{\gamma_2}(\ldots s_{\gamma_{K-2}}(y_{K-1} + s_{\gamma_{K-1}}(y_K - y_{K-1}))\ldots))\,. \tag{2.74}$$

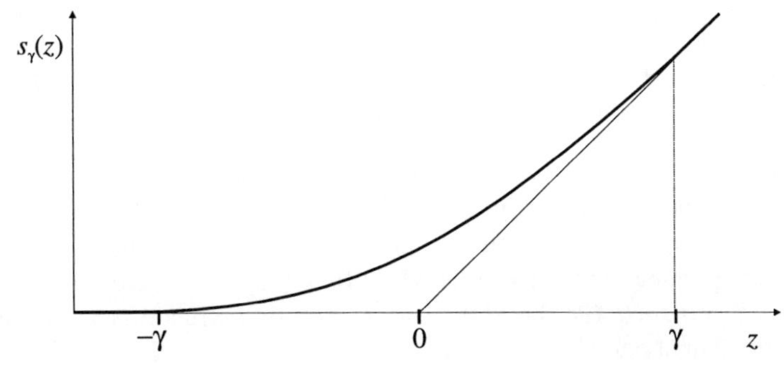

Figure 2.17. A smooth approximation of the maximum function. Reproduced from (Babuška, et al., 1996), ©1996 IEEE.

Determining γ. Rather than weighting by the degrees of fulfillment, the described interpolation scheme uses a set of smoothing parameters γ_i, $i = 1, 2, \ldots, K-1$, each for a pair of adjacent rules. From eq. (2.71) and (2.72) and from Figure 2.17, one can see that the interpolant starts to deviate from the consequent y_1 when $y_2 - y_1 = \gamma$. When using the antecedent membership functions and eq. (2.54), the same happens when x leaves the core of the membership function. Thus in the univariate case, for a pair of rules (R_i, R_{i+1}), γ_i can be directly obtained from the cores of the antecedent membership functions by:

$$\gamma_i = \frac{1}{2}(|(a_i^T c_{i,2} + b_i) - (a_{i+1}^T c_{i,2} + b_{i+1})| + |(a_i^T c_{i+1,1} + b_i) - (a_{i+1}^T c_{i+1,1} + b_{i+1})|), \qquad (2.75)$$

where $c_{i,1} = \min(C_i)$ and $c_{i,2} = \max(C_i)$, see also Figure 2.14a. Since eq. (2.72) always results in a symmetrical function, an average is taken from C_i and C_{i+1} in (2.75). γ can also be determined by optimization, especially when a high degree of accuracy is required, or when the antecedent membership functions are not reliable. This assumes, of course, that some data are available to measure the performance of the model.

For systems with multiple inputs, it is assumed that the rules are in the conjunctive form (2.50), and hence eq. (2.72) can be applied recursively for each component of x, as in the univariate case. Then, the partial results of the individual antecedent variables are combined. In a general multivariate case, the determination of the nesting order of rules in the recursion (2.74) is much more complicated, as it is not trivial to determine which rules are adjacent.

Example 2.5 This example demonstrates the smoothness and convexity properties of the described interpolation method, using two pre-defined TS rules. Figure 2.18a illustrates the difference between the conventional TS interpolation (2.54) and the interpolation (2.72), which uses the smoothing function (2.73). As shown in the bottom graph in Figure 2.18b, the gradient $\Delta y / \Delta x$ is not only bounded by the gradients of the consequents but is also a smooth function of x. The upper graph in Figure 2.18b shows that the gradient of the standard TS interpolant is a discontinuous function and is also not bounded by the consequents. The parameter $\gamma = 4$ was determined from the cores of the membership functions by using eq. (2.75). □

Example 2.6 A TS model is developed for the function $z = (x-1)^5 + (y-1)^5$ from 400 data points equidistantly spaced in $[-1,1] \times [-1,1]$). The membership functions were obtained by fuzzy clustering using the GK algorithm (see Chapter 4). The consequent parameters were estimated by fuzzy regression clustering, described in Section 3.5.3. Figure 2.19a shows the surface plot of the function and Figure 2.19b depicts the antecedent partition obtained.

As shown in Figure 2.20, the standard TS inference produces an undesired ridge on the approximated function surface, while the smoothing-maximum method results in a smooth surface with continuous second derivatives. Moreover, the RMS error is reduced from 3.4724 to 1.3927 (with the same consequent parameters). In this example, $\gamma = 7.7$ was obtained by numerical optimization. The same value of γ was used for all pairs of rules, since the overlap of the antecedent membership functions is approximately equal, as can be seen from Figure 2.19b. □

It can be seen that the interpolation based on the smoothing maximum is superior to the conventional TS interpolation formula (2.54), as it provides convex and smooth surfaces and it improves the numerical accuracy of the TS model. A limitation of this approach is that it can only be applied to a TS model where the rules are in the

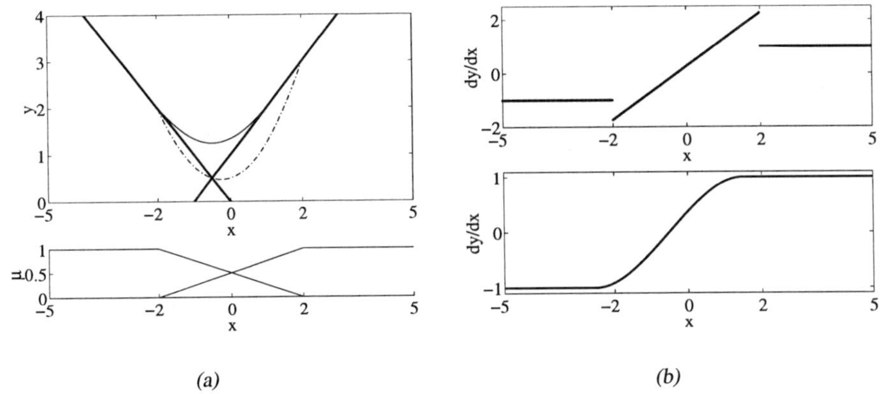

Figure 2.18. Convex interpolation of two TS rules: (a) shows the rule consequents (solid bold line), the conventional TS interpolation (dashed line) and the smoothing maximum interpolation (solid line). The bottom plot depicts the antecedent membership functions. Figure (b) shows the gradient $\Delta y/\Delta x$ for the original TS model (top) and of the smoothing maximum model (bottom). Reproduced from (Babuška, et al., 1996), ©1996 IEEE.

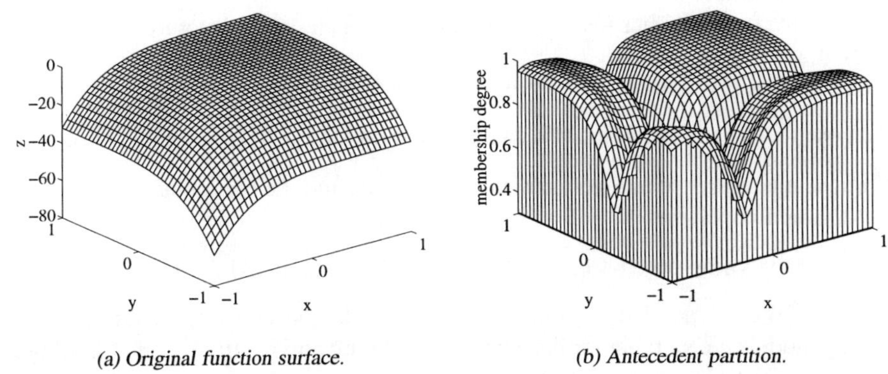

Figure 2.19. Function $z = (x-1)^5 + (y-1)^5$ and the antecedent partition of a TS fuzzy model obtained by fuzzy clustering. Reproduced from (Babuška, et al., 1996), ©1996 IEEE.

conjunctive form and where the consequents are facets to piece-wise convex or concave system surfaces.

2.4 Constructing Fuzzy Models

In the previous sections, the structures and the inference mechanisms of different classes of fuzzy models were reviewed. This section deals with the construction of

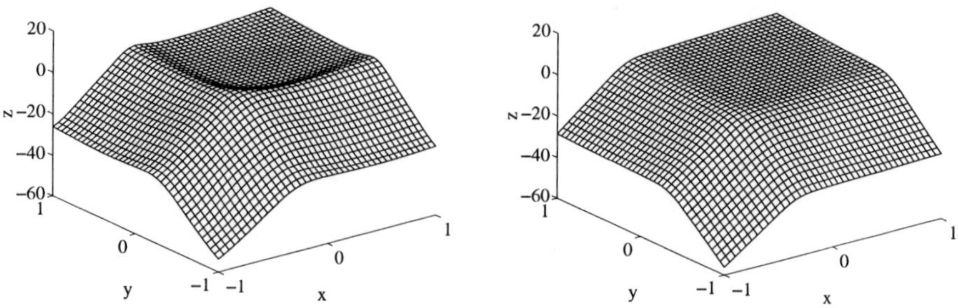

(a) Standard TS inference.

(b) Smoothing maximum.

Figure 2.20. The output of the TS model obtained with the standard fuzzy-mean method and the smoothing maximum method. Reproduced from (Babuška, et al., 1996), ©1996 IEEE.

fuzzy models. The aim is to outline the basic ideas and concepts, rather than give a complete survey of the different techniques. In this section, it is assumed that the structure of the model, i.e., the input and output variables have been determined. In the case of dynamic systems, it also means that the representation of the dynamics within the fuzzy model has been chosen. For more details on structure selection see Section 4.2.

To develop a fuzzy model, the following main decisions must made:

1. Choose the type of fuzzy model (Takagi–Sugeno, linguistic, relational, etc.).

2. Choose the inference and defuzzification methods and the particular fuzzy logic or set-theoretic operators.

3. Develop the knowledge base, i.e., the rules and the membership functions.

The choice of the type of fuzzy model depends on the particular application. While TS fuzzy models are suitable for the accurate modeling and identification of nonlinear systems, linguistic fuzzy models give a more qualitative description of the process under study. The level on which these two kinds of models can be interpreted is also different. A TS model with affine linear consequent can be interpreted in terms of local linear models of the system, and in terms of changes of the model parameters with respect to the antecedent variables. These relationships can give valuable insights into the nonlinearities, and can be exploited in the controller design. This information can be valuable for control engineers, but it may be difficult to understand for non-experts, such as process operators. Linguistic models, however, are interpreted by using natural language and qualitative terms. Such a description is more compatible with common sense knowledge, which is difficult to formalize mathematically, but may be less suitable for control design and mathematical analysis. Therefore, it might be useful to develop several models of different types for the same process, each of them serving a different purpose.

The choice of the inference and defuzzification methods in a given fuzzy model depends mainly on the techniques which are used to construct the model. Fuzzy models developed entirely from human knowledge are similar to fuzzy expert systems. In order to make an effective use of the available information, standard inference methods based on fuzzy implications or t-norms can be involved. When applying learning and adaptation techniques, however, inference mechanisms which make use of differentiable operators and membership functions are preferable, since the model can be tuned by gradient optimization methods.

The development of the rule base and of the membership functions can be based on prior knowledge, numerical measurements, or a combination of the two.

2.4.1 Knowledge-based Approach

Prior knowledge about the system under study can be in different forms. A main distinction can be made between formal knowledge in terms of a mechanistic description of the system and the (semi-)qualitative knowledge of experts, which is difficult to formalize. Both kinds of information can be incorporated in a fuzzy model. Section 7.1 gives an example of a linguistic model based primarily on expert knowledge, which is imprecise in its nature and hardly can be expressed in terms of mathematical equations.

Available process knowledge can also be used to characterize a complex nonlinear system as a collection of simpler, for instance, linear subsystems which are valid only in certain operating regimes. Such information can be encoded in the form of TS fuzzy rules. The variables which characterize the change of the operating regimes become antecedent variables, and membership functions are defined to specify the relevancy of each model for the given region (Murray-Smith and Johansen, 1997). The local models are then given by the consequents of the rules, and the global output is obtained as the weighted mean of the individual contributions. The consequent parameters can be estimated from data, or can be computed by the linearization of a known nonlinear white-box model. This approach can be illustrated using the well-known problem of balancing a pole on a cart (Wang, et al., 1995).

Example 2.7 The differential equations for the pole motion are:

$$\begin{aligned} \dot{x}_1 &= x_2 \\ \dot{x}_2 &= \frac{g\sin(x_1) - amlx_2^2\sin(2x_1)/2 - a\cos(x_1)u}{4l/3 - aml\cos^2(x_1)}, \end{aligned} \tag{2.76}$$

where x_1 and x_2 denote the angle and angular velocity of the pole respectively, g is the gravity constant, l is the distance from the pivot to the pole's center of mass, u is the force applied to the cart, and $a = 1/(m + M)$, where m is the mass of the pole and M the mass of the cart.

A simple homogeneous TS model for $x_1 \in (-\pi/2, \pi/2)$ can be designed by linearizing (2.76) around $|x_1| = 0$ and $|x_1| = \pi/2$. The corresponding TS rules are:

$$\begin{aligned} &\textbf{If } |x_1| \text{ is A}\text{ROUND }\text{Z}\text{ERO } \textbf{ then } \dot{\mathbf{x}} = \mathbf{A}_1\mathbf{x} + \mathbf{B}_1 u \\ &\textbf{If } |x_1| \text{ is } \textbf{not} \text{ A}\text{ROUND }\text{Z}\text{ERO } \textbf{ then } \dot{\mathbf{x}} = \mathbf{A}_2\mathbf{x} + \mathbf{B}_2 u \end{aligned} \tag{2.77}$$

The fuzzy set AROUND ZERO can be defined by a triangular membership function: $\mu(x_1) = 1 - 2|x_1|/\pi$. The membership degree of the complement **not** AROUND ZERO is $1 - \mu(x_1)$. Since $\pm\pi/2$ is not an equilibrium point, linearization of (2.76) around this point would give additional constant terms in (2.77), and hence an affine TS model would be obtained. Since this model is less suitable for controller design and analysis than the linear state-space TS model (2.77), Wang, et al. (1995) use the following approximation of the matrices $\mathbf{A}_1$, $\mathbf{B}_1$, $\mathbf{A}_2$ and $\mathbf{B}_2$:

$$\mathbf{A}_1 = \begin{bmatrix} 0 & 1 \\ g/(4l/3 - aml) & 0 \end{bmatrix}, \qquad \mathbf{B}_1 = \begin{bmatrix} 0 \\ a/(4l/3 - aml) \end{bmatrix},$$

$$\mathbf{A}_2 = \begin{bmatrix} 0 & 1 \\ 2g/(\pi(4l/3 - amlc^2)) & 0 \end{bmatrix}, \quad \mathbf{B}_2 = \begin{bmatrix} 0 \\ ac/(4l/3 - amlc^2) \end{bmatrix},$$

where $c = \cos(\pi/1.95)$ instead of $c = \cos(\pi/2)$, in order to obtain a controllable model. Derivation of matrices $\mathbf{B}_1$ and $\mathbf{B}_2$ follows directly from the linearization of (2.76) around $|x_1| = 0$ and $|x_1| = \pi/2$, respectively. Element $(a_1)_{2,1}$ is obtained by neglecting the second term in the numerator of (2.76). This is clearly an approximation, since this term can be neglected only for small angular velocities. Linearization of (2.76) around $\pi/2$ results in a constant term only, say c_2, where $c_2 = g/(4l/3 - amlc^2)$. For $|x_1| = \pi/2$ we can substitute c_2 and x_1 into the state equation $\dot{x}_2 = (a_2)_{2,1}x_1$ to obtain $c_2 = (a_2)_{2,1}\pi/2$ and thus $(a_2)_{2,1} = 2c_2/\pi$. This is again an approximation valid in the close neighborhood of $|x_1| = \pi/2$. As shown by Wang, et al. (1995), this model can be extended to the entire working range $x_1 \in [-\pi, \pi]$ by using four rules. A feedback Takagi-Sugeno controller with guaranteed stability that uses the same antecedents as the model can be designed. The consequents are control laws of the form $u = -\mathbf{P}_i\mathbf{x}$, $i = 1, \ldots, 4$, where the $\mathbf{P}_i$s are designed by pole placement. □

2.4.2 Data-driven Methods

Building fuzzy models from data involves methods based on fuzzy logic and approximate reasoning, but also ideas originating from the areas of neural networks, data analysis and conventional system identification. The strong potential of fuzzy models lies in their ability to combine information obtained from measured data with heuristic knowledge expressed in the form of rules. Two main approaches to the integration of knowledge and data in a fuzzy model can be distinguished:

1. The expert knowledge formulated as a collection of if–then rules and the associated membership functions. This initial model can be fine-tuned by using numerical data. The particular tuning algorithms exploit the fact that, at the computational level, a fuzzy model can be seen as a layered structure (network), similar to artificial neural networks, to which standard learning algorithms can be applied. This approach is usually called neuro-fuzzy modeling (Jang, 1993; Jang and Sun, 1993; Brown and Harris, 1994; Pedrycz, 1995; Jang, et al., 1997).

2. An initial rule base is constructed from numerical data. The extracted rules and membership functions provide an a posteriori interpretation of the system's behavior

and can be combined with additional knowledge. This approach is adopted in this book.

Templates for Membership Functions. The first approach assumes that the expert can initially provide the membership functions and the rules. For many systems, however, no particular knowledge is available about the partition of the operating space. In such a case, the domains of the antecedent variables can be simply partitioned into a specified number of equally spaced and shaped membership functions. The rule base can be established to cover all the combinations of the antecedent terms. In the literature, this approach is called template-based fuzzy modeling. The task of identification is then to estimate the remaining parameters in the fuzzy model from data. With regard to the types of fuzzy models presented in this chapter, these parameters include the consequents in the TS or singleton fuzzy model, and the relation defining the rule base in relational and linguistic fuzzy models.

Example 2.8 Consider a nonlinear dynamic system described by a first-order difference equation:

$$y(k+1) = y(k) + u(k)e^{-3|y(k)|}\,. \tag{2.78}$$

A set of 300 simulated input–output data shown in Figure 2.22a has been generated for identification. Suppose that it is known that the system is first order and that the nonlinearity is only caused by y, the following rule structure is chosen:

$$\textbf{If } y(k) \text{ is } A_i \textbf{ then } y_i(k+1) = a_i y(k) + b_i u(k)\,. \tag{2.79}$$

Assuming that no further prior knowledge is available, seven equally spaced triangular membership functions, A_1 to A_7, are defined in the domain of $y(k)$, as shown in Figure 2.21a.

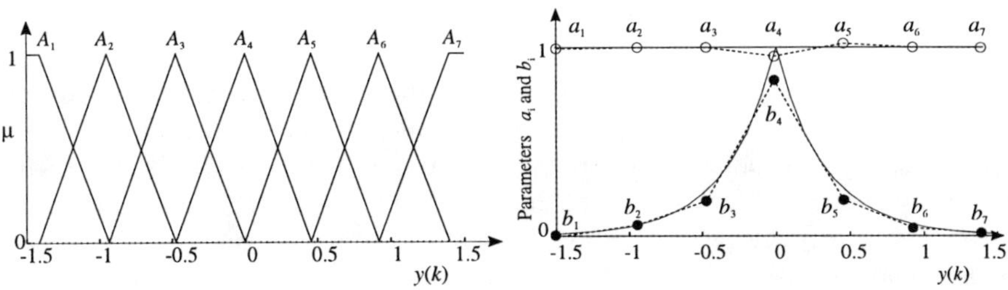

(a) Membership functions. *(b) Estimated consequent parameters.*

Figure 2.21. (a) Equidistant triangular membership functions designed for the antecedent $y(k)$; (b) comparison of the true system nonlinearity (solid line) and its approximation in terms of the estimated consequent parameters (dashed line).

The consequent parameters were estimated by the least-squares method as described in Section 5.5. Figure 2.21b gives a plot of the parameters a_i, b_i against the cores of the antecedent fuzzy sets A_i. Also plotted is the linear interpolation between the parameters (dashed line) and the true system nonlinearity (solid line). The interpolation between a_i and b_i is linear, since the membership functions are piece-wise linear (triangular). One can observe that the dependence of the consequent parameters on the antecedent variable approximates quite accurately the system's nonlinearity, which gives the model a certain transparency. The values of the parameters

$$\begin{aligned} \mathbf{a}^T &= [1.00, 1.00, 1.00, 0.97, 1.01, 1.00, 1.00], \\ \mathbf{b}^T &= [0.01, 0.05, 0.20, 0.81, 0.20, 0.05, 0.01] \end{aligned}$$

indicate the strong input nonlinearity and the linear dynamics in (2.78). Validation of the model in simulation using a different data set is given in Figure 2.22b. □

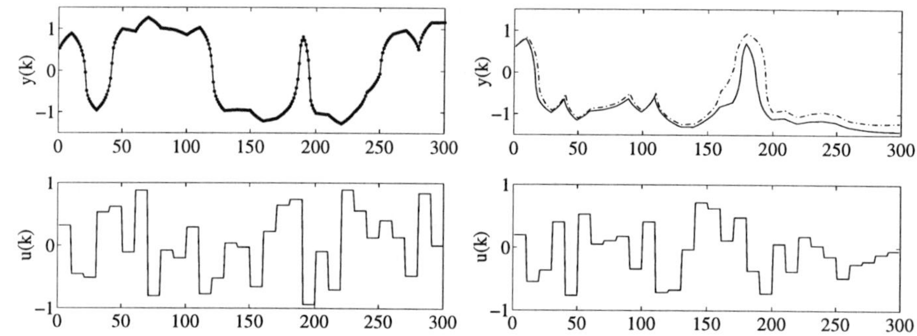

(a) Identification data.

(b) Model validation.

Figure 2.22. Identification and validation data. Solid line: process, dashed-dotted line: model.

A drawback of the template-based approach is that the number of rules in the model may grow very fast. If no knowledge is available as to which variables cause the nonlinearity of the system, all the antecedent variables are usually partitioned uniformly, which leads to an exponential increase of the number of rules, $K = \Pi_{j=1}^{p} N_j$, where N_j is the number of fuzzy sets per antecedent variable x_j and p is the number of antecedent variables. Having a large number of rules requires many data points for the estimation of the consequents. Typically, the data cover only a small fraction of the entire space; this entails that many parameters cannot be estimated, or that their estimates are inaccurate.

The complexity of the system's behavior is typically not uniform, which means, for instance, that certain regions can be well approximated by a single model, while other regions require rather fine partitioning. In order to obtain an efficient representation with as few rules as possible, the membership functions must be placed such that they capture the non-uniform behavior of the system. This often requires that system measurements are also used to extract the membership functions, as described in the following section.

Acquisition of Membership Functions from Data. A number of different approaches have been proposed in the literature to generate membership functions from data. They can be broadly classified into:

1. Methods that construct membership functions for the different variables independently of each other. These methods include, for instance, entropy equalization (Pedrycz, 1995), mass assignment (Baldwin, et al., 1995), or fuzzy clustering of the antecedent data (Pedrycz, 1993; Berenji and Khedar, 1993). A common drawback of these approaches is that they do not aim at the modeling of the entire system, but rather extract the membership functions, based on the data distributions in the individual variables, while disregarding the fact that a mapping between the variables is sought.

2. Methods that build the membership functions as a part of the function approximation problem, such as neuro-fuzzy gradient-based algorithms (Jang, 1992; de Oliveira, 1993; Jang, et al., 1997), orthogonal least-squares methods (Wang, 1994), successive partitioning (Takagi and Sugeno, 1985; Sugeno and Tanaka, 1991), spline construction algorithms (Brown and Harris, 1994; Kavli, 1993), or fuzzy clustering in the product space of input and output variables (Yoshinari, et al., 1993; Babuška and Verbruggen, 1994a; Zhao, et al., 1994; Nakamori and Ryoke, 1994). All these methods are based on nonlinear optimization techniques, and they aim at constructing the membership functions as a part of the input–output mapping, by minimizing an objective function. Some of these approaches are discussed in more detail below.

Neuro-fuzzy Identification. Recently, much research has been devoted to neuro-fuzzy modeling and identification. A number of different schemes have been studied, such as fuzzy-logic-based neurons (Pedrycz, 1995), neural nets with fuzzy weights (Buckley and Hayashi, 1994), or neuro-fuzzy adaptive models (Brown and Harris, 1994; Jang, et al., 1997). The most common approach to neuro-fuzzy identification is to use differentiable operators in the inference mechanism, differentiable (e.g., Gaussian) membership functions, and the weighted mean defuzzification. The inference scheme can then be represented in a network structure.

Figure 2.23 depicts an example of a singleton fuzzy model with two rules represented as a network. The rules are:

$$\textbf{If } x_1 \text{ is } A_{11} \textbf{ and } x_2 \text{ is } A_{21} \textbf{ then } y = b_1 .$$

$$\textbf{If } x_1 \text{ is } A_{12} \textbf{ and } x_2 \text{ is } A_{22} \textbf{ then } y = b_2 .$$

The nodes in the first layer compute the membership degree of the inputs in the antecedent fuzzy sets. The product nodes Π in the second layer represent the antecedent conjunction operator. The normalization node N and the summation node Σ realize the fuzzy-mean operator (2.54). Gradient-descent learning algorithms, such as backpropagation, can be applied to adjust the antecedent membership functions. Also other techniques, such as orthogonal least-squares learning (Wang, 1994), or splines construction algorithms (Brown and Harris, 1994; Kavli, 1993) are used. There are also clear relationships between neuro-fuzzy modeling and more conventional mathematical techniques introduced in the field of nonlinear regression analysis such as MARS (Friedman, 1991), or projection pursuit methods (Huber, 1985).

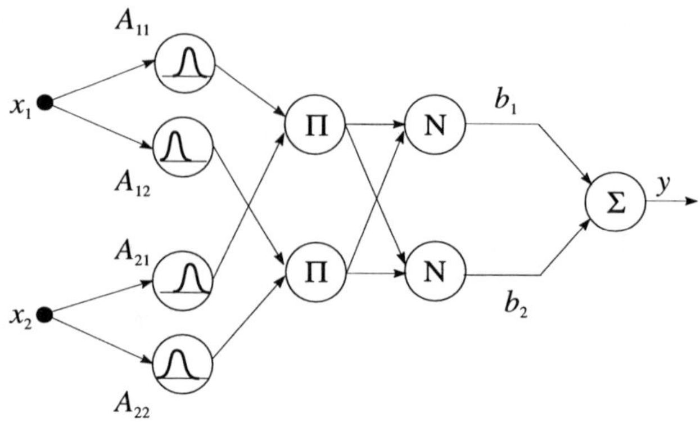

Figure 2.23. An example of a singleton fuzzy model with two rules represented as a (neuro-fuzzy) network.

Product-space Clustering. Identification methods based on fuzzy clustering originate from data analysis and pattern recognition, where the concept of graded membership is used to represent the degree to which a given object, represented as a vector of features, is similar to some prototype. The degree of similarity is usually calculated by using a suitable distance measure. Based on the similarity, feature vectors can be clustered such that the vectors within a cluster are as similar as possible, and vectors from different clusters are as dissimilar as possible. Cluster prototypes can be defined as linear subspaces (lines, planes, hyperplanes) and the distance measure quantifies the distance of data points from that linear subspace (Bezdek, 1981). These algorithms can be used to extract Takagi–Sugeno fuzzy models. Also, fuzzy clustering with an adaptive distance measure (Gustafson and Kessel, 1979) have been applied to the fuzzy identification of Takagi–Sugeno models for nonlinear dynamic systems (Babuška and Verbruggen, 1994a; Zhao, et al., 1994; Babuška and Verbruggen, 1995c; Babuška and Verbruggen, 1996e).

Each obtained cluster corresponds to one rule of the TS fuzzy model. The antecedent membership functions are generated by projecting the clusters onto the antecedent variables, and the consequent parameters are estimated by the least-squares methods. This approach is illustrated in Figure 2.24 and described in Chapters 4 and 5.

2.5 Summary and Concluding Remarks

This chapter has reviewed three different types of rule-based fuzzy models: linguistic (Mamdani-type) models, fuzzy relational models, and Takagi–Sugeno models. A major distinction can be made between the linguistic model, which has fuzzy sets in both the rule antecedents and consequents of the rules, and the TS model, where the consequents are (crisp) functions of the input variables. Fuzzy relational models can be regarded as an extension of linguistic models, which allow for different degrees of association between the antecedent and the consequent linguistic terms. All three types can be used to approximate static and dynamic nonlinear systems. There are,

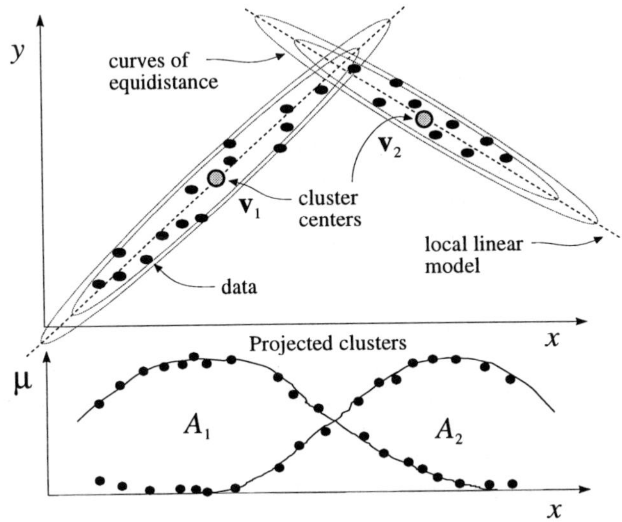

Figure 2.24. Hyperellipsoidal fuzzy clusters in the input–output product space.

however, differences between these models in terms of the approximation properties, the kind of knowledge they can represent, and the related identification methods.

Linguistic and relational fuzzy models are well suited for encoding imprecise knowledge and heuristics expressed in terms of if–then rules. Such rules are quite easily understandable to users without mathematical knowledge, as they do not involve any formulas. The type of interpolation between the individual rule consequents in linguistic models depends on the form of the membership functions, on the choice of the fuzzy logic operators and the defuzzification method. With algebraic operators, the fuzzy-mean defuzzification, and piece-wise linear membership functions, multilinear interpolation between the consequents is obtained. In such a case, the consequent parameters can easily be adjusted by means of linear estimation techniques.

Takagi–Sugeno fuzzy models provide a suitable framework for modeling by decomposition of a nonlinear system into a collection of local linear models. The individual consequents are linear systems which can be analyzed by using the standard tools of linear systems theory. A TS model can also more efficiently represent systems, since the rules are linear subsystems, rather than constants as in the case of the linguistic model. TS rules with a common structure of the consequent functions can be regarded as a mapping from the antecedent space to the space of the consequent parameters. Since this mapping is of the same type as that of the linguistic model, the TS fuzzy model can be interpreted linguistically at the level of system parameters. As an example, consider rules like

If *output is large* **then** *process gain is large* **and** *time constant is small.*

This sort of information can be useful for control engineers, as it is related to the methods for the analysis and synthesis of linear control systems. Moreover, the possible values of the parameters are always within a polytope (convex set) bounded by the individual consequent parameters. This property facilitates the analysis of TS fuzzy systems in a quasi-linear fashion.

The standard weighted mean inference in the TS model is suitable for the linear estimation of the consequent parameters and for the analysis of the model, but it results in a nonconvex and generally not smooth output of the global model, and an approximation error occurs. An alternative scheme described in this chapter yields convex and smooth surfaces and improves the numerical accuracy of the TS model.

A brief overview of methods to construct fuzzy models from prior knowledge and data has also been given. A detailed presentation of an identification technique based on fuzzy clustering is given in Chapters 4 and 5.

Finally, it should be noted that the use of membership functions (validity functions, basis functions) for partitioning the operating space of the system into overlapping regions is not used only in fuzzy modeling. Similar approaches can be found in operating regime-based modeling (Murray-Smith and Johansen, 1997), nonlinear phase regression (Seber and Wild, 1989), spline techniques (de Boor, 1978; Friedman, 1991), or radial basis functions (Jang and Sun, 1993). Despite different names, these techniques have a common mathematical basis.

3 FUZZY CLUSTERING ALGORITHMS

An effective approach to the identification of complex nonlinear systems is to partition the available data into subsets and approximate each subset by a simple model. Fuzzy clustering can be used as a tool to obtain a partitioning of data where the transitions between the subsets are gradual rather than abrupt. This chapter gives an introduction to the basic concepts of fuzzy clustering, and simultaneously serves as a reference to clustering algorithms that can be used to construct fuzzy models from data. The basic notions of clustering and the different types of partitions are defined in Sections 3.1 and 3.2. Section 3.3 presents the basic idea of fuzzy clustering with objective function and the fuzzy c-means algorithm. Sections 3.4 and 3.5 address algorithms which can detect clusters contained in linear subspaces of the data space. These methods include clustering with an adaptive distance measure, clustering with linear prototypes, and fuzzy regression clustering. Section 3.6 presents the approach known as possibilistic clustering. Section 3.7 addresses the determination of an appropriate number of clusters and Section 3.8 deals with pre-processing of the data. The aim of this chapter is to explain clustering at a level necessary to understand the subsequent chapters. For a more detailed treatment of the subject, the reader may refer to the classical monographs by Duda and Hart (1973), Bezdek (1981) and Jain and Dubes (1988). A more recent overview can be found in a collection of Bezdek and Pal (1992), and the monograph by Backer (1995). The notation and terminology in this chapter closely follows Bezdek (1981).

3.1 Cluster Analysis

The objective of cluster analysis is the classification of objects according to similarities among them, and the organizing of data into groups. Clustering techniques are among the *unsupervised* (learning) methods, since they do not use prior class identifiers. Most clustering algorithms also do not rely on assumptions common to conventional statistical methods, such as the underlying statistical distribution of data, and therefore they are useful in situations where little prior knowledge exists. The potential of clustering algorithms to reveal the underlying structures in data can be exploited, not only for classification and pattern recognition, but also for the reduction of complexity in modeling and optimization.

3.1.1 The Data

Clustering techniques can be applied to data that is quantitative (numerical), qualitative (categoric), or a mixture of both. In this book, the clustering of quantitative data is considered. The data are typically observations of some physical process. Each observation consists of n measured variables, grouped into an n-dimensional column vector $\mathbf{z}_k = [z_{1k}, \ldots, z_{nk}]^T$, $\mathbf{z}_k \in \mathbb{R}^n$. A set of N observations is denoted by $\mathbf{Z} = \{\mathbf{z}_k | k = 1, 2, \ldots, N\}$, and is represented as an $n \times N$ matrix:

$$\mathbf{Z} = \begin{bmatrix} z_{11} & z_{12} & \cdots & z_{1N} \\ z_{21} & z_{22} & \cdots & z_{2N} \\ \vdots & \vdots & \vdots & \vdots \\ z_{n1} & z_{n2} & \cdots & z_{nN} \end{bmatrix} . \tag{3.1}$$

In the pattern recognition terminology, the columns of this matrix are called *patterns* or objects, the rows are called the *features* or attributes, and $\mathbf{Z}$ is called the *pattern* or *data matrix*. The meaning of the columns and rows of $\mathbf{Z}$ depends on the context. In medical diagnosis, for instance, the columns of $\mathbf{Z}$ may represent patients, and the rows are then symptoms, or laboratory measurements for these patients. When clustering is applied to the modeling and identification of dynamic systems, the columns of $\mathbf{Z}$ contain samples of time signals, and the rows are, for instance, physical variables observed in the system (position, velocity, temperature, etc.). In order to represent the system's dynamics, past values of the variables are typically included in $\mathbf{Z}$ as well. More details on the choice of an appropriate representation for dynamic systems are given in Section 4.2.

3.1.2 What Are Clusters?

Various definitions of a cluster can be formulated, depending on the objective of clustering. Generally, one may accept the view that a cluster is a group of objects that are more similar to one another than to members of other clusters (Bezdek, 1981; Jain and Dubes, 1988). The term "similarity" should be understood as mathematical similarity, measured in some well-defined sense. In metric spaces, similarity is often defined by means of a *distance norm*. Distance can be measured among the data vectors themselves, or as a distance from a data vector to some prototypical object of the cluster. The prototypes are usually not known beforehand, and are sought by the clustering algorithms simultaneously with the partitioning of the data. The prototypes may be vectors of the same dimension as the data objects, but they can also be defined as "higher-level" geometrical objects, such as linear or nonlinear subspaces or functions, see Section 3.5.

Data can reveal clusters of different geometrical shapes, sizes and densities as demonstrated in Figure 3.1. While clusters (a) are spherical, clusters (b) to (d) can be characterized as linear and nonlinear subspaces of the data space. Algorithms that can detect subspaces of the data space are of particular interest for identification and will be discussed in detail later on. The performance of most clustering algorithms is influenced not only by the geometrical shapes and densities of the individual clusters but also by the spatial relations and distances among the clusters. Clusters can be well-separated, continuously connected to each other, or overlapping each other. The separation of clusters is influenced by the scaling and normalization of the data, as discussed in Section 3.8.

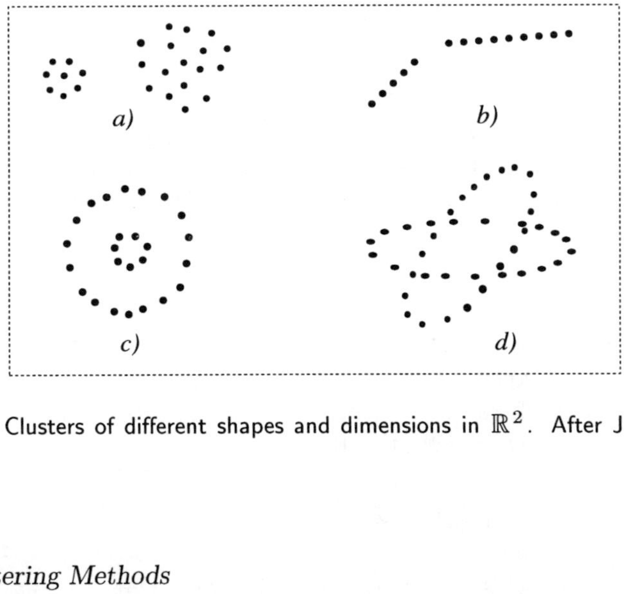

Figure 3.1. Clusters of different shapes and dimensions in $\mathbb{R}^2$. After Jain and Dubes (1988).

3.1.3 *Clustering Methods*

Many clustering algorithms have been introduced in the literature. Since clusters can formally be seen as subsets of the data set, one possible classification of clustering methods can be according to whether the subsets are fuzzy or crisp (hard). Hard clustering methods are based on classical set theory, and require that an object either does or does not belong to a cluster. Hard clustering means partitioning the data into a specified number of mutually exclusive subsets. Fuzzy clustering methods, however, allow the objects to belong to several clusters simultaneously, with different degrees of membership. In many situations, fuzzy clustering is more natural than hard clustering, as objects on the boundaries between several classes are not forced to fully belong to one of the classes, but rather are assigned membership degrees between 0 and 1 indicating their partial memberships. The discrete nature of the hard partitioning also causes analytical and algorithmic intractability of algorithms based on analytic functionals, since these functionals are not differentiable.

Another classification can be related to the algorithmic approach of the different techniques (Bezdek, 1981). *Agglomerative hierarchical methods* and *splitting hierarchical methods* form new clusters by reallocating memberships of one point at a time, based on some suitable measure of similarity. With *graph-theoretic methods*, $\mathbf{Z}$ is regarded as a set of nodes. Edge weights between pairs of nodes are based on a measure of similarity between these nodes. The third class of clustering algorithms uses an *objective function* to measure the desirability of partitions. Nonlinear optimization algorithms are used to search for local extrema of the objective function.

The remainder of this chapter focuses on fuzzy clustering with objective function. These methods lead to least-squares optimization, and hence there are close relationships between clustering with fuzzy objective function and statistical regression and systems identification methods. In fuzzy clustering, the objective function is differentiable, which is a useful property for optimization. The objective function methods are also relatively well understood, and mathematical results are available concerning the convergence properties and cluster validity assessment.

3.2 Hard and Fuzzy Partitions

The concept of *fuzzy partition* is essential for cluster analysis, and consequently also for the identification techniques that are based on fuzzy clustering. Fuzzy and possibilistic partitions can be seen as a generalization of *hard partition* which is formulated in terms of classical subsets.

3.2.1 Hard Partition

The objective of clustering is to partition the data set $\mathbf{Z}$ into c clusters. For the time being, assume that c is known, based on prior knowledge, for instance. Using classical sets, a *hard partition* of $\mathbf{Z}$ can be defined as a family of subsets $\{A_i | 1 \le i \le c\} \subset P(\mathbf{Z})$ with the following properties (Bezdek, 1981):

$$\bigcup_{i=1}^{c} A_i = \mathbf{Z}, \tag{3.2a}$$

$$A_i \cap A_j = \emptyset, \qquad 1 \le i \ne j \le c, \tag{3.2b}$$

$$\emptyset \subset A_i \subset \mathbf{Z}, \qquad 1 \le i \le c. \tag{3.2c}$$

Equation (3.2a) means that the subsets A_i collectively contain all the data in $\mathbf{Z}$. The subsets must be disjoint, as stated by (3.2b), and none of them is empty nor contains all the data in $\mathbf{Z}$ (3.2c). In terms of *membership functions*, equations (3.2) can be expressed as:

$$\bigvee_{i=1}^{c} \mu_{A_i} = \mathbf{1}, \tag{3.3a}$$

$$\mu_{A_i} \wedge \mu_{A_j} = \mathbf{0}, \qquad 1 \le i \ne j \le c, \tag{3.3b}$$

$$\mathbf{0} < \mu_{A_i} < \mathbf{1}, \qquad 1 \le i \le c. \tag{3.3c}$$

Here $\mathbf{0}$ and $\mathbf{1}$ denote zero and one function, respectively, and μ_{A_i} is the membership function of A_i. To simplify the notation, in this chapter we use μ_i instead of the usual μ_{A_i}. Further, by denoting $\mu_i(\mathbf{z}_k)$ by μ_{ik}, partitions can be conveniently represented in a matrix notation. A $c \times N$ matrix $\mathbf{U} = [\mu_{ik}]$ represents a hard partition if and only if its elements satisfy the conditions:

$$\mu_{ik} \in \{0, 1\}, \qquad 1 \leq i \leq c, \quad 1 \leq k \leq N, \tag{3.4a}$$

$$\sum_{i=1}^{c} \mu_{ik} = 1, \qquad 1 \leq k \leq N, \tag{3.4b}$$

$$0 < \sum_{k=1}^{N} \mu_{ik} < N, \qquad 1 \leq i \leq c, \tag{3.4c}$$

which directly follow from equations (3.3). The ith row of the hard partition matrix $\mathbf{U}$ contains values of the characteristic function of the ith subset A_i of $\mathbf{Z}$. The above discussion can be summarized in the following definition of hard partitioning space (Bezdek, 1981).

Definition 3.1 (Hard partitioning space) *Let $\mathbf{Z} = \{\mathbf{z}_1, \mathbf{z}_2, \ldots, \mathbf{z}_N\}$ be a finite set and $2 \leq c < N$ be an integer. The hard partitioning space for $\mathbf{Z}$ is the set*

$$M_{hc} = \left\{ \mathbf{U} \in \mathbb{R}^{c \times N} \middle| \mu_{ik} \in \{0, 1\}, \forall i, k; \sum_{i=1}^{c} \mu_{ik} = 1, \forall k; 0 < \sum_{k=1}^{N} \mu_{ik} < N, \forall i \right\}.$$

Example 3.1 Consider a data set $\mathbf{Z} = \{\mathbf{z}_1, \mathbf{z}_2, \ldots, \mathbf{z}_{10}\}$, shown in Figure 3.2.

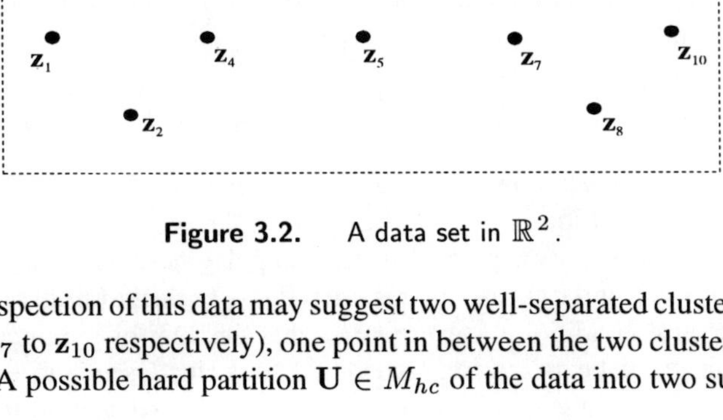

Figure 3.2. A data set in $\mathbb{R}^2$.

A visual inspection of this data may suggest two well-separated clusters (data points $\mathbf{z}_1$ to $\mathbf{z}_4$ and $\mathbf{z}_7$ to $\mathbf{z}_{10}$ respectively), one point in between the two clusters ($\mathbf{z}_5$), and an "outlier" $\mathbf{z}_6$. A possible hard partition $\mathbf{U} \in M_{hc}$ of the data into two subsets is given by:

$$\mathbf{U} = \begin{bmatrix} 1 & 1 & 1 & 1 & 1 & 1 & 0 & 0 & 0 & 0 \\ 0 & 0 & 0 & 0 & 0 & 0 & 1 & 1 & 1 & 1 \end{bmatrix}.$$

The first row of $\mathbf{U}$ defines point-wise the characteristic function for the first subset of $\mathbf{Z}$, A_1, and the second row defines the characteristic function of the second subset of $\mathbf{Z}$, A_2. Each sample must be assigned exclusively to one subset (cluster) of the partition. In this case, both the boundary point $\mathbf{z}_5$ and the outlier $\mathbf{z}_6$ have been assigned to A_1. It is clear that a hard partitioning may not give a realistic picture of the underlying data. Boundary data points may represent patterns with a mixture of properties of data in A_1 and A_2, and therefore cannot be fully assigned to either of these classes, or do they constitute a separate class. This shortcoming can be alleviated by using fuzzy and possibilistic partitions as shown in the following sections. □

3.2.2 Fuzzy Partition

Generalization of the hard partition to the fuzzy case follows directly by allowing μ_{ik} to attain real values in $[0, 1]$ (Ruspini, 1970). Conditions for a fuzzy partition matrix, analogical to (3.4) then are given by:

$$\mu_{ik} \in [0, 1], \qquad 1 \le i \le c, \quad 1 \le k \le N, \tag{3.5a}$$

$$\sum_{i=1}^{c} \mu_{ik} = 1, \qquad 1 \le k \le N, \tag{3.5b}$$

$$0 < \sum_{k=1}^{N} \mu_{ik} < N, \qquad 1 \le i \le c. \tag{3.5c}$$

Definition 3.2 (Fuzzy partitioning space) *Let $\mathbf{Z} = \{\mathbf{z}_1, \mathbf{z}_2, \ldots, \mathbf{z}_N\}$ be a finite set and let $2 \le c < N$ be an integer. The fuzzy partitioning space for $\mathbf{Z}$ is the set*

$$M_{fc} = \left\{ \mathbf{U} \in \mathbb{R}^{c \times N} \middle| \mu_{ik} \in [0, 1], \forall i, k; \sum_{i=1}^{c} \mu_{ik} = 1, \forall k; 0 < \sum_{k=1}^{N} \mu_{ik} < N, \forall i \right\}.$$

The ith row of the fuzzy partition matrix $\mathbf{U}$ contains values of the ith *membership function* of the fuzzy subset A_i of $\mathbf{Z}$. Equation (3.5b) constrains the sum of each column to 1, and thus the total membership of each $\mathbf{z}_k$ in $\mathbf{Z}$ equals one.

Example 3.2 Consider the data set from Example 3.1. One of the infinitely many fuzzy partitions in $\mathbf{Z}$ is:

$$\mathbf{U} = \begin{bmatrix} 1.0 & 1.0 & 1.0 & 0.8 & 0.5 & 0.5 & 0.2 & 0.0 & 0.0 & 0.0 \\ 0.0 & 0.0 & 0.0 & 0.2 & 0.5 & 0.5 & 0.8 & 1.0 & 1.0 & 1.0 \end{bmatrix}.$$

The boundary point $\mathbf{z}_5$ has now a membership degree of 0.5 in both classes, which correctly reflects its position in the middle between the two clusters. Note, however, that the outlier $\mathbf{z}_6$ has the same pair of membership degrees, even though it is further from the two clusters, and thus can be considered less typical of both A_1 and A_2 than $\mathbf{z}_5$. This is because condition (3.5b) requires that the sum of memberships of each point equals one. It can be, of course, argued that three clusters are more appropriate in this example than two. In general, however, it is difficult to detect outliers and assign them to extra clusters. The use of possibilistic partition, presented in the next section, overcomes this drawback of fuzzy partitions. □

3.2.3 Possibilistic Partition

A more general form of fuzzy partition, the *possibilistic partition*,[1] can be obtained by relaxing the constraint (3.5b). This constraint, however, cannot be completely removed, in order to ensure that each point is assigned to at least one of the fuzzy subsets with a membership greater than zero. Equation (3.5b) can be replaced by a less restrictive constraint $\forall k,\ \exists i,\ \mu_{ik} > 0$. The conditions for a possibilistic fuzzy partition matrix, analogical to (3.5) are:

$$\mu_{ik} \in [0,1], \qquad 1 \le i \le c, \quad 1 \le k \le N, \tag{3.6a}$$

$$\exists i,\ \mu_{ik} > 0, \qquad \forall k, \tag{3.6b}$$

$$0 < \sum_{k=1}^{N} \mu_{ik} < N, \qquad 1 \le i \le c. \tag{3.6c}$$

Definition 3.3 (Possibilistic partitioning space) *Let $\mathbf{Z} = \{\mathbf{z}_1, \mathbf{z}_2, \ldots, \mathbf{z}_N\}$ be a finite set and $2 \le c < N$ be an integer. The possibilistic partition space for $\mathbf{Z}$ is the set*

$$M_{pc} = \left\{ \mathbf{U} \in \mathbb{R}^{c \times N} \middle| \mu_{ik} \in [0,1], \forall i,k; \forall k,\ \exists i,\ \mu_{ik} > 0; 0 < \sum_{k=1}^{N} \mu_{ik} < N, \forall i \right\}.$$

Example 3.3 An example of a possibilistic partition matrix for our data set is:

$$\mathbf{U} = \begin{bmatrix} 1.0 & 1.0 & 1.0 & 1.0 & 0.5 & 0.2 & 0.0 & 0.0 & 0.0 & 0.0 \\ 0.0 & 0.0 & 0.0 & 0.0 & 0.5 & 0.2 & 1.0 & 1.0 & 1.0 & 1.0 \end{bmatrix}.$$

As the sum of elements in each column of $\mathbf{U} \in M_{fc}$ is no longer constrained, the outlier has a membership of 0.2 in both clusters, which is lower than the membership of the boundary point $\mathbf{z}_5$, reflecting a lower degree of typicality of this point for both sets. □

3.3 Fuzzy c-Means Clustering

Most analytical fuzzy clustering algorithms (and also all the algorithms presented in this chapter) are based on optimization of the basic *c*-means objective function, or some modification of it. Hence we start our discussion with presenting the fuzzy *c*-means functional (Dunn, 1974a).

---

[1] The term "possibilistic" (partition, clustering, etc.) has been introduced by Krishnapuram and Keller (1993). In the literature, the terms "constrained fuzzy partition" and "unconstrained fuzzy partition" are also used to denote partitions (3.5) and (3.6), respectively.

3.3.1 The Fuzzy c-Means Functional

A large family of fuzzy clustering algorithms is based on minimization of the *fuzzy c-means* functional formulated as (Bezdek, 1981):

$$J(\mathbf{Z}; \mathbf{U}, \mathbf{V}) = \sum_{i=1}^{c} \sum_{k=1}^{N} (\mu_{ik})^m \|\mathbf{z}_k - \mathbf{v}_i\|_{\mathbf{A}}^2 \tag{3.7a}$$

where

$$\mathbf{U} = [\mu_{ik}] \in M_{fc} \tag{3.7b}$$

is a fuzzy partition matrix of $\mathbf{Z}$,

$$\mathbf{V} = [\mathbf{v}_1, \mathbf{v}_2, \ldots, \mathbf{v}_c], \quad \mathbf{v}_i \in \mathbb{R}^n \tag{3.7c}$$

is a vector of *cluster prototypes* (centers), which have to be determined,

$$D_{ik\mathbf{A}}^2 = \|\mathbf{z}_k - \mathbf{v}_i\|_{\mathbf{A}}^2 = (\mathbf{z}_k - \mathbf{v}_i)^T \mathbf{A} (\mathbf{z}_k - \mathbf{v}_i) \tag{3.7d}$$

is a squared inner-product distance norm, and

$$m \in [1, \infty) \tag{3.7e}$$

is a weighting exponent which determines the fuzziness of the resulting clusters. The measure of dissimilarity in (3.7a) is the squared distance between each data point $\mathbf{z}_k$ and the cluster prototype $\mathbf{v}_i$. This distance is weighted by the power of the membership degree of that point $(\mu_{ik})^m$. The value of the cost function (3.7a) can be seen as a measure of the total variance of $\mathbf{z}_k$ from $\mathbf{v}_i$.

3.3.2 The Fuzzy c-Means Algorithm

The minimization of the c-means functional (3.7a) represents a nonlinear optimization problem that can be solved by using a variety of available methods, ranging from grouped coordinate minimization (Bezdek, et al., 1987; Hathaway and Bezdek, 1991a), over simulated annealing (DeSarbo, 1982), to genetic algorithms (Babu and Murty, 1994). The most popular method, however, is a simple Picard iteration through the first-order conditions for stationary points of (3.7a), known as the fuzzy c-means (FCM) algorithm, which is given in Algorithm 3.1.

The stationary points of the objective function (3.7a) can be found by adjoining the constraint (3.5b) to J by means of Lagrange multipliers:

$$\bar{J}(\mathbf{Z}; \mathbf{U}, \mathbf{V}, \boldsymbol{\lambda}) = \sum_{i=1}^{c} \sum_{k=1}^{N} (\mu_{ik})^m D_{ik\mathbf{A}}^2 + \sum_{k=1}^{N} \lambda_k \left[\sum_{i=1}^{c} \mu_{ik} - 1 \right], \tag{3.8}$$

and by setting the gradients of $\bar{J}$ with respect to $\mathbf{U}$, $\mathbf{V}$ and λ to zero. If $D_{ik\mathbf{A}}^2 > 0, \forall i, k$ and $m > 1$, then $(\mathbf{U}, \mathbf{V}) \in M_{fc} \times \mathbb{R}^{n \times c}$ may minimize (3.7a) only if

$$\mu_{ik} = \frac{1}{\sum_{j=1}^{c} (D_{ik\mathbf{A}}/D_{jk\mathbf{A}})^{2/(m-1)}}, \quad 1 \le i \le c \quad 1 \le k \le N, \tag{3.9a}$$

and

$$\mathbf{v}_i = \frac{\sum_{k=1}^{N} (\mu_{ik})^m \mathbf{z}_k}{\sum_{k=1}^{N} (\mu_{ik})^m}; \quad 1 \le i \le c. \tag{3.9b}$$

This solution also satisfies the remaining constraints (3.5a) and (3.5c). Note that eq. (3.9b) gives $\mathbf{v}_i$ as the weighted mean of the data items that belong to a cluster, where the weights are the membership degrees. That is why the algorithm is called "c-means". The FCM algorithm iterates through (3.9a) and (3.9b).

Algorithm 3.1 (Fuzzy c-means (FCM))

Given the data set $\mathbf{Z}$, choose the number of clusters $1 < c < N$, the weighting exponent $m > 1$, the termination tolerance $\epsilon > 0$ and the norm-inducing matrix $\mathbf{A}$. Initialize the partition matrix randomly, such that $\mathbf{U}^{(0)} \in M_{fc}$.

Repeat for $l = 1, 2, \ldots$

Step 1: **Compute the cluster prototypes (means):**

$$\mathbf{v}_i^{(l)} = \frac{\sum_{k=1}^{N} (\mu_{ik}^{(l-1)})^m \mathbf{z}_k}{\sum_{k=1}^{N} (\mu_{ik}^{(l-1)})^m}, \quad 1 \le i \le c.$$

Step 2: **Compute the distances:**

$$D_{ik\mathbf{A}}^2 = (\mathbf{z}_k - \mathbf{v}_i^{(l)})^T \mathbf{A} (\mathbf{z}_k - \mathbf{v}_i^{(l)}), \quad 1 \le i \le c, \quad 1 \le k \le N.$$

Step 3: **Update the partition matrix:**

if $D_{ik\mathbf{A}} > 0$ for $1 \le i \le c, \quad 1 \le k \le N$,

$$\mu_{ik}^{(l)} = \frac{1}{\sum_{j=1}^{c} (D_{ik\mathbf{A}}/D_{jk\mathbf{A}})^{2/(m-1)}},$$

otherwise

$$\mu_{ik}^{(l)} = 0 \text{ if } D_{ik\mathbf{A}} > 0, \text{ and } \mu_{ik}^{(l)} \in [0,1] \text{ with } \sum_{i=1}^{c} \mu_{ik}^{(l)} = 1.$$

until $||\mathbf{U}^{(l)} - \mathbf{U}^{(l-1)}|| < \epsilon$.

Remark 1. A singularity in FCM occurs at Step 3 when $D_{is\mathbf{A}} = 0$ for some $\mathbf{z}_k$ and one or more cluster prototypes $\mathbf{v}_s$, $s \in S \subset \{1, 2, \ldots, c\}$. In this case, the membership degree in (3.9a) cannot be computed. When this happens, 0 is assigned to each μ_{ik},

$i \in \bar{S}$ and the membership is distributed arbitrarily among μ_{sj} subject to the constraint $\sum_{s \in S} \mu_{sj} = 1, \forall k$.

Remark 2. Equations (3.9) are only first-order necessary conditions for stationary points of the functional (3.7a). Sufficiency of these conditions and convergence of the algorithm has been proven by Bezdek (1980).

Remark 3. The alternating optimization scheme used by FCM loops through the estimates $\mathbf{U}^{(l-1)} \to V^{(l)} \to \mathbf{U}^{(l)}$ and terminates as soon as $||\mathbf{U}^{(l)} - \mathbf{U}^{(l-1)}|| < \epsilon$. Alternatively, the algorithm can be initialized with $V^{(0)}$, loop through $V^{(l-1)} \to \mathbf{U}^{(l)} \to V^{(l)}$, and terminate on $||V^{(l)} - V^{(l-1)}|| < \epsilon$. The error norm in the termination criterion is usually chosen as $\max_{ik}(|\mu_{ik}^{(l)} - \mu_{ik}^{(l-1)}|)$. Different results may be obtained with the same values of ϵ, since the termination criterion used in Algorithm 3.1 requires that more parameters become close to one another. The usual setting of the termination criterion is $\epsilon = 0.001$, even though $\epsilon = 0.01$ works well in most cases.

Remark 4. The weighting exponent m is a rather important parameter, as it significantly influences the resulting partition. As m approaches one from above, the partition becomes hard ($\mu_{ik} \in \{0, 1\}$) and $\mathbf{v}_i$ are ordinary means of the clusters. As $m \to \infty$, the partition becomes maximally fuzzy ($\mu_{ik} = 1/c$) and the cluster means are all equal to the grand mean of $\mathbf{Z}$. These limit properties of (3.7) are independent of the optimization method used (Pal and Bezdek, 1995). Usually, $m = 2$ is chosen.

Remark 5. The number of clusters c is the most important parameter, in the sense that the remaining parameters have secondary effects on $\mathbf{U}$, compared to the effects of the number of clusters. The choice of the number of clusters is discussed in Sections 3.7 and 4.5.

3.3.3 Inner-product Norms

The shape of the clusters is determined by the choice of the matrix $\mathbf{A}$ in the distance measure (3.7d). A common choice is $\mathbf{A} = \mathbf{I}$, which induces the standard Euclidean norm:

$$D_{ik}^2 = (\mathbf{z}_k - \mathbf{v}_i)^T (\mathbf{z}_k - \mathbf{v}_i). \tag{3.10}$$

$\mathbf{A}$ can be chosen as an $n \times n$ diagonal matrix that accounts for different variances in the directions of the coordinate axes of $\mathbf{Z}$:

$$\mathbf{A}_D = \begin{bmatrix} (1/\sigma_1)^2 & 0 & \cdots & 0 \\ 0 & (1/\sigma_2)^2 & \cdots & 0 \\ \vdots & \vdots & \ddots & \vdots \\ 0 & 0 & \cdots & (1/\sigma_n)^2 \end{bmatrix}. \tag{3.11}$$

This matrix induces a diagonal norm on $\mathbb{R}^n$. Finally, $\mathbf{A}$ can be defined as the inverse of the $n \times n$ sample covariance matrix of $\mathbf{Z}$: $\mathbf{A} = \mathbf{R}^{-1}$, with

$$\mathbf{R} = \frac{1}{N} \sum_{k=1}^{N} (\mathbf{z}_k - \bar{\mathbf{z}})(\mathbf{z}_k - \bar{\mathbf{z}})^T. \tag{3.12}$$

Here $\bar{\mathbf{z}}$ denotes the sample mean of the data. In this case, $\mathbf{A}$ induces the Mahalanobis norm on $\mathbb{R}^n$ (Bezdek, 1981).

The norm metric influences the clustering criterion by changing the measure of dissimilarity. The Euclidean norm induces hyperspherical clusters, i.e., clusters whose surfaces of constant membership are hyperspheres. Both the diagonal and the Mahalanobis norm generate hyperellipsoidal clusters, the difference is that with the diagonal norm, the axes of the hyperellipsoids are parallel to the coordinate axes while with the Mahalanobis norm the orientation of the hyperellipsoid is arbitrary, as shown in Figure 3.3. A common limitation of clustering algorithms based on a fixed distance norm is that such a norm induces a fixed topological structure on $\mathbb{R}^n$ and forces the objective function to prefer clusters of that shape even if they are not present (see Example 3.4). In Section 3.4, we will see that the norm-inducing matrix $\mathbf{A}$ can be adapted by using estimates of the data covariance, and can be used to estimate the dependence of the data in each cluster.

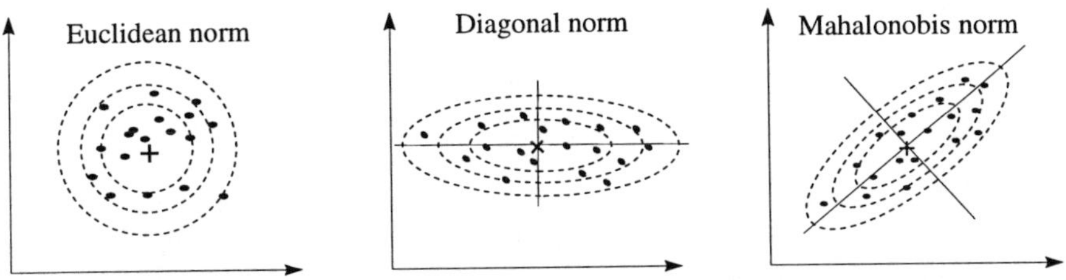

Figure 3.3. Different distance norms used in fuzzy clustering.

Example 3.4 *Fuzzy c-means clustering.* Consider a synthetic data set in $\mathbb{R}^2$, which contains two well-separated clusters of different shapes, as depicted in Figure 3.4. The samples in both clusters are drawn from the normal distribution. The standard deviation for the upper cluster is 0.2 for both axes, whereas in the lower cluster it is 0.2 for the horizontal axis and 0.05 for the vertical axis. From the membership level curves, one can see that the FCM algorithm strictly imposes a circular shape on both clusters, even though the lower cluster is rather elongated. The norm-inducing matrix was set to $\mathbf{A} = \mathbf{I}$ for both clusters, the weighting exponent was $m = 2$, and the termination criterion $\epsilon = 0.01$. The algorithm was initialized with a random partition matrix and converged after 4 iterations.

Note that it is of no help to use another $\mathbf{A}$, since the clusters may differ both in shape and orientation. Generally, different matrices $\mathbf{A}_i$ are required for the different clusters, but there is no guideline as to how to choose them a priori. Section 3.4.1 presents the partition obtained with the Gustafson–Kessel algorithm based on an adaptive distance norm. □

The following sections present several extensions of the basic c-means algorithm. A common feature of these algorithms is that they can detect clusters which lie in subspaces of the data space. The methods can be broadly classified into two groups:

- Algorithms using an adaptive distance measure, such as the Gustafson–Kessel algorithm (Gustafson and Kessel, 1979), or the fuzzy maximum likelihood estimation algorithm (Gath and Geva, 1989). These algorithms are presented in Section 3.4.

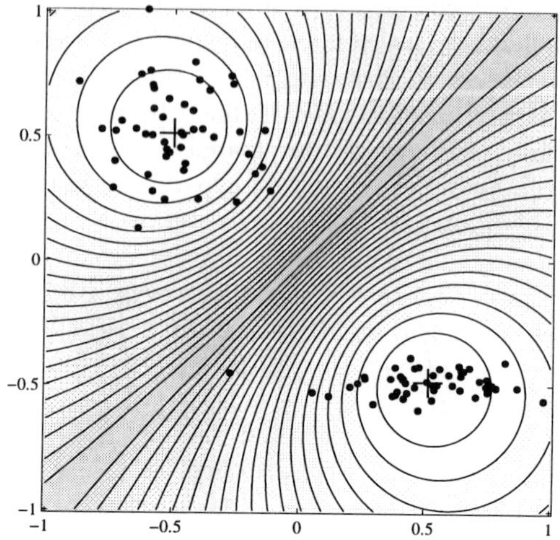

Figure 3.4. The fuzzy c-means algorithm imposes a spherical shape on the clusters, regardless of the actual data distribution. The dots represent the data points, '+' are the cluster means. Also shown are level curves of the clusters. Dark shading corresponds to membership degrees around 0.5.

- Algorithms based on hyperplanar or functional prototypes, or prototypes defined by functions. They include the fuzzy c-varieties (Bezdek, 1981), fuzzy c-elliptotypes (Bezdek, et al., 1981a), and fuzzy regression models (Hathaway and Bezdek, 1993b). These methods are presented in Section 3.5.

In addition, Section 3.6 presents a class of possibilistic clustering algorithms, which search for possibilistic partitions in the data, i.e., partitions where the constraint (3.5b) is relaxed.

3.4 Clustering with Fuzzy Covariance Matrix

A family of algorithms can be derived from the basic FCM scheme by adapting the inner-product norm (3.7d). Two of them are presented in this section: the Gustafson–Kessel algorithm and the algorithm based on fuzzy maximum likelihood estimates.

3.4.1 Gustafson–Kessel Algorithm

Gustafson and Kessel (1979) extended the standard fuzzy c-means algorithm by employing an adaptive distance norm, in order to detect clusters of different geometrical shapes in one data set. Each cluster has its own norm-inducing matrix $\mathbf{A}_i$, which yields the following inner-product norm:

$$D^2_{ik\mathbf{A}_i} = (\mathbf{z}_k - \mathbf{v}_i)^T \mathbf{A}_i (\mathbf{z}_k - \mathbf{v}_i)\,. \tag{3.13}$$

The matrices $\mathbf{A}_i$ are used as optimization variables in the c-means functional, thus allowing each cluster to adapt the distance norm to the local topological structure of the data. Let $\mathbf{A}$ denote a c-tuple of the norm-inducing matrices: $\mathbf{A} = (\mathbf{A}_1, \mathbf{A}_2, \ldots, \mathbf{A}_c)$. The objective functional of the GK algorithm is defined by:

$$J(\mathbf{Z}; \mathbf{U}, \mathbf{V}, \mathbf{A}) = \sum_{i=1}^{c} \sum_{k=1}^{N} (\mu_{ik})^m D^2_{ik\mathbf{A}_i} \tag{3.14}$$

where $\mathbf{U} \in M_{fc}$, $\mathbf{V} \in \mathbb{R}^{n \times c}$ and $m > 1$. The solutions,

$$(\mathbf{U}, \mathbf{V}, \mathbf{A}) = \underset{M_{fc} \times \mathbb{R}^{n \times c} \times \mathrm{PD}^n}{\arg\min} \; J(\mathbf{Z}; \mathbf{U}, \mathbf{V}, \mathbf{A}), \tag{3.15}$$

are stationary points of J, where PD^n denotes a space of $n \times n$ positive definite matrices. For a fixed $\mathbf{A}$, conditions (3.9) can be directly applied. However, the objective function (3.14) cannot be directly minimized with respect to $\mathbf{A}_i$, since it is linear in $\mathbf{A}_i$. J could be made as small as desired by making $\mathbf{A}_i$ less positive definite. To obtain a feasible solution, $\mathbf{A}_i$ must be constrained in some way. The usual way of accomplishing this is to constrain the determinant of $\mathbf{A}_i$. Allowing the matrix $\mathbf{A}_i$ to vary with its determinant fixed corresponds to optimizing the cluster's shape while its volume remains constant:

$$|\mathbf{A}_i| = \rho_i, \quad \rho_i > 0, \quad \forall i\,. \tag{3.16}$$

Using the Lagrange multiplier method, the following expression for $\mathbf{A}_i$ is obtained:

$$\mathbf{A}_i = [\rho_i \det(\mathbf{F}_i)]^{1/n} \mathbf{F}_i^{-1}, \tag{3.17}$$

where $\mathbf{F}_i$ is the *fuzzy covariance matrix* of the ith cluster defined by:

$$\mathbf{F}_i = \frac{\sum_{k=1}^{N} (\mu_{ik})^m (\mathbf{z}_k - \mathbf{v}_i)(\mathbf{z}_k - \mathbf{v}_i)^T}{\sum_{k=1}^{N} (\mu_{ik})^m}\,. \tag{3.18}$$

Note that the substitution of equations (3.17) and (3.18) into (3.13) gives a generalized squared Mahalanobis distance norm between $\mathbf{z}_k$ and the cluster mean $\mathbf{v}_i$, where the covariance is weighted by the membership degrees in $\mathbf{U}$. The GK algorithm is given in Algorithm 3.2.

Remark 1. The same applies to the choice of m as in the case of the FCM algorithm, see Section 3.3.2.

Remark 2. Without any prior knowledge, the cluster volumes ρ_i are simply fixed at 1 for each cluster. A drawback of the GK algorithm is that due to the constraint (3.16), it only can find clusters of approximately equal volumes. Repetitive application of GK clustering with varying volumes is mentioned in Section 4.5.2 in combination with a compatible cluster merging procedure.

Remark 3. The eigenstructure of the cluster covariance matrix provides information about the shape and orientation of the cluster. The ratio of the lengths of the cluster's hyperellipsoid axes is given by the ratio of the square roots of the eigenvalues of $\mathbf{F}_i$. The directions of the axes are given by the eigenvectors of $\mathbf{F}_i$, as shown in Figure 3.5. Linear subspaces of the data space are represented by flat hyperellipsoids, which can

Algorithm 3.2 (Gustafson–Kessel (GK) algorithm)

Given the data set $\mathbf{Z}$, choose the number of clusters $1 < c < N$, the weighting exponent $m > 1$ and the termination tolerance $\epsilon > 0$. Initialize the partition matrix randomly, such that $\mathbf{U}^{(0)} \in M_{fc}$.

Repeat for $l = 1, 2, \ldots$

Step 1: **Compute cluster prototypes (means):**

$$\mathbf{v}_i^{(l)} = \frac{\sum_{k=1}^{N} (\mu_{ik}^{(l-1)})^m \mathbf{z}_k}{\sum_{k=1}^{N} (\mu_{ik}^{(l-1)})^m}, \quad 1 \le i \le c.$$

Step 2: **Compute the cluster covariance matrices:**

$$\mathbf{F}_i = \frac{\sum_{k=1}^{N} (\mu_{ik}^{(l-1)})^m (\mathbf{z}_k - \mathbf{v}_i^{(l)})(\mathbf{z}_k - \mathbf{v}_i^{(l)})^T}{\sum_{k=1}^{N} (\mu_{ik}^{(l-1)})^m}, \quad 1 \le i \le c.$$

Step 3: **Compute the distances:**

$$D_{ik\mathbf{A}_i}^2 = (\mathbf{z}_k - \mathbf{v}_i^{(l)})^T \left[(\rho_i \det(\mathbf{F}_i))^{1/n} \mathbf{F}_i^{-1} \right] (\mathbf{z}_k - \mathbf{v}_i^{(l)}), \quad 1 \le i \le c, \quad 1 \le k \le N.$$

Step 4: **Update the partition matrix:**

if $D_{ik\mathbf{A}_i} > 0$ for $1 \le i \le c, \quad 1 \le k \le N$,

$$\mu_{ik}^{(l)} = \frac{1}{\sum_{j=1}^{c} (D_{ik\mathbf{A}_i} / D_{jk\mathbf{A}_i})^{2/(m-1)}},$$

otherwise

$$\mu_{ik}^{(l)} = 0 \text{ if } D_{ik\mathbf{A}_i} > 0, \text{ and } \mu_{ik}^{(l)} \in [0, 1] \text{ with } \sum_{i=1}^{c} \mu_{ik}^{(l)} = 1.$$

until $\|\mathbf{U}^{(l)} - \mathbf{U}^{(l-1)}\| < \epsilon$.

be seen as hyperplanes. The eigenvector corresponding to the smallest eigenvalue determines the normal to the hyperplane, and can be used to compute optimal local linear models from the covariance matrix, as shown in Lemma 5.1.

Remark 4. An advantage of the GK algorithm over FCM is that GK can detect clusters of different shape and orientation in one data set, as demonstrated in Ex-

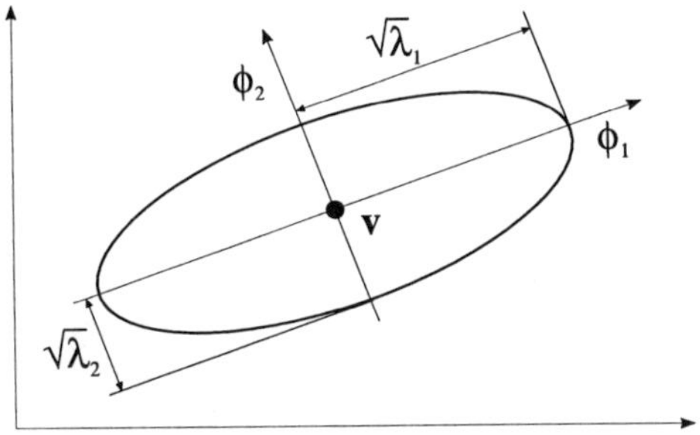

Figure 3.5. Equation $(\mathbf{z} - \mathbf{v})^T \mathbf{F}^{-1} (\mathbf{x} - \mathbf{v}) = 1$ defines a hyperellipsoid. The length of the jth axis of this hyperellipsoid is given by $\sqrt{\lambda_j}$ and its direction is spanned by $\boldsymbol{\phi}_j$, where λ_j and $\boldsymbol{\phi}_j$ are the jth eigenvalue and the corresponding eigenvector of $\mathbf{F}$, respectively.

ample 3.5. It is, however, computationally more involved than FCM, since the inverse and determinant of the cluster covariance matrix must be calculated in each iteration.

Example 3.5 The GK algorithm was applied to the data set from Example 3.4, using the same initial settings as the FCM algorithm. Figure 3.4 shows that the GK algorithm can adapt the distance norm to the underlying distribution of the data. One nearly circular cluster and one elongated ellipsoidal cluster are obtained. The shape of the clusters can be determined from the eigenstructure of the resulting covariance matrices $\mathbf{F}_i$. The eigenvalues of the clusters are:

cluster	λ_1	λ_2	$\sqrt{\lambda_1}/\sqrt{\lambda_2}$
upper	0.0352	0.0310	1.0666
lower	0.0482	0.0028	4.1490

One can see that the ratios given in the last column reflect quite accurately the ratio of the standard deviations in each data group (1 and 4 respectively). For the lower cluster, the unitary eigenvector corresponding to λ_2, $\boldsymbol{\phi}_2 = [0.0134, 0.9999]^T$, can be seen as a normal to a line representing the second cluster's direction, and it is, indeed, nearly parallel to the vertical axis. □

3.4.2 Fuzzy Maximum Likelihood Estimates Clustering

The fuzzy maximum likelihood estimates (FMLE) clustering algorithm employs a distance norm based on the fuzzy maximum likelihood estimates, proposed by Bezdek and Dunn (1975):

$$D_{ik\Sigma_i} = \frac{[\det \Sigma_i]^{1/2}}{P_i} \exp\left[\frac{1}{2}(\mathbf{z}_k - \mathbf{v}_i)^T \Sigma_i^{-1} (\mathbf{z}_k - \mathbf{v}_i)\right]. \tag{3.19}$$

Note that, contrary to the GK algorithm, this distance norm involves an exponential term and thus decreases faster than the inner-product norm. Σ_i denotes the fuzzy

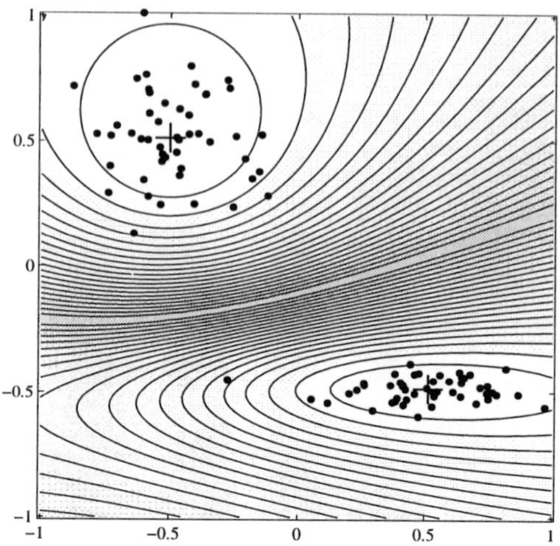

Figure 3.6. The Gustafson–Kessel algorithm can detect clusters of different shape and orientation. The points represent the data, '+' are the cluster means. Also shown are level curves of the clusters. Dark shading corresponds to membership degrees around 0.5.

covariance matrix of the ith cluster, given by:

$$\Sigma_i = \frac{\sum_{k=1}^{N} \mu_{ik}(\mathbf{z}_k - \mathbf{v}_i)(\mathbf{z}_k - \mathbf{v}_i)^T}{\sum_{k=1}^{N} \mu_{ik}} . \tag{3.20}$$

The difference between the matrix $\mathbf{F}_i$ in (3.18) and the Σ_i defined above is that the latter does not involve the weighting exponent m. This is because the two weighted covariance matrices arise as generalizations of the classical covariance from two different concepts. P_i is the prior probability of selecting cluster i, given by:

$$P_i = \frac{1}{N} \sum_{k=1}^{N} \mu_{ik} . \tag{3.21}$$

The membership degrees μ_{ik} are interpreted as the posterior probabilities, $\mu_{ik} \approx h(i|\mathbf{z}_k)$, of selecting the ith cluster given the data point $\mathbf{z}_k$. The iterative scheme of the FMLE algorithm (3.3) is very similar to that of the GK algorithm. Gath and Geva (1989) reported that the FMLE algorithm is able to detect clusters of varying shapes, sizes and densities. This is because the cluster covariance matrix is used in conjunction with an "exponential" distance, and the clusters are not constrained in volume. However, FMLE needs a good initialization, as due to the exponential distance norm, it tends to converge to a near local optimum.

3.5 Clustering with Linear Prototypes

In the algorithms described so far, the clusters are represented by their *prototypical points* (centers), $\mathbf{v}_i \in \mathbb{R}^n$, i.e., geometrical structures of the same "type" as the data.

Algorithm 3.3 (Fuzzy maximum likelihood estimate clustering)

Given the data set $\mathbf{Z}$ and a good initial partition matrix $\mathbf{U}^{(0)} \in M_{fc}$, choose the termination tolerance $\epsilon > 0$.

Repeat for $l = 1, 2, \ldots$

Step 1: **Compute cluster prototypes (means):**

$$\mathbf{v}_i^{(l)} = \frac{\sum_{k=1}^{N} \mu_{ik}^{(l-1)} \mathbf{z}_k}{\sum_{k=1}^{N} \mu_{ik}^{(l-1)}}, \quad 1 \le i \le c\,.$$

Step 2: **Compute cluster covariance matrices and prior probabilities:**

$$\Sigma_i = \frac{\sum_{k=1}^{N} \mu_{ik}^{(l-1)} (\mathbf{z}_k - \mathbf{v}_i^{(l)})(\mathbf{z}_k - \mathbf{v}_i^{(l)})^T}{\sum_{k=1}^{N} \mu_{ik}^{(l-1)}}, \quad 1 \le i \le c,$$

$$P_i = \frac{1}{N} \sum_{k=1}^{N} \mu_{ik}^{(l-1)}, \quad 1 \le i \le c\,.$$

Step 3: **Compute the distances:**

$$D_{ik\Sigma_i} = \frac{[\det(\Sigma_i)]^{1/2}}{P_i} \exp[\frac{1}{2}(\mathbf{z}_k - \mathbf{v}_i^{(l)})^T \Sigma_i^{-1} (\mathbf{z}_k - \mathbf{v}_i^{(l)})], \quad 1 \le i \le c,\ 1 \le k \le N\,.$$

Step 4: **Update the partition matrix:**

if $D_{ik\Sigma_i} > 0$ for $1 \le i \le c, \quad 1 \le k \le N$,

$$\mu_{ik}^{(l)} = \frac{1}{\sum_{j=1}^{c} (D_{ik\Sigma_i}/D_{jk\Sigma_i})^{2/(m-1)}},$$

otherwise

$$\mu_{ik}^{(l)} = 0 \text{ if } D_{ik\Sigma_i} > 0, \text{ and } \mu_{ik}^{(l)} \in [0,1] \text{ with } \sum_{i=1}^{c} \mu_{ik}^{(l)} = 1\,.$$

until $\|\mathbf{U}^{(l)} - \mathbf{U}^{(l-1)}\| < \epsilon$.

The FCM algorithm uses a fixed distance norm, and thus strongly prefers clusters of a geometrical shape induced by that norm. The GK and FMLE algorithms remedy this drawback by locally adapting the distance norm. A conceptually different approach is to define the prototypes as r-dimensional linear or nonlinear subspaces of the data space, where $0 \leq r \leq n-1$. Algorithms based on this approach are reviewed in the following sections.

3.5.1 *Fuzzy c-Varieties*

The main idea of the fuzzy c-varieties (FCV) algorithm (Bezdek, et al., 1981a) is to measure the distances of data from r-dimensional linear varieties, i.e., lines ($r = 1$), planes ($r = 2$) or hyperplanes ($2 < r < n$). This family of algorithms can detect clusters lying in r-dimensional linear subspaces of $\mathbb{R}^n$. The corresponding objective functional is given by:

$$J_V(\mathbf{Z}; \mathbf{U}, V_r) = \sum_{i=1}^{c} \sum_{k=1}^{N} (\mu_{ik})^m D_{rik}^2, \tag{3.22}$$

where V_r is a set of c linear r-dimensional varieties $V_r = (v_{r1}, v_{r2}, \ldots, v_{rc})$, and D_{rik}^2 is a squared orthogonal distance from $\mathbf{z}_k$ to the linear variety v_{ri}:

$$D_{rik}^2 = ||\mathbf{z}_k - \mathbf{v}_i||^2 - \sum_{j=1}^{r} \langle \mathbf{z}_k - \mathbf{v}_i, \mathbf{s}_{ij} \rangle^2 . \tag{3.23}$$

$(\mathbf{s}_{i1}, \mathbf{s}_{i2}, \ldots, \mathbf{s}_{ir})$ is an r-tuple of linearly independent vectors spanning the variety v_{ir}, $\mathbf{v}_i$ is a point through which the variety passes, and $\langle \cdot, \cdot \rangle$ denotes the scalar product. The functional $J_V(\mathbf{Z}; \mathbf{U}, V_r)$ attains its local minimum with respect to $\mathbf{U}$ if and only if the conditions (3.9) hold and $\mathbf{s}_{ij}$ is the unit eigenvector of the cluster covariance matrix (3.18) corresponding to the jth largest eigenvalue (Bezdek, et al., 1981a). By substituting the distance measure (3.23) into eq. (3.9a), the FCV algorithm (3.4) follows as a straightforward generalization of FCM. A major drawback of the FCV algorithm is that the linear variety is not limited in size, and thus the algorithm tends to connect collinear clusters that may be well separated. Moreover, FCV also does not partition correctly when the sizes of the varieties vary from cluster to cluster. In such case, the algorithm usually gets stuck in poor local minima (Dave, 1992).

3.5.2 *Fuzzy c-Elliptotypes*

The fuzzy c-elliptotypes (FCE) algorithm (Bezdek, et al., 1981b) attempts to alleviate some of the drawbacks of the fuzzy c-varieties algorithm by forcing each cluster to have a center of gravity $\mathbf{v}_i$, and by measuring the distance as a convex combination of the FCM and FCV distances:

$$D_{eik} = \alpha D_{ik} + (1 - \alpha) D_{rik}, \tag{3.24}$$

Algorithm 3.4 (Fuzzy c-varieties)

Given the data set $\mathbf{Z}$, choose the number of clusters $1 < c < N$, the weighting exponent $m > 1$, the dimension of the prototypical varieties $0 < r < n$ and the termination tolerance $\epsilon > 0$. Initialize the partition matrix randomly, such that $\mathbf{U}^{(0)} \in M_{fc}$.

Repeat for $l = 1, 2, \ldots$

Step 1: **Compute cluster centers (means):**

$$\mathbf{v}_i^{(l)} = \frac{\sum_{k=1}^{N} \mu_{ik}^{(l-1)} \mathbf{z}_k}{\sum_{k=1}^{N} \mu_{ik}^{(l-1)}}, \quad 1 \le i \le c\,.$$

Step 2: **Compute cluster covariance matrices:**

$$\mathbf{F}_i^{(l)} = \frac{\sum_{k=1}^{N} (\mu_{ik}^{(l-1)})^m (\mathbf{z}_k - \mathbf{v}_i^{(l)})(\mathbf{z}_k - \mathbf{v}_i^{(l)})^T}{\sum_{k=1}^{N} (\mu_{ik}^{(l-1)})^m}, \quad 1 \le i \le c\,.$$

Step 3: **Extract principal eigenvectors.** Extract from each $\mathbf{F}_i^{(l)}$ its r principal eigenvectors $\mathbf{s}_{ij}^{(l)}$, $j = 1, 2, \ldots, r$ (eigenvectors corresponding to the r largest eigenvalues).

Step 4: **Compute the distances:**

$$D_{rik}^2 = \|\mathbf{z}_k - \mathbf{v}_i^{(l)}\|^2 - \sum_{j=1}^{r} (\langle \mathbf{z}_k - \mathbf{v}_i^{(l)}, \mathbf{s}_j^{(l)} \rangle)^2,$$
$$1 \le i \le c, \quad 1 \le k \le N\,.$$

Step 5: **Update the partition matrix:**

if $D_{rik} > 0$ for $1 \le i \le c, \quad 1 \le k \le N$,

$$\mu_{ik}^{(l)} = \frac{1}{\sum_{j=1}^{c} (D_{rik}/D_{rjk})^{2/(m-1)}},$$

otherwise

$$\mu_{ik}^{(l)} = 0 \text{ if } D_{rik} > 0, \text{ and } \mu_{ik}^{(l)} \in [0, 1] \text{ with } \sum_{i=1}^{c} \mu_{ik}^{(l)} = 1\,.$$

until $\|\mathbf{U}^{(l)} - \mathbf{U}^{(l-1)}\| < \epsilon$.

where $\alpha \in [0, 1]$, D_{ik} is the Euclidean distance of $\mathbf{z}_k$ from $\mathbf{v}_i$ (3.10), and D_{rik} is given by eq. (3.23). Substituting equations (3.10) and (3.23) into (3.24) yields:

$$D_{eik} = (\mathbf{z}_k - \mathbf{v}_i)^T(\mathbf{z}_k - \mathbf{v}_i) - \alpha \sum_{j=1}^{r} \langle \mathbf{z}_k - \mathbf{v}_i, \mathbf{s}_{ij} \rangle^2 . \tag{3.25}$$

The first term in this expression measures the distance from the cluster center, and thus restricts the cluster size. The geometric interpretation of eq. (3.25) is that the level surfaces of the convex combinations D_{eik} are hyperellipsoids obtained by stretching a hypersphere defined by the Euclidean norm in the directions of vectors $\mathbf{s}_{ij}$. The FCE algorithm is identical to FCV, with the exception that (3.25) is used to compute the distances and that the mixing coefficient α must be defined.

As shown in Section 4.4.2, the FCE algorithm does not completely correct the problems of FCV. Moreover the value for α must be chosen carefully. If α is common to all clusters, the algorithm will seek clusters of the same elliptical shape. Techniques have also been proposed to adaptively select the mixing coefficient α for each cluster (Gunderson, 1983).

3.5.3 *Fuzzy c-Regression Models*

The last fuzzy clustering algorithm presented in this chapter is the fuzzy c-regression models (FCRM) algorithm proposed by Hathaway and Bezdek (1993b). This algorithm estimates parameters of c regression models together with a fuzzy c-partitioning of the data. The regression models take the general form

$$y_k = f_i(\mathbf{x}_k; \boldsymbol{\theta}_i), \tag{3.26}$$

where the functions f_i are parameterized by $\boldsymbol{\theta}_i \in \mathbb{R}^{p_i}$. The membership degree $\mu_{ik} \in \mathbf{U}$ is interpreted as a weight representing the extent to which the value predicted by the model $f_i(\mathbf{x}_k; \boldsymbol{\theta}_i)$ matches y_k. The prediction error is defined by:

$$E_{ik}(\boldsymbol{\theta}_i) = \left[y_k - f_i(\mathbf{x}_k; \boldsymbol{\theta}_i)\right]^2 \tag{3.27}$$

but other measures can be applied as well, provided they fulfill the minimizer property stated by Hathaway and Bezdek (1993b). The family of objective functions for fuzzy c-regression models is defined for $\mathbf{U} \in M_{fc}$ and $(\boldsymbol{\theta}_1, \ldots \boldsymbol{\theta}_c) \in \mathbb{R}^{p_1} \times \mathbb{R}^{p_2} \times \ldots \times \mathbb{R}^{p_c}$ by:

$$E_m(\mathbf{U}, \{\boldsymbol{\theta}_i\}) = \sum_{i=1}^{c} \sum_{k=1}^{N} (\mu_{ik})^m E_{ik}(\boldsymbol{\theta}_i) . \tag{3.28}$$

One possible approach to minimize the objective function (3.28) is the grouped coordinate minimization method (Hathaway and Bezdek, 1991a), given in Algorithm 3.5.

A specific situation for Step 1 of the algorithm arises when the regression functions f_i in (3.26) are linear in the parameters $\boldsymbol{\theta}_i$. In such a case, the parameters can be obtained as a solution of a weighted least-squares problem where the membership

Algorithm 3.5 (Fuzzy c-regression models)

Given a set of data $\mathbf{Z} = \{(\mathbf{x}_1, y_1), \ldots, (\mathbf{x}_N, y_N)\}$, specify c, the structure of the regression models (3.26) and the error measure (3.27). Choose the weighting exponent $m > 1$ and the termination tolerance $\epsilon > 0$. Initialize the partition matrix randomly, such that $\mathbf{U}^{(0)} \in M_{fc}$.

Repeat for $l = 1, 2, \ldots$

Step 1: **Calculate values for the model parameters** $\boldsymbol{\theta}_i^*$ that globally minimize the function $E_m(\mathbf{U}^{(l)}, \{\boldsymbol{\theta}_i\})$.

Step 2: **Update the partition matrix:**

$$\mu_{ik}^{(l)} = \frac{1}{\sum_{j=1}^{c} (E_{ik}/E_{jk})^{2/(m-1)}}, \quad 1 \leq i \leq c, \quad 1 \leq k \leq N .$$

if $E_{ik} = 0$ for some $i = s$, set $\mu_{sk} = 1$ and $\mu_{ik} = 0, \forall i \neq s$.

until $||\mathbf{U}^{(l)} - \mathbf{U}^{(l-1)}|| < \epsilon$.

degrees of the fuzzy partition matrix $\mathbf{U}$ serve as the weights. Define the matrix $\mathbf{X} \in \mathbb{R}^{N \times p}$, the vector $\mathbf{y} \in \mathbb{R}^N$, and the matrix $\mathbf{W}_i \in \mathbb{R}^{N \times N}$, as follows:

$$\mathbf{X} = \begin{bmatrix} \mathbf{x}_1^T \\ \mathbf{x}_2^T \\ \vdots \\ \mathbf{x}_N^T \end{bmatrix}, \quad \mathbf{y} = \begin{bmatrix} y_1 \\ y_2 \\ \vdots \\ y_N \end{bmatrix}, \quad \mathbf{W}_i = \begin{bmatrix} \mu_{i1} & 0 & \cdots & 0 \\ 0 & \mu_{i2} & \cdots & 0 \\ \vdots & \vdots & \ddots & \vdots \\ 0 & 0 & \cdots & \mu_{iN} \end{bmatrix} . \tag{3.29}$$

The optimal parameters $\boldsymbol{\theta}_i$ are then computed by:

$$\boldsymbol{\theta}_i = \left[\mathbf{X}_e^T \mathbf{W}_i \mathbf{X}_e\right]^{-1} \mathbf{X}_e^T \mathbf{W}_i \mathbf{y} . \tag{3.30}$$

The FCRM algorithm suffers from the same drawback as the FCV algorithms, as the clusters are not limited in size. The advantage of the algorithm is that is can also fit locally nonlinear models to data, such as polynomials, which are still linear in their parameters and hence lead to a linear estimation problem in Step 1 of Algorithm 3.5.

3.6 Possibilistic Clustering

The clustering approaches derived from the FCM functionals use the "probabilistic" constraint (3.5b), which states that the sum of membership degrees of each data point equals one. It has been recognized that the membership degrees generated by FCM-based algorithms do not always correspond to the degree of typicality. These problems arise in situations, where the total membership of a data point to all the clusters does not equal one, as in the presence of outliers, see Example 3.3. Several approaches have been suggested to replace (3.5b) by a less restrictive constraint. The method proposed by Krishnapuram and Keller (1993) uses the following objective function:

$$J(\mathbf{Z}, \eta; \mathbf{U}, \mathbf{V}) = \sum_{i=1}^{c} \sum_{k=1}^{N} (\mu_{ik})^m \|\mathbf{z}_k - \mathbf{v}_i\|_A^2 + \sum_{i=1}^{c} \eta_i \sum_{k=1}^{N} (1 - \mu_{ik})^m, \tag{3.31}$$

where η_i are positive constants. The first term is identical to the FCM objective function (3.7a). The second term forces the memberships μ_{ik} to be as large as possible, thus avoiding the trivial solution of the optimization problem, $\mathbf{U} = 0$, which would occur after simply removing constraint (3.5b). Note that the columns in $\mathbf{U}$ are now independent, which makes it possible to decompose the global objective function (3.31) into c individual objective functions for the c clusters. Differentiating with respect to $\mathbf{U}$ and setting to zero leads to the following necessary condition for μ_{ik}:

$$\mu_{ik} = \frac{1}{1 + \left(\frac{D_{ik\mathbf{A}_i}}{\eta_i}\right)^{2/(m-1)}}, \tag{3.32}$$

where $D_{ik\mathbf{A}}^2 = \|\mathbf{z}_k - \mathbf{v}_i\|_{\mathbf{A}}^2$ is the squared inner-product norm. The necessary conditions for the prototypes $\mathbf{v}_i$ are identical to the corresponding conditions for FCM and its derivatives. The value of η_i determines the width of the resulting possibility distribution and simultaneously specifies the relative weighting of the second term in (3.31). The same value may be chosen for all clusters, if they all are expected to be similar, or it can be chosen based on the initial partition, in proportion to the average intra-cluster distance:

$$\eta_i = \frac{\sum_{k=1}^{N} (\mu_{ik})^m D_{ik\mathbf{A}}^2}{\sum_{k=1}^{N} (\mu_{ik})^m}. \tag{3.33}$$

The basic "possibilistic" c-means (PCM) algorithm, which follows as a straightforward generalization of the FCM iteration, is given in Algorithm 3.6.

As mentioned above, by removing the constraint (3.5b), the membership functions of the c clusters become independent of each other. This makes PCM more sensitive to initialization, since nothing prevents the algorithm from converging to degenerate possibilistic partitions where all clusters are identical, or very similar to each other. Typically, FCM may be used to find an initial partition for PCM. The concept of possibilistic clustering has also been applied to the GK algorithm and other FCM derivatives (Krishnapuram and Keller, 1993).

Example 3.6 To illustrate the difference between FCM and PCM, these two algorithms are applied to an artificial data set, similar to the set in Example 3.1. In both cases, the settings of the parameters are: $\mathbf{A}_i = \mathbf{I}$ for all clusters, $m = 2$ and $\epsilon = 0.01$. The FCM algorithm is initialized with a random partition matrix, and the PCM with the partition generated by FCM. In the following, we refer to the left cluster as cluster 1 and to the right cluster as cluster 2.

By comparing the level curves in Figure 3.7a and Figure 3.7b, one can see that the membership degrees generated by the possibilistic algorithm correspond to the

Algorithm 3.6 (Possibilistic c-means (PCM))

Given the data set $\mathbf{Z}$ and a good initial partition $\mathbf{U}^{(0)} \in M_{pc}$, choose the number of clusters $1 < c < N$, the weighting exponent $m > 1$, the termination tolerance $\epsilon > 0$ and the norm-inducing matrix $\mathbf{A}$. Estimate η_i using (3.33).

Repeat for $l = 1, 2, \ldots$

Step 1: **Compute cluster prototypes (means):**

$$\mathbf{v}_i^{(l)} = \frac{\sum_{k=1}^{N} (\mu_{ik}^{(l-1)})^m \mathbf{z}_k}{\sum_{k=1}^{N} (\mu_{ik}^{(l-1)})^m}, \quad 1 \leq i \leq c.$$

Step 2: **Compute the distances:**

$$D_{ik\mathbf{A}}^2 = (\mathbf{z}_k - \mathbf{v}_i^{(l)})^T \mathbf{A} (\mathbf{z}_k - \mathbf{v}_i^{(l)}), \quad 1 \leq i \leq c, \quad 1 \leq k \leq N.$$

Step 3: **Update the partition matrix:**

$$\mu_{ik}^{(l)} = \frac{1}{1 + \left(\frac{D_{ik\mathbf{A}}}{\eta_i}\right)^{2/(m-1)}} \quad 1 \leq i \leq c, \quad 1 \leq k \leq N.$$

until $||\mathbf{U}^{(l)} - \mathbf{U}^{(l-1)}|| < \epsilon$.

distances from the cluster prototype, and that they are not influenced by the neighboring cluster (the level curves generated by PCM are almost circular while in the case of FCM they are distorted near the cluster boundary).

Note also that the performance of PCM is not influenced by the presence of the outlier A. As expected, the membership degrees assigned to this point by FCM are $\mu(A) = [0.4973, 0.5027]^T$, thus approximately equal to the membership degrees of B, $\mu(B) = [0.4931, 0.5069]^T$. It is obvious that point B is much closer to the prototypes of both clusters than A, and thus should have a greater degree of membership. The PCM algorithm accounts for this difference by assigning much lower membership degrees to A than to B ($\mu(A) = [0.0215, 0.0244]^T$ and $\mu(B) = [0.1147, 0.1263]^T$, respectively).

Further, it is interesting to note that the possibilistic partition correctly reflects the symmetrical form of the clusters (the level curves in Figure 3.7b are almost circular). This observation can be confirmed by examining the memberships of points C and D that are both approximately at the same distance from the center of cluster 1. Since, in the fuzzy partition, the membership degrees are relative to the distance from a point to *all* the clusters, point C receives greater membership in the cluster 1,

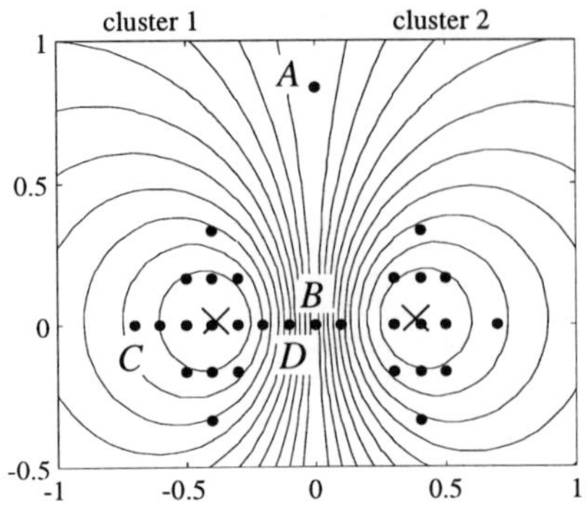

(a) Fuzzy clustering.

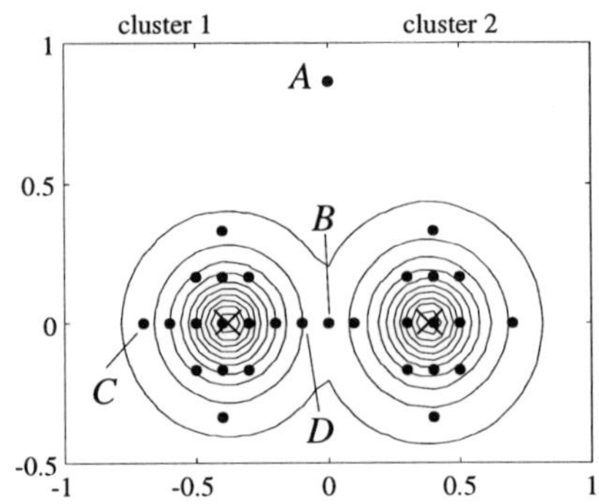

(b) Possibilistic clustering.

Figure 3.7. Comparison of the partitions generated by fuzzy and possibilistic c-means algorithms.

$\mu(C) = [0.9234, 0.0766]^T$, than point D with $\mu(D) = [0.7367, 0.2633]^T$ which is closer to cluster 2. The PCM algorithm considers points C and D equally typical for cluster 1 and assigns them similar membership degrees $\mu(C) = [0.1500, 0.0177]^T$ and $\mu(D) = [0.1935, 0.0832]^T$ respectively. □

3.7 Determining the Number of Clusters

When clustering real data without any a priori information about the data structure, one usually has to make assumptions about the number of underlying subgroups (clusters) c in the data. The chosen clustering algorithm then searches for c clusters, regardless of whether they are really present in the data or not. Two main approaches to determining the appropriate number of clusters in data can be distinguished:

- Clustering data for different values of c, and using *validity measures* to assess the goodness of the obtained partitions. Different scalar validity measures have been proposed in the literature. Section 4.5.1 gives an overview of validity measures used with the adaptive distance clustering algorithms, and demonstrates their performance on several examples.

- Starting with a sufficiently large number of clusters, and successively reducing this number by merging clusters that are similar (compatible) with respect to some predefined criteria. This approach, called *compatible cluster merging*, is presented in Section 4.5.2.

3.8 Data Normalization

Distance norms are sensitive to variations in the numerical ranges of the different features. The Euclidean distance, for example, assigns more weighting to features with wide ranges than to those with narrow ranges. The result of clustering can thus be negatively influenced by, for instance, choosing different measurement units. In pattern recognition literature, it is often suggested that the data should be appropriately normalized before clustering (Jain and Dubes, 1988). The simplest type of normalization is the subtraction of the feature means $\bar{z}_j$:

$$z_{jk} = z^*_{jk} - \bar{z}_j, \tag{3.34}$$

which makes the feature values invariant to rigid displacements of the coordinates. The asterisk denotes the raw (unscaled) data. Another type of normalization translates and scales the axes so that all the features have zero mean and unit variance:

$$z_{jk} = \frac{z^*_{jk} - \bar{z}_j}{\sigma_j} . \tag{3.35}$$

However, normalization is not always desirable, as it may alter the separation between clusters and negatively influence the results of clustering. It turns out that clustering algorithms based on adaptive distance measure, see Section 3.4, are less sensitive to data scaling, since the adaptation of the distance measure automatically compensates for the differences in scale. The following example illustrates this property.

Example 3.7 An artificial data set contains two well-separated clusters which are relatively far apart along the horizontal axis. The clusters are practically linear and parallel to the vertical axis. Figure 3.8a shows level curves of a fuzzy partition obtained with the FCM algorithm on the original (unscaled data).

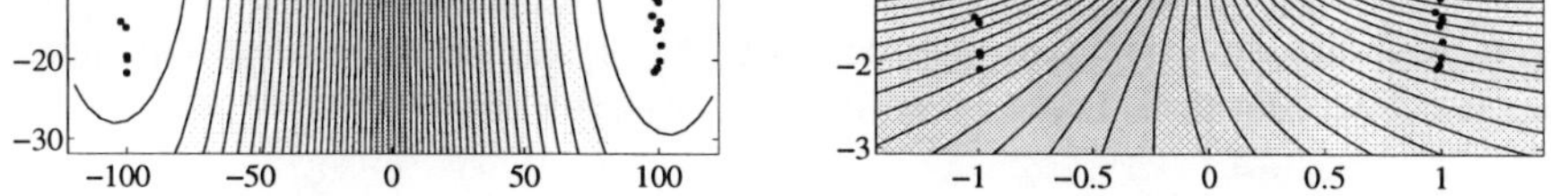

(a) Original data. *(b) Normalized data.*

Figure 3.8. The fuzzy *c*-means algorithm is sensitive to the scaling (normalization) of data. The dots represent the data points, the circles are the cluster means.

Because of the large distance between the two clusters, FCM is able to pick the cluster shapes quite correctly. The same algorithm was applied to data normalized by eq. (3.35). Note that the scales in Figure 3.8a and Figure 3.8b are different. Since the distance between the clusters becomes considerably smaller due to scaling, the partition reflects the influence of the two clusters on each other, and the geometrical shape of the clusters no longer corresponds to the underlying data structure. Figure 3.9 gives the results for the GK algorithm. Note that almost identical partitions are obtained for both the raw and the normalized data sets. □

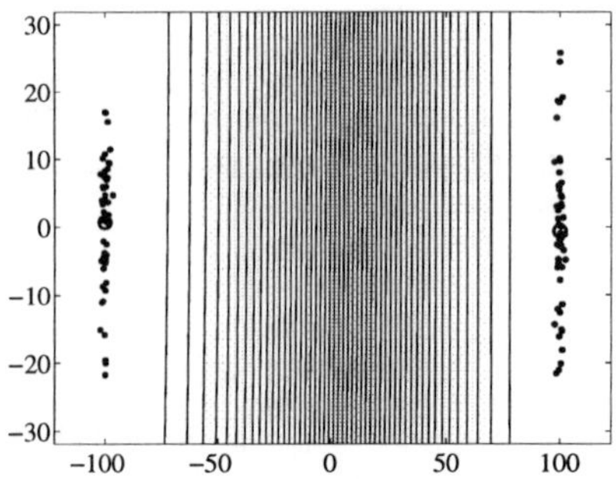

(a) Original data.

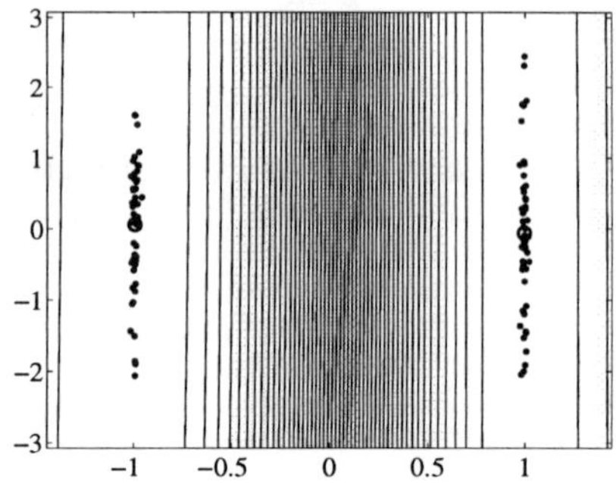

(b) Normalized data.

Figure 3.9. The Gustafson-Kessel algorithm is less sensitive to the data scale. The dots represent the data points, the circles are the cluster means.

3.9 Summary and Concluding Remarks

Fuzzy clustering is a powerful unsupervised method for data analysis. A large number of clustering algorithms have been proposed in the literature, and applied to a variety of real-world problems. In this chapter, methods that can be used to detect clusters contained in subspaces of the data space have been presented. These methods can be applied to the approximation of nonlinear systems, and can facilitate the task of building and analyzing models of complex systems based on numerical data, as shown in Chapter 5. This particular aim imposes some requirements on the performance and validation of the clustering algorithms that may be quite different from those usually considered in the pattern recognition literature. A discussion of this issue and analysis of the selected algorithms is presented in the following chapter.

4 PRODUCT-SPACE CLUSTERING FOR IDENTIFICATION

This chapter addresses the decomposition of a nonlinear identification problem into a set of locally linear models by means of product-space fuzzy clustering. The identification procedure is first outlined in Section 4.1. Structure selection and the choice of regressors in the modeling of dynamic systems are discussed in Section 4.2. Section 4.3 describes the principle of identification of nonlinear systems by product-space clustering. The choice of clustering algorithms is discussed in Section 4.4. Section 4.5 deals with the determination of the number of clusters by means of validity measures and compatible cluster merging.

4.1 Outline of the Approach

Figure 4.1 outlines the individual steps of the identification procedure, which is iterative in its nature. In a typical modeling session, some of the steps may be repeated for different choices of the various parameters. The purpose and the different steps and the related methods are outlined below.

Step 1: Design of identification experiments and data collection. This is an important initial step for any identification method, since it determines the information content of the identification data set. As opposed to linear techniques, pseudo-random binary excitation signals are not suitable for nonlinear identification in general, and for fuzzy clustering in particular. Although the choice of the excitation signal may be problem dependent, the input data should preferably excite the system

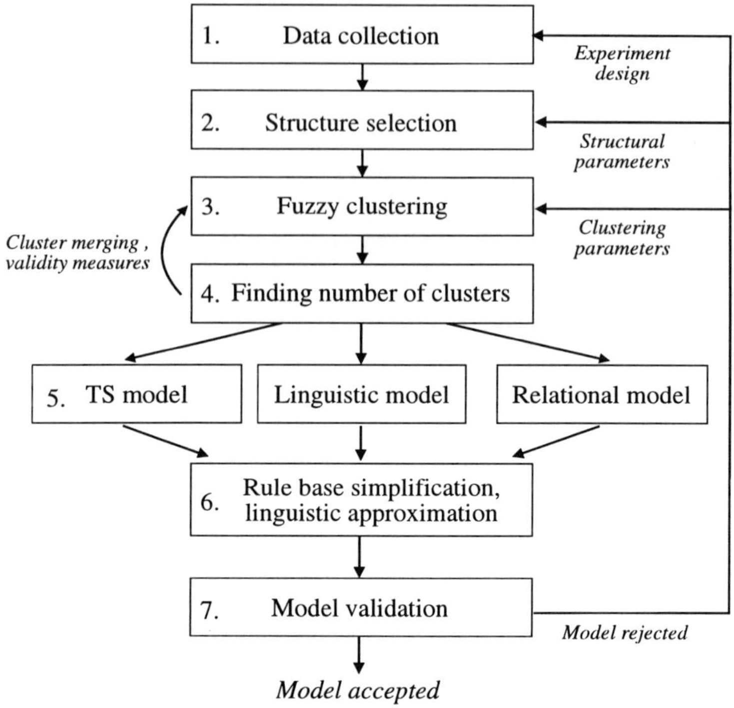

Figure 4.1. Overview of the identification approach based on fuzzy clustering.

in the entire range of the considered variables both in amplitude and in frequency. The pseudo-random binary signal is not suitable, since it only contains two amplitude levels. Typical choices are a multi-sinusoidal signal or a step-wise signal with random amplitude and random width (Godfrey, 1993). White noise of small amplitude is often added to these signals in order to guarantee proper excitation of the process dynamics. The choice of a suitable sampling period, the design of (anti-aliasing) filters, the duration of the experiments, etc., are other important issues of the experiment design.

Step 2: Structure selection. The purpose of this step is to determine the relevant input and output variables with respect to the aim of the modeling exercise. When identifying dynamic systems, the structure and the order of the model dynamics must be chosen. Structure selection allows us to translate the identification of a dynamic system into a regression problem that can be solved in a static or quasi-static manner. The structure can be selected in an automated way by comparing different candidate structures in terms of some performance measures. In most cases, a reasonable choice can be made by the user, based on the prior knowledge about the process.

Step 3: Clustering of the data. Structure selection leads to a nonlinear static regression problem, which is then approximated by a collection of local linear submodels. The location and the parameters of the submodels are found by partitioning the available data into hyperplanar or hyperellipsoidal clusters. Each of the clusters defines a fuzzy region in which the system can be approximated locally by a linear submodel.

Step 4: Selection of the number of clusters. By applying cluster validity measures, compatible cluster merging, or a combination of the two techniques, an appropriate number of clusters can be found. This step typically involves several repetitions of Step 3 for a different number of clusters and a different initial partition matrix.

Step 5: Generation of an initial fuzzy model. Fuzzy clustering divides the available data into groups in which local linear relations exists between the inputs and the output. In order to obtain a model suitable for prediction or controller design, a rule-based fuzzy model of a selected structure is derived from the available fuzzy partition matrix and from the cluster prototypes. The rules, the membership functions and other parameters that constitute the fuzzy model are extracted in an automated way. The exact procedure applied at this step depends on the type of fuzzy model required and on the purpose of modeling (prediction, analysis, control design, etc.), as discussed in more detail in Chapter 5.

Step 6: Simplification and reduction of the initial model. Initial fuzzy models obtained from data may be redundant in the sense that they contain more membership functions than are necessary to describe the system. Fuzzy similarity measures can be applied to simplify or reduce the initial fuzzy rule base and to obtain linguistic interpretation of the membership functions. Chapter 5 presents the corresponding techniques.

Step 7: Model validation. By means of validation, the final model is either accepted as appropriate for the given purpose, or it is rejected. In the latter case, some steps of the identification loop shown in Figure 4.1 may be repeated with a different setting, as it is usual also in other approaches to linear and nonlinear system identification (Ljung, 1987; Johansen, 1994). In addition to the usual numerical validation by means of simulation, interpretation of fuzzy models plays an important role in the validation step. The coverage of the input space by the rules can be analyzed, and, for an incomplete rule base, additional rules can be provided based on prior knowledge, local linearization, or first-principle models.

4.2 Structure Selection

In fuzzy modeling, the problem of structure selection[1] can be divided into three subproblems: 1) choice of input and output variables; 2) representation of the system's dynamics, and 3) choice of the fuzzy model's granularity.

[1] Structure selection is also called structure identification in the literature, see, for instance, (Sugeno and Yasukawa, 1993; Johansen, 1994).

Choice of input and output variables. Although most identification methods assume that the input and output variables of the process are known (Ljung, 1987), in reality, especially for multivariable and closed-loop systems, it is often not clear which variables should be considered as the model inputs. The selection of the input and output variables is based on the aim of the modeling exercise, on the prior knowledge related to the (expected) process dynamics, and on additional variables that may cause the nonlinearity of the system. Statistical techniques, such as correlation analysis, can be used in combination with prior knowledge. This step can also be partially automated. Several candidate models with different input variables can be compared in terms of some performance measure, and the best one is then selected.

Representation of the system's dynamics. A common approach is to transform the identification of a dynamic system into a static regression problem (Leonaritis and Billings, 1985; Chen and Billings, 1989; Sjöberg, et al., 1995). The choice of this particular transformation is usually based on a combination of a priori knowledge with intuition, insights, and understanding of the process behavior. Mechanistic (physical, first-principle) modeling of the well-understood relationships and physical laws can guide the selection of the relevant variables, and of the model's order. This transformation can be regarded as a mapping from the domain of time signals into a space of variables that fully determine the state of the system. These variables are called the regressors. The system's behavior can be predicted by means of a static mapping from the space of regressors to the space of the model output (regressand). A major distinction can be made between input–output models, state-space models, and hybrid (semi-mechanistic) models. The choice of the regressors is a crucial step, as an inappropriate choice may hamper the modeling effort. Choosing too poor a structure (too few regressors) results in inaccurate modeling of the process dynamics and nonlinearities. Choosing a structure richer than necessary (too many regressors) leads to badly conditioned estimation problems and to "overfitting" the data.

Granularity of the fuzzy model. This choice is related to the number of linguistic terms defined for each variable and therefore also to the number of rules in the model. In the fuzzy modeling literature, the term "structure identification" often pertains to this step (Sugeno and Kang, 1988). When fuzzy clustering is applied to generate fuzzy models from data, the number of clusters is the primary parameter that must be chosen. This choice can be based on prior knowledge, or can be done automatically, by techniques which are presented in Section 4.5. It should be noted that the granularity of the final model is also related to the particular type of the model used (e.g., Takagi–Sugeno, linguistic, etc.). The ways in which these models approximate functions differ, which means that the granularity needed to achieve a required level of accuracy may differ as well, see Sections 2.1 to 2.3. In practice, a trade-off is usually sought between the accuracy and the complexity of the model.

4.2.1 The Nonlinear Regression Problem

Fuzzy systems are general function approximators (Wang, 1992; Kosko, 1994; Zeng and Singh, 1994a; 1995b), and as such can be applied to general nonlinear regression problems. Nonlinear regression is the modeling of the static dependence of a response variable, called the regressand, $y \in Y \subset \mathbb{R}$ on the regression vector $\mathbf{x} = [x_1, \ldots, x_p]^T$ over some domain $X \subset \mathbb{R}^p$. The elements of the regression vector will be called the *regressors* and the domain X the *regressor space*. The system that generated the data is presumed to be described by:

$$y \approx f(\mathbf{x}) . \tag{4.1}$$

The deterministic function $f(\cdot)$ captures the dependence of y on $\mathbf{x}$, and $\approx$ reflects the fact that y will not be an exact function of $\mathbf{x}$[2]. The aim of regression is to use the data to construct a function $F(\mathbf{x})$ that can serve as a reasonable approximation of $f(\mathbf{x})$ not only for the given data, but over the entire domain X. The definition of "reasonable approximation" depends on the purpose for which the model is constructed. If the aim of modeling is to obtain predictions of y, accuracy is the most relevant criterion. Lack of accuracy is usually defined by the integral error

$$I = \int_X \|f(\mathbf{x}) - F(\mathbf{x})\| \mathrm{d}\mathbf{x}, \tag{4.2}$$

over the entire domain X. In general, this error cannot be computed, since the value of f is known only at the available data points. Therefore, the average prediction error over the available data is often used

$$J = \frac{1}{N} \sum_{i=1}^{N} \|f(\mathbf{x}_i) - F(\mathbf{x}_i)\|, \tag{4.3}$$

where N denotes the number of data samples. The attainment of the minimum of I in (4.2) implies the best model possible with the selected structure. This is, however, not the case with the criterion J, which only guarantees that the model fits the available data with the least error. A separate validation step is hence necessary, in order to assess the goodness of the model over the entire region of interest X.

Apart from accurate predictions, the goal may be to obtain a model that can be used to analyze and understand the properties of the real system that generated the data. A strong potential of fuzzy models is that they describe systems as a collection of simple local submodels that are expressed as rules. The rules can be formulated using natural language, and as such may be more understandable than complicated mathematical expressions. Rules can also be combined with analytical models commonly used by control engineers, such as the local linear models in the Takagi–Sugeno structure described in Section 2.3.

4.2.2 Input–output Black-box Models

There are a number of possibilities for the choice of regressors in nonlinear black-box identification. Since extensive literature is available on this topic (Leonaritis and Billings, 1985; Chen and Billings, 1989; Sjöberg, et al., 1994), only a brief review

[2] In connection with the identification of dynamic systems and time series, also more general regression structures can be defined, as discussed in Section 4.2.2.

is given here with respect to the use of the particular structures for clustering-based identification.

The NARX (Nonlinear AutoRegressive with eXogenous input) model is frequently used with many nonlinear identification methods, such as neural networks (Hunt, et al., 1992), radial basis functions (Chen, et al., 1991), CMAC (Brown and Harris, 1994), and fuzzy models (Yager and Filev, 1994; Wang, 1994). The NARX model establishes a relation between the past input–output data and the predicted output:

$$\hat{y}(k+1) = F\left(y(k), \ldots, y(k-n_y+1), u(k), \ldots, u(k-n_u+1)\right), \tag{4.4}$$

where k denotes discrete time samples, n_u and n_y are integers related to the system's order, and F denotes a fuzzy model. In the NARX model, the regression vector is a collection of a finite number of past inputs and outputs, $\mathbf{x}(k) = [y(k), \ldots, y(k-n_y+1), u(k), \ldots, u(k-n_u+1)]^T$. The regressand is the predicted output $\hat{y}(k+1)$. Hence, from a set of observed inputs and outputs of an unknown dynamic system, $S = \{(u(j), y(j)) \mid j = 1, 2, \ldots, N\}$, the function $F(\cdot)$ in (4.4) can be approximated by using static nonlinear regression. Pure delays from the input to the output can be directly incorporated in the regression vector $\mathbf{x}(k) = [y(k), \ldots, y(k-n_y+1), u(k-n_d+1), \ldots, u(k-n_d-n_u+2)]^T$, where n_d is the delay in samples. For the sake of brevity, the models will be given for $n_d = 1$. Nonlinear counterparts of other linear black-box structures can be used as well. The Nonlinear Finite Impulse Response (NFIR) model employs only the past values of $u(k)$ as regressors:

$$\hat{y}(k+1) = F\left(u(k), u(k-1), \ldots, u(k-n_u+1)\right). \tag{4.5}$$

Also here the unknown function $F(\cdot)$ can be directly inferred from the data by using static regression. Note, however, that n_u may be quite large and is not directly related to the system's order.

The Nonlinear Output Error (NOE) model involves the past model predictions instead of the data:

$$\hat{y}(k+1) = F\left(\hat{y}(k), \ldots, \hat{y}(k-n_y+1), u(k), \ldots, u(k-n_u+1)\right). \tag{4.6}$$

The "innovations" forms offer additional freedom in describing the effect of disturbances. In the NARMAX model, for instance, the prediction error $e(k) = y(k) - \hat{y}(k)$ and its past values are included in the regression vector as well:

$$\begin{aligned} \hat{y}(k+1) = \; & F\big(y(k), \ldots, y(k-n_y+1), u(k), \ldots, u(k-n_u+1), \\ & e(k), \ldots, e(k-n_e)\big). \end{aligned} \tag{4.7}$$

The regression vector in the NOE and innovations model forms cannot be constructed directly from the data, since it includes the past values of the model output. The corresponding regression problem must be solved iteratively. In principle, clustering methods can be applied to these structures, but as fuzzy clustering is iterative in itself, it leads to a rather complex optimization problem. It is, however, reasonable to expect that the partition of the operating space of a nonlinear system into regions described by linear models can be correctly determined regardless of the exact noise model. Each of the local models in the TS model structure can then use an appropriate noise model in order to improve the prediction and to obtain more accurate estimates of the model parameters.

The choice of the regression vector can be seen as a projection from the time domain into the space with orthonormal basis $\{z^{-k}|k = 0, 1, 2, \ldots\}$. Besides this usual choice of the basis, other projections may be employed in order to reduce the dimension of the regression problem, and to provide numerically better conditioned problems (van den Hof, et al., 1995). Application of this approach to nonlinear identification was proposed by Schram, et al. (1996).

Determination of the input–output model structure boils down to the choice of the model type and of the related structural parameters n_y, n_u and of the pure delays n_d. A priori knowledge is typically used to make a first guess of the range of these parameters and a structure is then sought within this range that minimizes a certain criterion. The most straightforward approach is to directly use the mean-square prediction error,

$$J = \frac{1}{N} \sum_{i=1}^{N} (y(i) - \hat{y}(i))^2, \tag{4.8}$$

evaluated on a different data set than was used to identify the system, in order to avoid fitting the noise. In connection with fuzzy models, Sugeno and Yasukawa (1993) also applied the *regularity criterion*. The available data set is first divided into two subsets, say A and B. For each subset, one model is built and the regularity criterion RC is computed by

$$RC = \frac{1}{2} \left[\frac{1}{N_A} \sum_{i=1}^{N_A} [y^A(i) - \hat{y}^A_B(i)]^2 + \frac{1}{N_B} \sum_{i=1}^{N_B} [y^B(i) - \hat{y}^B_A(i)]^2 \right], \tag{4.9}$$

where N_A (N_B) is the number of data points in A (B), y^A (y^B) is the systems output in the subset A (B), $\hat{y}^A_B$ is the output of model B simulated with data from A, and $\hat{y}^B_A$ is the output of model A simulated with data from B. More general v-fold cross-validation techniques can be applied as well (Weiss and Kulikowski, 1991).

Since linear identification techniques are much simpler and numerically more robust than nonlinear methods, it is usually worthwhile to start with a linear model to determine the structure. A variety of tools can be used, such as the singular value test (Verhaegen and Dewilde, 1992) or the information-theoretic criteria (Akaike, 1974; Rissanen, 1978). The structure of the best linear model is then used as a starting point for nonlinear modeling. Also cluster validity measures can guide the selection of the model's order and the number of clusters within the given structure.

So far, it has been assumed that y is a scalar, i.e., the system under study is a MISO system. With input–output models, MIMO systems can be represented in two ways: the function F is a vector-valued function, or the MIMO system is decomposed into a set of coupled MISO systems. While the former approach is typically used with neural networks, in fuzzy modeling the decomposition approach is mostly adopted. The reason is that the it is more flexible if each output is associated with a different sort of nonlinearity. One output may contain a complex nonlinearity in some region, while another output may be linear in the same region. By decomposing the MIMO mapping into several MISO mappings, the number of membership functions and rules can be reduced. In this book we use the latter approach.

4.2.3 State-space Framework

In addition to the most frequently used input–output structures, fuzzy models can also represent nonlinear systems in the state-space form:

$$\begin{aligned} \mathbf{x}(k+1) &= g(\mathbf{x}(k), \mathbf{u}(k)) \\ \mathbf{y}(k) &= h(\mathbf{x}(k)) \end{aligned}$$

where state transition function g maps the current state $\mathbf{x}(k)$ and the input $\mathbf{u}(k)$ into a new state $\mathbf{x}(k+1)$. The output function h maps the state $\mathbf{x}(k)$ into the output $\mathbf{y}(k)$. An example of a rule-based representation of a state-space model is the following Takagi–Sugeno model:

$$\textbf{If } \mathbf{x}(k) \text{ is } A_i \textbf{ and } \mathbf{u}(k) \text{ is } B_i \textbf{ then } \begin{cases} \mathbf{x}(k+1) = \mathbf{A}_i\mathbf{x}(k) + \mathbf{B}_i\mathbf{u}(k) \\ \mathbf{y}(k) \quad\;\; = \mathbf{C}_i\mathbf{x}(k) \end{cases} \tag{4.10}$$

for $i = 1, \ldots, K$. Here $\mathbf{A}_i$, $\mathbf{B}_i$, $\mathbf{C}_i$ are matrices of appropriate dimensions, associated with the ith rule.

The state-space representation is useful when the prior knowledge allows us to model the system from first principles such as mass and energy balances. In literature, this approach is called white-box state-space modeling (Ljung, 1987). If the state is directly measured on the system, or can be reconstructed from other measured variables, both g and h can be approximated by using nonlinear regression techniques. An advantage of the state-space modeling approach is that the structure of the model is related to the structure of the real system, and, consequently, also the model parameters are often physically relevant. This is usually not the case in the input-output models. In addition, the dimension of the regression problem in state-space modeling is often smaller than with input–output models, since the state of the system can be represented with a vector of a lower dimension than the regression (4.4). For these reasons, subspace identification techniques are gaining increasing popularity for linear systems as well as for special classes of nonlinear systems (Westwick and Verhaegen, 1996).

4.2.4 Semi-mechanistic Modeling

With physical insights in the system, nonlinear transformations of the measured signals can be involved. When modeling, for instance, the relation between the room temperature and the voltage applied to an electric heater, the power signal is computed by squaring the voltage, since it is the heater power rather than the voltage that causes the temperature to change (Lindskog and Ljung, 1994). This new variable is then used in a linear black-box model instead of the voltage itself. The motivation for using nonlinear regressors in nonlinear models is not to waste effort (rules, parameters, etc.) on estimating facts that are already known.

Another approach is based on a combination of white-box and black-box models. In many systems, such as chemical and biochemical processes, the modeling task can be divided into two subtasks: modeling of well-understood mechanisms based on mass and energy balances (first-principle modeling), and approximation of partially known relationships such as specific reaction rates. A number of hybrid modeling approaches have been proposed that combine first principles with nonlinear black-box models, e.g., neural networks (Psichogios and Ungar, 1992; Thompson and Kramer, 1994) or fuzzy models (Babuška, et al., 1996). A neural network or a fuzzy model is typically used as a general nonlinear function approximator that "learns" the unknown relationships from data and serves as a predictor of unmeasured process quantities that are difficult to model from first principles.

As an example, consider the modeling of a fed-batch stirred bioreactor described by the following equations derived from the mass balances (Psichogios and Ungar, 1992):

$$\frac{\mathrm{d}X}{\mathrm{d}t} = \eta(\cdot)X - \frac{F}{V}X \tag{4.11a}$$

$$\frac{\mathrm{d}S}{\mathrm{d}t} = -k_1\eta(\cdot)X + \frac{F}{V}[S_i - S] \tag{4.11b}$$

$$\frac{\mathrm{d}V}{\mathrm{d}t} = F \tag{4.11c}$$

where X is the biomass concentration, S is the substrate concentration, V is the reactor's volume, F is the inlet flow rate, k_1 is the substrate to cell conversion coefficient, and S_i is the inlet feed concentration. These mass balances provide a partial model. The kinetics of the process are represented by the specific growth rate $\eta(\cdot)$ which accounts for the conversion of the substrate to biomass, and it is typically a complex nonlinear function of the process variables. Many different models have been proposed to describe this function, but choosing the right model for a given process may not be straightforward. The hybrid approach is based on an approximation of $\eta(\cdot)$ by a nonlinear (black-box) model from process measurements and incorporates the identified nonlinear relation in the white-box model. The data can be obtained from batch experiments, for which $F = 0$, and eq. (4.11a) reduces to the expression:

$$\frac{\mathrm{d}X}{\mathrm{d}t} = \eta(\cdot)X, \tag{4.12}$$

where $\eta(\cdot)$ appears explicitly. This model is then used in the white-box model given by equations (4.11) for both batch and fed-batch regimes. An application of the semi-mechanistic approach to modeling the kinetics of enzymatic Penicillin G conversion is given in Section 7.3.

4.3 Identification by Product-space Clustering

The principle of identification by product-space clustering is to approximate a nonlinear regression problem by decomposing it into several local linear subproblems. This approach has a number of advantages in comparison with global nonlinear models, such as neural networks. The model structure is easy to understand and interpret, both qualitatively and quantitatively. Various types of knowledge can be integrated in the model, including empirical knowledge, measured data and available mathematical models. In addition, the approach has computational advantages and lends itself to straightforward adaptive and learning algorithms (Murray-Smith and Johansen, 1997).

Fuzzy clustering is applied in the product space of the regressors and the regressand: $X \times Y$. Let $\mathbf{X}$ denote the matrix in $\mathbb{R}^{N \times p}$, having the regression vectors $\mathbf{x}_k^T$ in its rows, and let $\mathbf{y}$ denote the column vector in $\mathbb{R}^N$, containing the regressands y_k:

$$\mathbf{X} = \begin{bmatrix} \mathbf{x}_1^T \\ \mathbf{x}_2^T \\ \vdots \\ \mathbf{x}_N^T \end{bmatrix}, \qquad \mathbf{y} = \begin{bmatrix} y_1 \\ y_2 \\ \vdots \\ y_N \end{bmatrix}. \tag{4.13}$$

N denotes the number of data samples, p is the dimension of the regression vector. For an input–output model of a dynamic system, the matrix $\mathbf{X}$ contains shifted versions of the input and output data. As an example, assume a second-order NARX model $y(k+1) = F\left(y(k), y(k-1), u(k), u(k-1)\right)$. With the set of available measurements, $S = \{(u(j), y(j)) \mid j = 1, 2, \ldots, N_d\}$, the regressor matrix and the regressand vector are:

$$\mathbf{X} = \begin{bmatrix} y(2) & y(1) & u(2) & u(1) \\ y(3) & y(2) & u(3) & u(2) \\ \vdots & \vdots & \vdots & \vdots \\ y(N_d-1) & y(N_d-2) & u(N_d-1) & y(N_d-2) \end{bmatrix}, \qquad \mathbf{y} = \begin{bmatrix} y(3) \\ y(4) \\ \vdots \\ y(N_d) \end{bmatrix}.$$

In this example, $N = N_d - 2$. The decomposition of a global nonlinear mapping into a set of locally linear models is based on a geometrical interpretation of the regression problem. The unknown nonlinear function $y = f(\mathbf{x})$ represents a nonlinear (hyper)surface in the product space: $(X \times Y) \subset \mathbb{R}^{p+1}$. This surface is called the *regression surface*.

Example 4.1 For low-order systems, the regression surface can be visualized. As an example, consider a series connection of a static dead-zone/saturation nonlinearity with a first-order linear dynamic system (a Hammerstein system):

$$y(k+1) = 0.6y(k) + w(k), \tag{4.14a}$$

where $w = f(u)$ is given by:

$$w = \begin{cases} 0, & -0.3 \le u \le 0.3, \\ u, & 0.3 \le |u| \le 0.8, \\ 0.8\,\mathrm{sign}(u), & 0.8 \le |u|. \end{cases} \tag{4.14b}$$

The input-output description of the system using the NARX model (4.4) can be seen as a surface in the space $(U \times Y \times Y) \subset \mathbb{R}^3$, as shown in Figure 4.2a. As another example, consider a state-space system (Chen and Billings, 1989):

$$\begin{aligned} x(k+1) &= x(k) + u(k), \\ y(k) &= \exp(-x(k)). \end{aligned} \tag{4.15}$$

For this system, an input–output regression model $y(k+1) = y(k)\exp(-u(k))$ can be derived. The corresponding regression surface is shown in Figure 4.2b. Note that if the measurements of the state of this system are available, the state and output mappings in (4.15) can be approximated separately, yielding one two-variate linear and one univariate nonlinear problem, which can be solved more easily. □

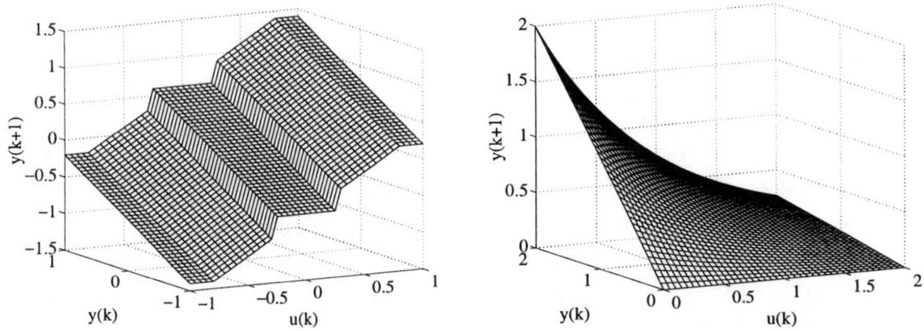

(a) *System with a dead zone and saturation.* (b) *System* $y(k+1) = y(k)\exp(-u(k))$.

Figure 4.2. Regression surfaces of two nonlinear dynamic systems.

The available data represents a sample from the regression surface. By clustering the data, local linear models can be found that approximate the regression surface in an optimal way. The set (matrix) of data to be clustered, denoted $\mathbf{Z}$, is constructed by concatenating the regressor data matrix $\mathbf{X}$ and the regressand vector $\mathbf{y}$:

$$\mathbf{Z}^T = [\mathbf{X}, \mathbf{y}] \,. \tag{4.16}$$

This data set is a subset of the Cartesian product space $X \times Y$ defined by the nonlinear functional relationship (4.1):

$$\mathbf{Z} \subset X \times Y \quad \text{such that} \quad \mathbf{y} \approx f(\mathbf{X})\,. \tag{4.17}$$

The data set $\mathbf{Z}$ is partitioned into fuzzy subsets by applying fuzzy clustering algorithms capable of detecting linear substructures in data, see Sections 3.4 and 3.5. The choice of an appropriate clustering algorithm is addressed in Section 4.4.

The membership of the data samples in the clusters is described by the fuzzy partition matrix. Each cluster is characterized by its center and covariance matrix which represents the variance of the data in the cluster. A fuzzy clustering algorithm $\mathcal{C}$ can be regarded as a mapping $\mathcal{C}$: $(Z \times \mathbb{N}) \rightarrow (M_{fc} \times \mathbb{R}^{n \times c} \times \mathrm{PD}^n)$:

$$(\mathbf{U}, \mathbf{V}, \mathbf{F}) = \mathcal{C}(\mathbf{Z}, c; \mathbf{U}^0, m, \epsilon), \tag{4.18}$$

where c is the number of clusters, $\mathbf{U}^0$ is the initial partition matrix and m, ϵ are the parameters of the clustering algorithm, see Chapter 3. The partition matrix $\mathbf{U}$ contains the membership degrees of the data points in the clusters with prototypes $\mathbf{V}$.

The cluster covariance matrix $\mathbf{F}_i$ conveys information about the shape and orientation of the ith cluster. The jth eigenvalue and the jth unit eigenvector of $\mathbf{F}_i$ are denoted λ_{ij} and ϕ_{ij} respectively. The eigenvalues of the cluster covariance matrix $\mathbf{F}_i$ are arranged such that:

$$\lambda_{i1} \geq \lambda_{i2} \geq \cdots \geq \lambda_{in}, \tag{4.19}$$

and the eigenvectors are labeled accordingly. The eigenvectors ϕ_{i1} to ϕ_{in-1} span the ith cluster's linear subspace and the nth eigenvector ϕ_{in} is the normal to this linear subspace. Since λ_{in} is the smallest eigenvalue, ϕ_{in} is called the smallest eigenvector.

When the intrinsic dimension of the data is p, λ_{in} is in orders of magnitude smaller than the remaining eigenvalues. This means that the n-dimensional data can be locally represented by a linear combination of $n-1$ variables. Hence, the proportions between the eigenvalues can be used to check whether an appropriate structure has been chosen. If all the eigenvalues are in the same order of magnitude, then no functional relationship between the regressors and the regressand has been detected (the chosen structure may not be rich enough). However, when several relatively small eigenvalues are found, the regression problem may be (locally) of a lower dimension than was assumed, or the data might not be rich enough.

Example 4.2 This example illustrates the partition of a nonlinear function into linear submodels by product-space clustering. Consider the univariate nonlinear function given by:

$$y = 1 \cdot 10^{-4} \sin(0.001x^2)x^3 + \epsilon, \quad x \in [0, 100], \tag{4.20}$$

where $\epsilon \sim N(0, 25)$ is a normally distributed random noise. Figure 4.3a shows 200 data points generated by (4.20).

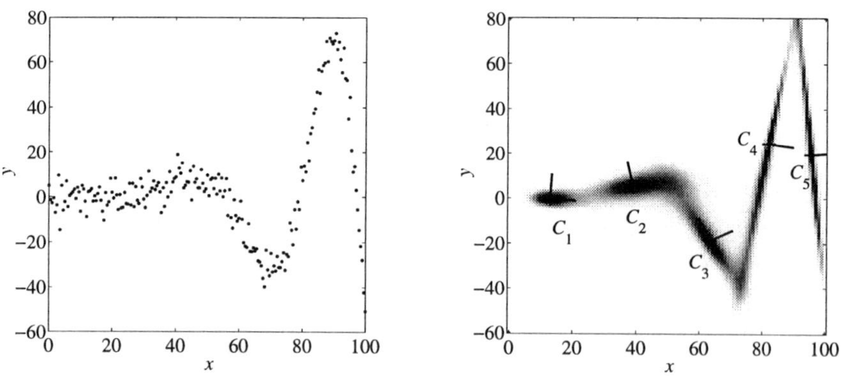

(a) Identification data.

(b) Obtained clusters.

Figure 4.3. Sinusoidal function and its piece-wise linear approximation by clusters. Dark regions in plot (b) correspond to high membership degrees. Also shown is the smallest eigenvector of each cluster.

The available data pairs $(x_i, y_i)^T$, $i = 1, 2, \ldots, 200$ are organized in a 2 by 200 data matrix:

$$\mathbf{Z} = \begin{bmatrix} x_1, x_2, \ldots, x_{200} \\ y_1, y_2, \ldots, y_{200} \end{bmatrix}, \tag{4.21}$$

and are clustered using the GK algorithm. In this example, the number of clusters was set a priori to five (see Section 4.5 for methods to determine the number of clusters). The fuzziness coefficient was $m = 2$. The GK algorithm converged after 26 iterations and formed fuzzy clusters depicted in Figure 4.3b.

By comparing the smallest eigenvalue of the second cluster ($\lambda_{12} = 19.1$) with the smallest eigenvalue of the fifth cluster ($\lambda_{52} = 0.4$) one can see that the data in cluster C_1 has much greater variance than the data in cluster C_5. In other words, C_5 is more planar (linear) than C_1. This observation is confirmed by examining Figure 4.3.

The orientation of the clusters is given by the directions of the hyperellipsoid axes, i.e., the eigenvectors ϕ_{ij}. In our example, the smallest eigenvector of the cluster C_1 is $\phi_{1,2} = [0.063, 0.998]^T$. One can see that this vector is nearly perpendicular to the x-axis, i.e., the cluster is nearly parallel with x. The eigenvector $\phi_{5,2} = [0.997, 0.071]^T$ indicates that the cluster C_5 is nearly parallel with the y-axis, as can also be seen from Figure 4.3b.

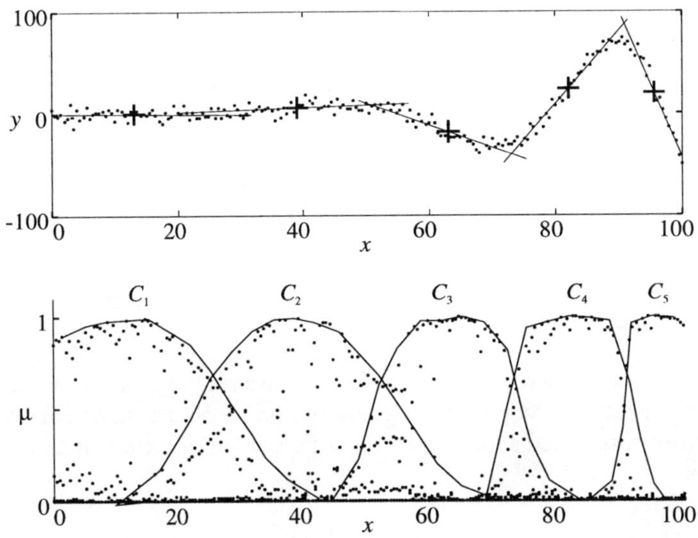

Figure 4.4. The upper graph shows the data and the local linear models given by the largest eigenvector of the cluster covariance matrices. Cluster centers are denoted by the '+' markers. The bottom graph depicts membership functions for x obtained as approximate envelopes of the projection of the fuzzy partition matrix onto x.

The individual partitioning subsets in $\mathbf{U}$ projected on the regressor x define directly fuzzy regions, in which the data can be reasonably approximated by the linear submodels defined by the clusters, see Figure 4.4. Additional processing is required to obtain membership functions suitable for prediction and interpretation of the model is described in Section 5.1.1. The overlap of the adjacent membership functions already gives an indication of the local character of the nonlinearity. For instance, the membership functions of clusters C_1 and C_2 overlap more than those of clusters C_4 and C_5, which corresponds to the form of the approximated function. □

4.4 Choice of Clustering Algorithms

Clustering algorithms suitable for locating clusters that are linear subspaces of the data have been reviewed in Sections 3.4 and 3.5. These algorithms are based on extensions of the c-means functional, which is a least-squares criterion for minimizing the variance of the data from the cluster means. These algorithms differ in the definition of the distance measure and of the prototypical structure for the clusters. Because of these differences, each algorithm performs in a different way for the same data set. With regard to the choice of a suitable clustering algorithm for system identification, it is required that:

- The clusters represent local linear models of the function being approximated. This facilitates the analysis and control design based on the obtained model. In this respect, fuzzy models generated by product space clustering differ from the majority of other local approximation methods, like radial basis functions or splines.
- The projection of the partition matrix onto the regressors results in a semantically interpretable partition with unimodal fuzzy sets. An example of such a projection is shown in Figure 4.4.
- The algorithm is robust with respect to the initialization, and does not suffer from convergence to local optima representing unsatisfactory solutions of the approximation problem.
- The algorithm is able to reveal clusters of different sizes, since some regions in the regression space can be easily represented as a single linear model, while other regions may require finer partitioning.

The advantages and drawbacks of the different clustering algorithms are discussed in the following sections. The presented results are based on the statistical evaluation of a large number of repeated experiments, in order to eliminate the influence of random initialization.

4.4.1 Clustering with Adaptive Distance Measure

Two algorithms are considered in this section, the Gustafson–Kessel (GK) algorithm and the fuzzy maximum likelihood estimates (FMLE) algorithm.

Gustafson–Kessel Algorithm. The GK algorithm appears to be a suitable method for identification purposes, because of the following properties:

- The size of the clusters is limited by the definition of the distance measure (3.13). The fuzzy sets induced by the partition matrix are compact, have typically one distinct extreme, and hence are easy to interpret.

- In comparison with the other considered algorithms, the GK algorithm is relatively insensitive to the initialization of the partition matrix (or cluster prototypes).
- As the GK algorithm is based on an adaptive distance measure, it is not so sensitive to scaling (normalization, standardization) of the data, see Example 3.7.
- The GK algorithm can detect clusters of different shapes, not only linear subspaces, see Figure 3.6.

The GK algorithm has, however, also some drawbacks:

- The calculation of the inverse and of the determinant of the covariance matrix in each iteration slows down the algorithm considerably for a large data dimension n and a large number of clusters c.
- When only a small number of data samples is available, or when the data are linearly dependent, numerical problems occur when the covariance matrix becomes close to singular. For this reason, the GK algorithm cannot be applied to purely linear problems, in the ideal noise-free case.
- Without any prior knowledge, the volumes, ρ_i, of the clusters are set equal to each other. The GK algorithm then cannot detect clusters that differ largely in their volumes. Hence, for problems where the regions approximated by linear submodels significantly differ in size, the locations and the parameters of the submodels are biased, as illustrated by the following example.

Example 4.3 Consider a piece-wise linear function

$$y = \begin{cases} a_1 x + b_1 + \epsilon, & 0 \le x \le 0.2, \\ a_2 x + b_2 + \epsilon, & 0.2 < x \le 1, \end{cases} \quad \text{with} \quad \begin{array}{ll} a_1 = 5, & b_1 = 0, \\ a_2 = 0.2, & b_2 = 0.96, \end{array} \tag{4.22}$$

where $\epsilon \sim N(0, \sigma^2)$ with $\sigma = 0.08$. A hundred data pairs (x, y) generated by (4.22) for $x \in [0, 1]$ were clustered in two clusters by the GK algorithm. The volume constraints were set equal, $\rho_1 = \rho_2 = 1$. The algorithm converged in 14 iterations to a solution depicted in Figure 4.5.

Due to the equal cluster volumes, the GK algorithm arrives at a biased solution, where the boundary between the two clusters is shifted away from the true value 0.2. By comparing the estimated[3] parameters,

$$\begin{array}{ll} \hat{a}_1 = 2.2784 \pm 0.1662, & \hat{b}_1 = 0.2783 \pm 0.0357, \\ \hat{a}_2 = 0.2409 \pm 0.0383, & \hat{b}_2 = 0.9225 \pm 0.0260, \end{array}$$

with the true values in (4.22), one can see the bias in $\hat{a}_1$ and $\hat{b}_1$. When there is some prior knowledge about the volumes, it can be translated into an appropriate setting for the parameters ρ_i. This setting can also be sought by means of compatible cluster merging, as described in Section 4.5.2. Another possibility is to use the FMLE algorithm as shown below. □

[3] The parameters are estimated by the weighted least-squares method, see Section 5.1.2.

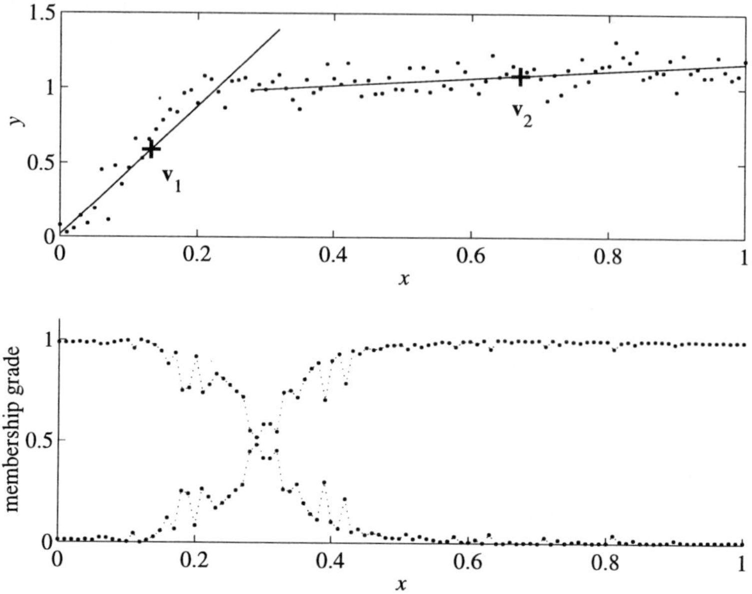

Figure 4.5. The GK algorithm with equal volume constraints gives a biased estimate when the clusters differ considerably in size. The upper plot shows the data and the estimated local models. The bottom plot gives the membership functions.

Fuzzy Maximum Likelihood Clustering. As suggested by Gath and Geva (1989), the FMLE clustering algorithm should be able to automatically detect clusters of varying volumes, contrary to the GK algorithm. A drawback of the FMLE algorithm is that it generates almost crisp partitions due to the exponential distance measure (3.19), and, consequently, it is also more sensitive to the initial conditions. It is thus useful to generate the initial partition, for instance, by the GK algorithm, and then initialize the FMLE algorithm with that partition.

Example 4.4 Consider the data in Example 4.3. Taking the fuzzy partition generated by the GK algorithm as the initial partition and running the FMLE algorithm resulted in a solution shown in Figure 4.6. The number of iterations needed for the FMLE algorithm to converge was 11 (with the termination criterion $\epsilon = 0.01$).

The boundary between the two clusters is now correctly located at 0.2. The numerical values of the parameter estimates and their standard deviations for the FMLE algorithm are:

$$\hat{a}_1 = 5.1443 \pm 0.1382, \qquad \hat{b}_1 = -0.0222 \pm 0.0161,$$
$$\hat{a}_2 = 0.2141 \pm 0.0322, \qquad \hat{b}_2 = 0.9424 \pm 0.0208\,.$$

By comparing these values with the true parameters in (4.22), one can see that the estimates obtained by the FMLE algorithm are more accurate than those of the GK method. Also the variance of the parameter estimates is smaller, since the clusters overlap less. □

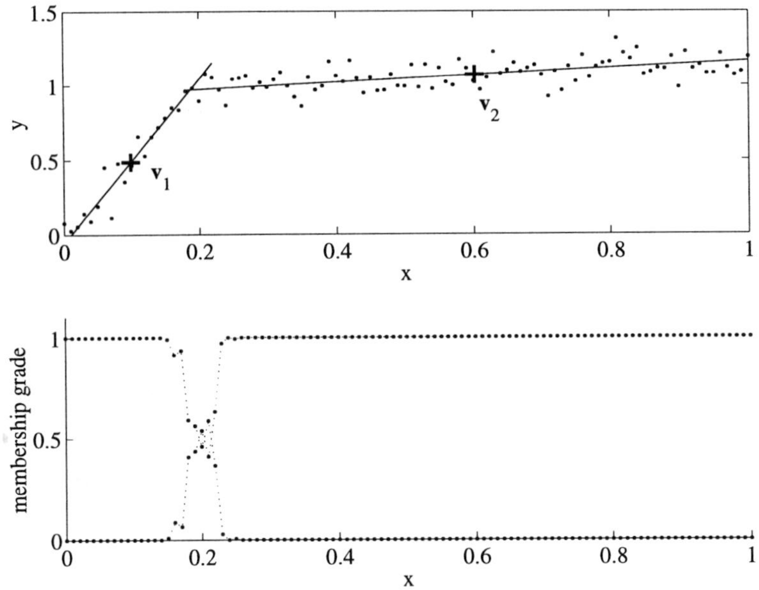

Figure 4.6. The FMLE algorithm gives a better estimate of the local models and also of the boundary between them. The upper plot shows the data and the estimated local models, the membership functions are given in the bottom plot.

Possibilistic Clustering. The fuzzy sets in the projected partition of Figure 4.5 are not convex. Data points which are quite distant from a cluster's center may belong to that cluster with a relatively high membership degree, because they happen to be in the direction of the ellipsoid's principal axis. The probabilistic constraint on the membership degrees causes that these data points automatically receive lower membership in the other clusters, which results in a nonconvexity of the fuzzy sets induced by these clusters.

The possibilistic variant of the GK algorithm (PGK) is supposed to be able to solve this drawback, see Section 3.6. From the experiments carried out, it appears that the PGK algorithm converges to a degenerate partition with all clusters grouped near each other; see Figure 4.7. Note that convergence to such a solution is not possible with the algorithms derived from the FCM functional, because the probabilistic constraint (3.5b) prevents such solutions. It is of no help to start the PGK algorithm with a good initial partition. The result shown in Figure 4.7 was obtained for the initialization with the GK partition shown in Figure 4.4.

4.4.2 Fuzzy c-lines and c-elliptotypes

The fuzzy c-lines (FCL) and fuzzy c-elliptotypes (FCE) algorithms are designed to detect linear clusters, see Section 3.5.2. The cluster prototypes are defined as linear varieties (lines in $\mathbb{R}^2$, planes in $\mathbb{R}^3$, and hyperplanes in a general multidimensional

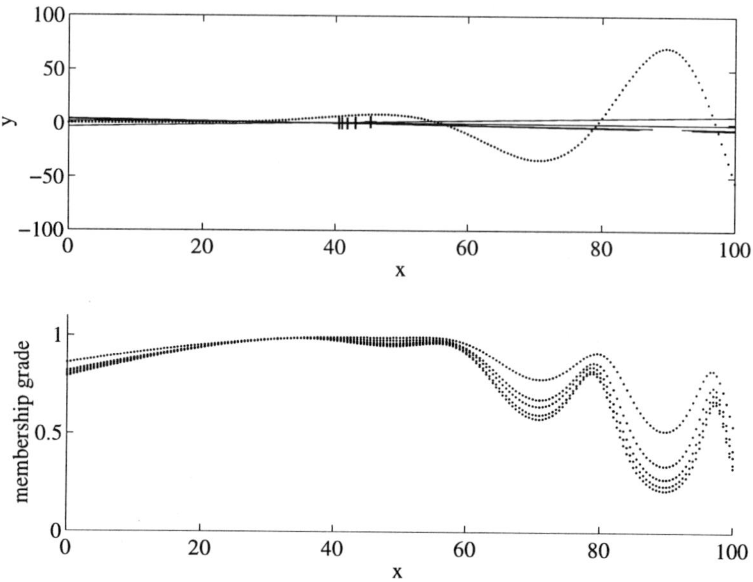

Figure 4.7. The possibilistic variant of the GK algorithm often converges to a useless partition with all the cluster centers grouped next to each other. The upper plot shows the data, the centers and the estimated local models. The membership functions are given in the bottom plot.

space) and the distance metric measures the distance of data points from the linear varieties. However, the FCL algorithm proves to be unsuitable for function approximation purposes.

An inherent drawback of the FCL method is that the cluster size is not limited, and thus the algorithm tends to connect well-separated clusters, and, consequently, generates nonconvex fuzzy sets. The FCE method attempts to alleviate this drawback by taking a convex sum of the distance from the linear variate and a distance from a point, in order to limit the size of the cluster, see eq. (3.24). The determination of the weighting coefficient α in this convex sum remains a significant problem. The exact value of α is problem dependent, and is also influenced by scaling of the data, since (3.24) involves the Euclidean distance measure, see Section 3.8.

The FCL and FCE algorithms appear to be more sensitive to initialization than the GK algorithm. For fitting more complex functions such as (4.20), starting from a random initial partition results in convergence to poor solutions, such as the one shown in Figure 4.8. Finally, the FCL and FCE algorithms are computationally more demanding than the algorithms based on adaptive distance measures. In MATLAB, for instance, eigenvector extraction requires about ten times more floating-point operations than the computation of a matrix inverse and determinant.

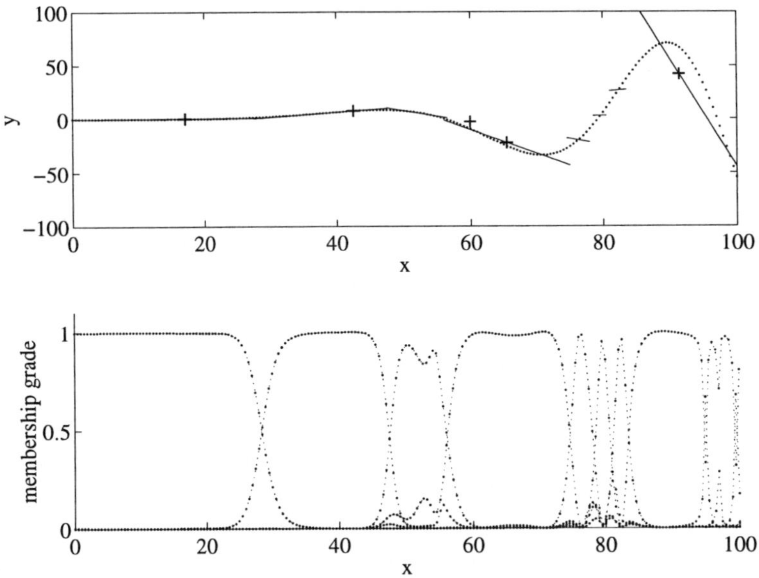

Figure 4.8. The FCE algorithm often converges to poor solutions. The upper plot shows the data and the estimated local models, the membership functions are given in the bottom plot.

4.4.3 Fuzzy c-regression Models

The fuzzy c-regression (FCR) algorithm yields simultaneous estimates for the parameters of the local regression models together with the partitioning of the data, see Section 3.5.3. Contrary to the previous methods, with this approach, the cluster prototypes are not geometrical objects in the data space, but are defined explicitly by functional relationships in terms of regression equations. For linear regression, the FCR algorithm suffers from the same drawback as the FCL method, i.e., the size of the clusters is not limited. Consequently, the algorithm is rather sensitive to initialization.

Also computationally, the FCR algorithm is very involved, as it requires solving a regression problem at each iteration. The computational costs depend on the nature of this regression problem (linear or nonlinear in parameters), and on the amount of data involved. For local linear regression such as that illustrated in Example 4.2, the computational costs are in orders of magnitude higher than for the GK algorithm. An advantage of the FCR method is that it can easily accommodate nonlinear local models such as polynomials, whose parameters still can be estimated by linear least-squares techniques. The use of locally nonlinear submodels can reduce the complexity of the global model, and can also improve the numerical accuracy of the fuzzy model.

4.5 Determining the Number of Clusters

Before fuzzy clustering can be applied, the number of clusters must be specified. Two methods to determine the number of clusters are considered in this section: cluster validity measures, and compatible cluster merging. Validity measures assess the goodness of the obtained partition by using criteria like the within-cluster distance, the partition density, the entropy, etc. Cluster merging approaches start with a higher number of clusters than are expected for the particular problem. The initial number of clusters is then reduced by successively merging compatible clusters until some threshold is reached and no more clusters can be merged.

4.5.1 Cluster Validity Measures

The use of cluster validity measures is a standard approach to determining an appropriate number of clusters in a data set. Clustering algorithms generally aim at locating well-separated and compact clusters. When the number of clusters is chosen equal to the number of groups that actually exist in the data, it can be expected that the clustering algorithm will identify them correctly. When this is not the case, misclassifications appear, and the clusters are not likely to be well separated and compact. Hence, most cluster validity measures are designed to quantify the separation and the compactness of the clusters. However, as Bezdek (1981) points out, the concept of cluster validity is open to interpretation and can be formulated in different ways. Consequently, many validity measures have been introduced in the literature, see (Bezdek, 1981; Gath and Geva, 1989; Backer, 1995; Pal and Bezdek, 1995) among others.

A widely accepted principle is that good clusters are not very fuzzy, i.e., do not exhibit too much overlap. Gath and Geva (1989) suggested assessing the goodness of the obtained partition by evaluating the separation between the clusters, the volume of the clusters, and the number of data points concentrated in the vicinity of the cluster prototype. Validity measures that account for these requirements are given below. *Fuzzy hypervolume* V_h is defined by:

$$V_h = \sum_{i=1}^{c} [\det(\mathbf{F}_i)]^{1/2}, \tag{4.23}$$

where $\mathbf{F}_i$'s are the cluster covariance matrices. Good partitions are indicated by small values of V_h. The *average partition density* D_A is defined by

$$D_A = \frac{1}{c} \sum_{i=1}^{c} \frac{S_i}{[\det(\mathbf{F}_i)]^{1/2}}, \tag{4.24}$$

where S_i is the sum of membership degrees of the data vectors that lie within a hyperellipsoid whose radii are the standard deviations of the cluster features:

$$S_i = \sum_k \mu_{ik}, \ \forall k, \text{ such that } (\mathbf{z}_k - \mathbf{v}_i)^T \mathbf{F}_i^{-1} (\mathbf{z}_k - \mathbf{v}_i) < 1 \,. \tag{4.25}$$

The *partition density* D_P is defined by:

$$D_P = \frac{\sum_{i=1}^{c} S_i}{V_h} \,. \tag{4.26}$$

Good partitions are indicated by large values of D_A and D_P. In addition to these criteria, Krishnapuram and Freg (1992) investigated the use of the *total within-cluster distance*, given by

$$W_T = \sum_{i=1}^{c} \sum_{k=1}^{N} \mu_{ik}^m D_{ik}^2, \tag{4.27}$$

and the *average within-cluster distance*

$$W_A = \frac{1}{c} \sum_{i=1}^{c} \frac{\sum_{k=1}^{N} \mu_{ik}^m D_{ik}^2}{\sum_{k=1}^{N} \mu_{ik}^m}. \tag{4.28}$$

As this measure monotonically decreases with c, a "knee" on the graph will indicate a suitable partition.

A measure specifically designed for approximating functions by means of fuzzy clustering is the *flatness index*, defined as the ratio between the smallest and largest eigenvalue of the cluster covariance matrix[4] (Babuška and Verbruggen, 1995d):

$$t_i = \frac{\lambda_{in}}{\lambda_{i1}}. \tag{4.29}$$

When clustering data which describe a functional relationship, the clusters are flat. Consequently, the smallest eigenvalue λ_{in} of the covariance matrix is considerably smaller than the remaining eigenvalues. The flatness index attains low values for clusters which are large and flat. For the entire partition, an aggregate measure called the *average cluster flatness* is given by:

$$t_A = \frac{1}{c} \sum_{i=1}^{c} \frac{\lambda_{in}}{\lambda_{i1}}. \tag{4.30}$$

Aiming simultaneously at a low prediction error of the model given by:

$$e = \frac{1}{N} \sum_{k=1}^{N} (y_k - \hat{y}_k)^2, \tag{4.31}$$

where y_k and $\hat{y}_k$ are the true data and the predicted output, respectively, and N is the number of data items, one can combine the average flatness index (4.30) with the prediction error (4.31) to obtain a performance criterion

$$\nu = t \cdot e, \tag{4.32}$$

which prefers a few flat clusters to a larger number of small ones, if both settings lead to approximately the same prediction error. This approach conceptually resembles the use of information criteria in linear system identification (Akaike, 1974).

Cluster validity analysis is performed by running the clustering algorithm for different values of c, and usually also several times for each c with a different initialization.

[4] The flatness index corresponds to the condition number of the cluster covariance matrix.

The validity measure is calculated for each run, and the number of clusters which minimizes (maximizes) the measure is chosen as the "correct" number of clusters in the data. Validity measures can also be evaluated for different structures of the model (different choice of input variables and model orders), as shown in Example 5.11. Note that the use of validity measures is computationally quite involved, as clustering must be repeated many times.

Example 4.5 The average within-cluster distance (AWCD), the fuzzy hypervolume (FHV), the average partition density (APD), and the average cluster flatness (ACF) were applied to the data set generated by eq. (4.20). For each $c = 2, 3, \ldots, 10$, the clustering algorithm was run ten times with different initializations. The results are shown in Figure 4.9. Recall that for the average partition density, a maximum is sought, while the remaining measures indicate a good partition by small values. The AWCD measure is, in this case, monotonic, with a knee at five clusters, which is a good partition as one can see from Figure 4.3b. The FHV and ACF measures have both first local minima at five clusters, as well. The first local maximum of APD indicates six clusters, which is also acceptable. □

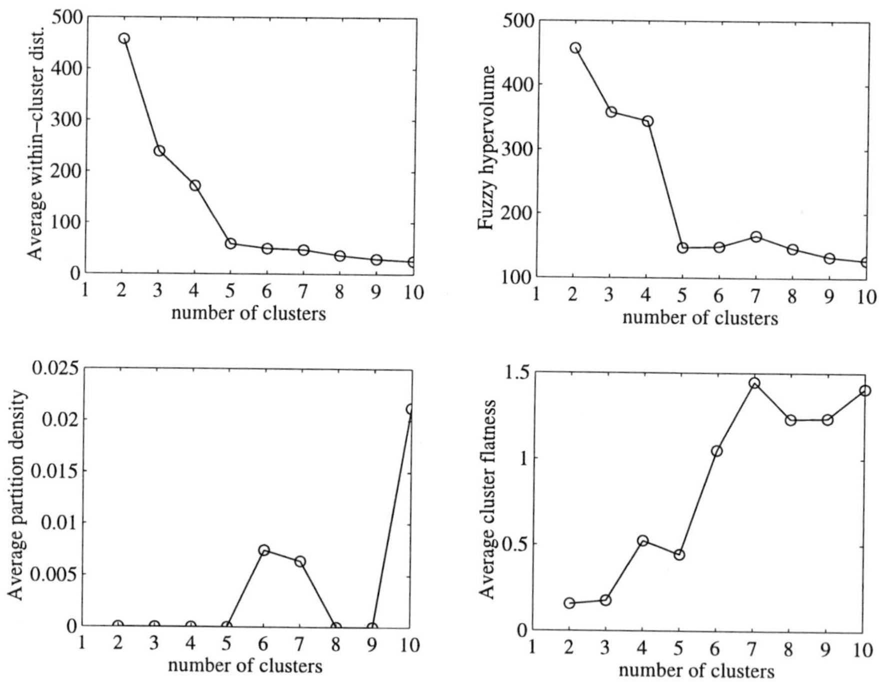

Figure 4.9. Performance of selected validity measures on the data set generated by eq. (4.20).

Example 4.6 Consider a three-dimensional data set generated by the equation $y = (x_1 - 1)^5 + (x_2 - 1)^5$ for 400 data points $(x_1, x_2) \in [-1, 1]^2$. The function surface is given in Figure 4.10a.

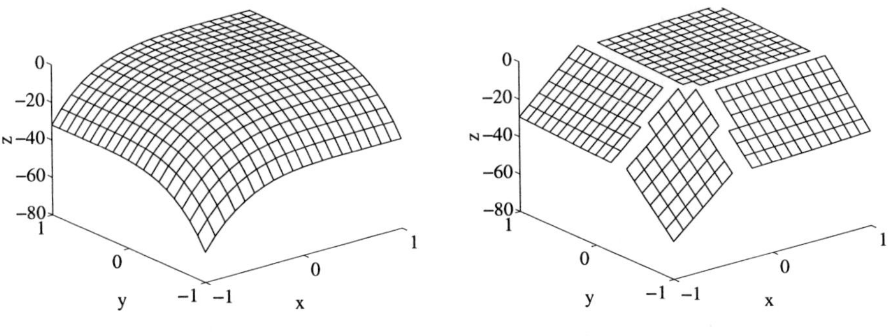

(a) Function surface.

(b) Local linear models.

Figure 4.10. Function $z = (x-1)^5 + (y-1)^5$ and its approximation by four local linear models. Only the parts of the local models corresponding to membership degrees greater than 0.5 are shown.

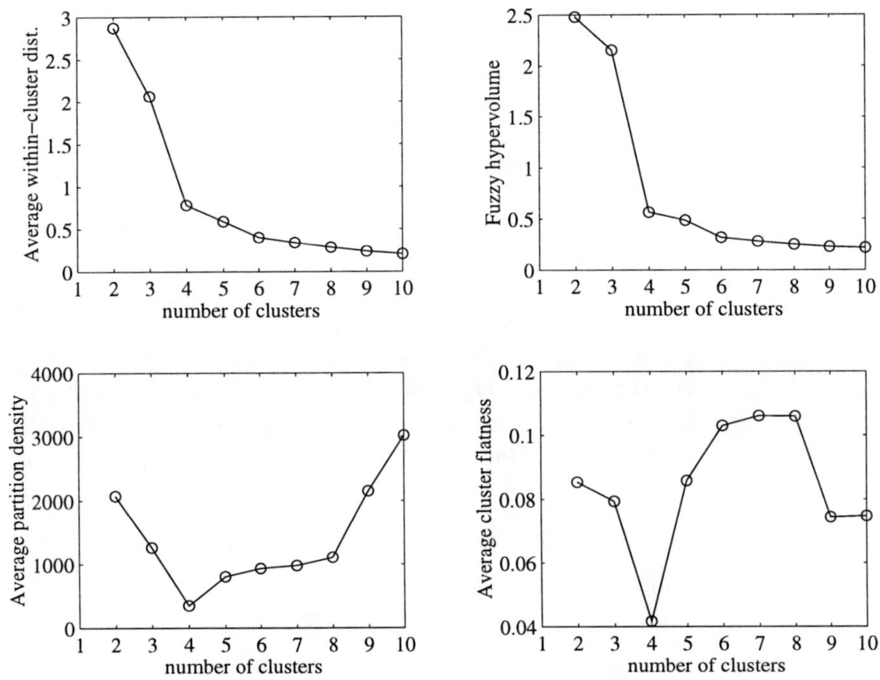

Figure 4.11. Performance of selected validity measures on the function shown in Figure 4.10a.

Figure 4.11 shows the performance of the validity measures. For this data set, all measures except APD indicate four clusters as an appropriate number. Partitioning the data into four clusters gives a reasonable piece-wise linear approximation of the surface, as shown in Figure 4.10b. A smooth approximation of this surface by a TS fuzzy model is shown in Example 2.6. □

4.5.2 Compatible Cluster Merging

A compatible cluster merging (CCM) algorithm was proposed by Krishnapuram and Freg (1992) for finding an appropriate number of linear or planar clusters in 2D or 3D image data. The algorithm starts with $c = c_{max}$, which is greater than the maximum number of clusters expected for the particular problem. The number of clusters is then reduced by successively merging compatible clusters until some threshold is reached and no more clusters can be merged. Figure 4.12 depicts the principal idea of compatible cluster merging.

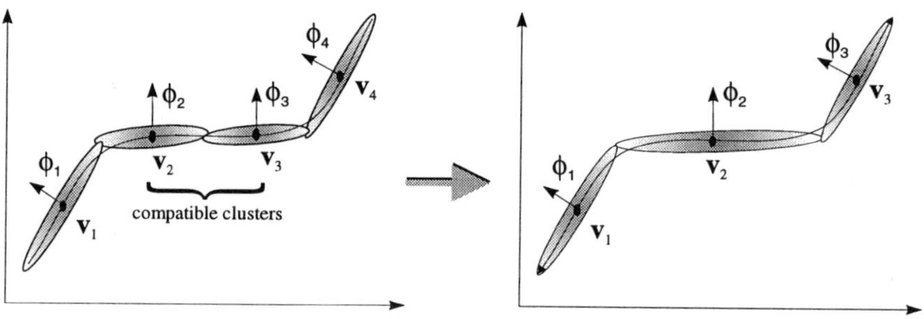

Figure 4.12. Merging of compatible clusters.

The key elements of the CCM algorithm are the criteria which measure the degree of compatibility between clusters. The degree of compatibility is determined on the basis of the geometrical properties of the clusters, by analyzing the eigenvalues and the unit eigenvectors of the cluster covariance matrices. Let the centers of two clusters be $\mathbf{v}_i$ and $\mathbf{v}_j$, the eigenvalues of the two clusters be $\{\lambda_{i1}, \ldots, \lambda_{in}\}$ and $\{\lambda_{j1}, \ldots, \lambda_{jn}\}$, and the unit eigenvectors be $\{\phi_{i1}, \ldots, \phi_{in}\}$ and $\{\phi_{j1}, \ldots, \phi_{jn}\}$. The eigenvalues and the corresponding eigenvectors are arranged in descending order of the eigenvalues. The cluster compatibility criteria proposed by Krishnapuram and Freg are stated as follows:

$$| \phi_{in}^T \phi_{jn} | \geq k_1, \quad k_1 \text{ close to } 1 \tag{4.33a}$$

$$\left| \frac{\phi_{in}^T + \phi_{jn}^T}{2} \frac{\mathbf{v}_i - \mathbf{v}_j}{||\mathbf{v}_i - \mathbf{v}_j||} \right| \leq k_2, \quad k_2 \text{ close to } 0 \tag{4.33b}$$

$$\frac{||\mathbf{v}_i - \mathbf{v}_j||}{\sqrt{\lambda_i} + \sqrt{\lambda_j}} \leq k_3, \quad k_3 \text{ between 2 and 4.} \tag{4.33c}$$

The first condition states that the clusters should be merged if they are parallel. The second condition states that the normals to the hyperplanes should be orthogonal to the line connecting the two cluster centers (the clusters should lie in the same hyperplane). The last condition states that the cluster centers should be sufficiently close to one another. The range of values for k_3 is derived for a 2D geometrical interpretation. Figure 4.13 gives an example where clusters 2 and 3 satisfy all the three conditions, cluster 4 violates condition (4.33b) and cluster 1 violates both conditions (4.33a) and (4.33b). The cluster merging criteria are evaluated pair-wise for all clusters.

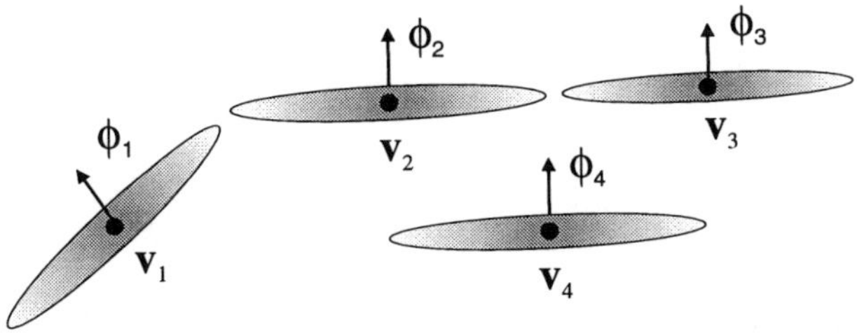

Figure 4.13. Compatible cluster merging of ellipsoidal clusters in $\mathbb{R}^2$.

Although good results are reported for the clustering of image data (Krishnapuram and Freg, 1992; Krishnapuram, 1994), direct application of the above criteria to modeling and function approximation yields undesirable results, especially if the data are noisy. A modified CCM algorithm has been proposed, which is based on less conservative criteria and uses an automated algorithm which replaces the three thresholds in (4.33) by a single user-defined parameter (Kaymak and Babuška, 1995).

Modified Compatibility Criteria. This algorithm is based on two compatibility criteria which are aggregated in one by using the geometric mean as a compensatory decision-making operator. The fuzzy relational clustering algorithm is applied to the aggregated result, to obtain groups of clusters that can be transitively merged. A heuristic check is performed at the final stage, to confirm whether the clusters should indeed be merged. This is necessary for preserving small clusters in the interesting regions, which may have been found when clustering with a large number of clusters. The following criteria are defined for merging clusters:

$$s_{ij}^1 = |\, \phi_{in}^T \phi_{jn} \,| \geq k_1, \quad k_1 \text{ close to } 1, \tag{4.34a}$$

$$s_{ij}^2 = \|\mathbf{v}_i^* - \mathbf{v}_j^*\| \leq k_2, \quad k_2 \text{ close to } 0, \tag{4.34b}$$

where $\mathbf{v}_i^*$ and $\mathbf{v}_j^*$ the cluster centers normalized by eq. (3.35) to have a zero mean and a unit standard deviation. Criterion (4.34a) assesses whether the clusters are parallel, and criterion (4.34b) measures the distance between the cluster centers. Criterion (4.33b) has been removed in order to allow for merging clusters in noisy data, in cases similar to the one shown in Figure 4.16. Scaling of the distance between the cluster centers by the largest eigenvalue in (4.33c) is not used, since it does not give good results for higher dimensional data, especially for elongated clusters that are parallel. Evaluating the criteria (4.34) for all pairs of clusters gives two matrices in $\mathbb{R}^{c\times c}$, $\mathbf{S}^1 = [s_{ij}^1]$ and $\mathbf{S}^2 = [s_{ij}^2]$, whose elements indicate the degree of compatibility (similarity) between the ith and the jth cluster measured according to the corresponding criterion.

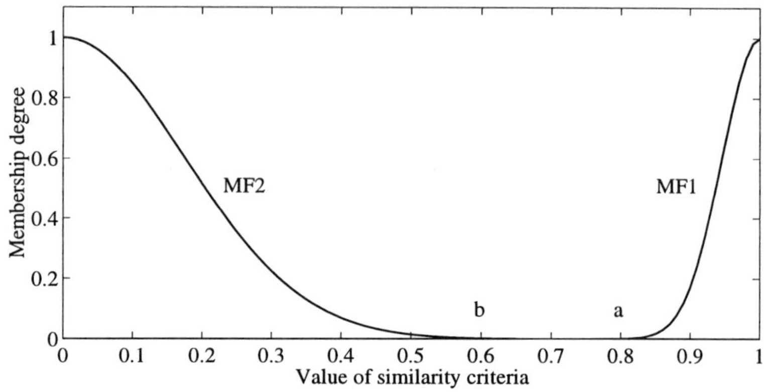

Figure 4.14. Membership functions which indicate the degree to which two clusters are compatible, based on their score in the criteria (4.34).

Aggregation of the Criteria. The criteria (4.34) by themselves do not provide a decision as to which clusters should be merged. In order to obtain such a decision, the evidence from the two criteria must be combined and some thresholding must be applied. It is still quite impractical to use k_1 and k_2 as two hard thresholds, since the two criteria may, to a certain degree, compensate for one another. Hence, an algorithm is given that first maps the elements of the matrices $\mathbf{S}^1$ and $\mathbf{S}^2$ onto a unit interval by using two membership functions MF1 and MF2, depicted in Figure 4.14. These membership functions give the degree of compatibility (similarity) between the pairs of clusters, based on the evidence from the two compatibility criteria.

The exponential membership functions (A.5) are used. The parameters of the membership functions are defined such that c_l for MF1 equals one, c_r for MF2 equals one, and $\mu_{\mathrm{MF1}}(a) = \mu_{\mathrm{MF2}}(b) = 0.01$. The support points a and b are calculated by:

$$a = \frac{1}{c(c-1)} \sum_{i=1}^{c} \sum_{\substack{j=1 \\ j \neq i}}^{c} s_{ij}^1, \quad b = \frac{1}{c(c-1)} \sum_{i=1}^{c} \sum_{\substack{j=1 \\ j \neq i}}^{c} s_{ij}^2 \,. \tag{4.35}$$

Note that these membership functions are not fixed but are determined adaptively. The support points a and b are calculated such that the cluster pairs with a score lower than the mean value have the membership degree equal to zero. As a result, matrices $\tilde{\mathbf{S}}^1$ and $\tilde{\mathbf{S}}^2$ are obtained where:

$$\tilde{s}_{ij}^1 = \mu_{\mathrm{MF1}}(s_{ij}^1), \quad \tilde{s}_{ij}^2 = \mu_{\mathrm{MF2}}(s_{ij}^2) \,. \tag{4.36}$$

By using these membership functions, the scores of the cluster pairs are mapped into the unit interval and the differences between the pairs are amplified. Moreover, making the membership functions variable alleviates problems with scaling of the data. Ideal candidates correspond to $\tilde{s}_{ij}^1 = 1$ and $\tilde{s}_{ij}^2 = 1$.

Note that the criteria (4.34) may partially compensate each other. Two clusters that are not completely parallel, but are close to each other, may be merged. The same applies to the clusters that are parallel but somewhat far from each other. Taking this fact into account, the elements of $\tilde{\mathbf{S}}^1$ and $\tilde{\mathbf{S}}^2$ are combined to form one similarity matrix $\mathbf{S}^0 = [s^0_{ij}]$, by using the geometric mean as a compensatory aggregation operator (Zimmermann, 1996; Kaymak and van Nauta Lemke, 1993):

$$s^0_{ij} = \sqrt{\tilde{s}^1_{ij}\tilde{s}^2_{ij}}\,. \tag{4.37}$$

The diagonal elements of $\mathbf{S}^0$ are by definition equal to 1.

Merging Clusters. The next step is determining which clusters are to be merged, given the similarity matrix $\mathbf{S}^0$. The merging of clusters is a crisp decision, which means that the fuzzy similarity matrix $\mathbf{S}^0$ must be converted into a binary one by thresholding it with a suitably chosen threshold. At each step of the algorithm, a group, or several groups of clusters can be merged. The merging is done transitively, i.e., if the values in $\mathbf{S}^0$ suggest merging cluster i with cluster j and cluster j with cluster k, all the three clusters are merged together. The transitivity in groups of compatible clusters is determined by means of fuzzy relational clustering (Dunn, 1974b; Yang, 1993). This algorithm is based on a successive application of the max-min composition:

$$\begin{aligned}
&\mathbf{S}^0 := \mathbf{S},\; i := 0,\\
&\textbf{repeat}\\
&\quad i := i+1,\\
&\quad \mathbf{S}_i := \mathbf{S}^0 \circ \mathbf{S}_{i-1}\\
&\textbf{until } \mathbf{S}_i = \mathbf{S}_{i-1}\\
&\mathbf{S} := \mathbf{S}_i
\end{aligned} \tag{4.38}$$

The matrix $\mathbf{S}$ is now thresholded with a user-defined threshold γ to obtain a binary matrix $\mathbf{S}^\gamma$ whose elements are given by:

$$s^\gamma_{ij} = \begin{cases} 1, & \text{if } s_{ij} \geq \gamma, \\ 0, & \text{otherwise.} \end{cases} \tag{4.39}$$

The new number of clusters c' is given by the rank of $\mathbf{S}^\gamma$. In order to maintain the partition, the merging of clusters is accomplished by adding up the corresponding rows of the fuzzy partition matrix $\mathbf{U}$. In order to provide a simple algebraic operation for computing the new partition matrix, $\mathbf{S}^\gamma$ is converted to its reduced row echelon form and only the first c' rows are retained in $\mathbf{S}^\gamma$. The meaning of this new matrix is that the ith cluster (i.e., the ith row of the new partition matrix $\mathbf{U}$) is created by adding up the rows of the original partition matrix with indices j corresponding to nonzero elements s^γ_{ij}. The new fuzzy partition matrix $\mathbf{U}'$ containing the merged clusters is thus computed by:

$$\mathbf{U}' = \mathbf{S}^\gamma \mathbf{U}\,. \tag{4.40}$$

Heuristic Step. Merging clusters directly after relational clustering may give undesired results if incompatible clusters are located "between" the compatible ones,

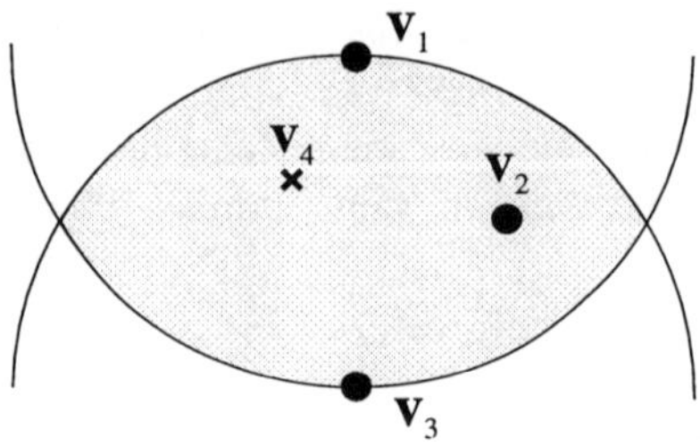

Figure 4.15. A group of compatible clusters $\{\mathbf{v}_1, \mathbf{v}_2, \mathbf{v}_3\}$ are not merged if an incompatible cluster center ($\mathbf{v}_4$) is within their mutual neighborhood (the shaded region) in the regressor space.

as shown schematically in Figure 4.15. This situation is remedied by a heuristic, which states that a set M of compatible clusters are only merged if there is no incompatible cluster center within the mutual neighborhood of the centers of the compatible clusters. This heuristic is expressed by the following condition:

$$\min_{\substack{\mathbf{v}_i^* \in M \\ \mathbf{v}_k^* \notin M}} \max d_{ik} > \max_{\mathbf{v}_i^*, \mathbf{v}_j^* \in M} d_{ij}, \tag{4.41}$$

which must be satisfied to merge the clusters in M. In this equation the distance d_{ij} is defined by

$$d_{ij} = \| \, \mathrm{proj}_{\mathbb{N}_p}(\mathbf{v}_i^*) - \mathrm{proj}_{\mathbb{N}_p}(\mathbf{v}_j^*) \|,$$

where $\mathrm{proj}_{\mathbb{N}_p}(\mathbf{v}_i^*)$ represents the projection of the cluster centers onto the regressor space. The condition (4.41) is checked for each row l of $\mathbf{S}^\gamma$. M is then the set: $M = \{m | s_{lm}^\gamma = 1\}$. If (4.41) is not met for some l, the lth row of $\mathbf{S}^\gamma$ is broken down into $|M|$ rows, each containing only one nonzero element (i.e., no merging is done when applying (4.40)). The procedure stops when $\mathbf{U}' = \mathbf{U}$, i.e., when no more clusters can be merged with the current value of γ. The modified CCM algorithm is summarized in Algorithm 4.1.

The CCM algorithm has been applied to a variety of simulated and real-world function approximation and modeling problems. Three examples of synthetic data are given in this section, to illustrate the features of the algorithm. An application to real data can be found in Section 7.2.

Example 4.7 Applying the cluster merging criteria (4.34) to the data from Figure 4.5 with an initial number of clusters $c_{\max} = 4$ leads to the result shown in Figure 4.16. The matrices $\mathbf{S}^1$ and $\mathbf{S}^2$ computed by evaluating the criteria (4.34) are:

$$\mathbf{S}^1 = \begin{bmatrix} 0 & 0.58 & 0.68 & 0.53 \\ 0.58 & 0 & 0.99 & 0.99 \\ 0.68 & 0.99 & 0 & 0.98 \\ 0.53 & 0.99 & 0.98 & 0 \end{bmatrix}, \quad \mathbf{S}^2 = \begin{bmatrix} 0 & 2.08 & 2.46 & 3.15 \\ 2.08 & 0 & 0.71 & 1.55 \\ 2.46 & 0.71 & 0 & 0.84 \\ 3.15 & 1.55 & 0.84 & 0 \end{bmatrix}.$$

By applying (4.35) to $\mathbf{S}^1$ and $\mathbf{S}^2$, the support points for the membership functions are calculated: $a = 0.46$ and $b = 1.80$. After evaluating the membership degrees, $\tilde{\mathbf{S}}^1$ and

Algorithm 4.1 (Compatible cluster merging.)

Choose the initial number of clusters $c = c_{max}$ and the threshold γ.

Repeat

Step 1: **Cluster the data into** c **clusters.** The partition matrix $\mathbf{U}$, the cluster centers $\mathbf{V}$ and the covariance matrices $\mathbf{F}_i$ are obtained.

Step 2: **Evaluate the compatibility criteria**. Using eq. (4.34), compute matrices $\mathbf{S}^1$ and $\mathbf{S}^2$.

Step 3: **Calculate the compatibility matrix.** By using eq. (4.35), determine the membership functions MF1 and MF2. Apply (4.36) to obtain $\tilde{\mathbf{S}}^1$ and $\tilde{\mathbf{S}}^2$. By using eq. (4.37), aggregate $\tilde{\mathbf{S}}^1$ and $\tilde{\mathbf{S}}^2$ into the compatibility matrix $\mathbf{S}_0$.

Step 4: **Determine groups of clusters for merging.** Perform fuzzy relational clustering on $\mathbf{S}_0$ using (4.38). Apply threshold γ (4.39), and compute the reduced row echelon form of $\mathbf{S}^\gamma$.

Step 5: **Check the heuristic.** For each row of $\mathbf{S}^\gamma$, evaluate (4.41). If not fulfilled, modify $\mathbf{S}^\gamma$ accordingly.

Step 6: **Compute the new partition matrix** $\mathbf{U}'$ **using eq. (4.40).**

until $\mathbf{U}' = \mathbf{U}$

$\tilde{\mathbf{S}}^2$ are combined by the geometric mean (4.37) to obtain the initial similarity matrix $\mathbf{S}^0$ shown below. To obtain a group of clusters to be transitively merged, the relational clustering algorithm (4.38) is applied to $\mathbf{S}^0$. In this simple example, the algorithm converges after the first iteration, yielding the matrix $\mathbf{S}$:

$$\mathbf{S}^0 = \begin{bmatrix} 1 & 0 & 0 & 0 \\ 0 & 1 & 0.55 & 0.28 \\ 0 & 0.55 & 1 & 0.34 \\ 0 & 0.28 & 0.34 & 1 \end{bmatrix} \xrightarrow{\text{Relational clustering}} \mathbf{S} = \begin{bmatrix} 1 & 0 & 0 & 0 \\ 0 & 1 & 0.55 & 0.34 \\ 0 & 0.55 & 1 & 0.34 \\ 0 & 0.34 & 0.34 & 1 \end{bmatrix}.$$

By thresholding $\mathbf{S}$ with $\gamma = 0.3$, a group of clusters 2, 3 and 4 is obtained as candidates for merging. Applying a threshold greater than 0.34 and lower than 0.55 results in the merging of only two clusters, 2 and 3. For thresholds above 0.55, no merging occurs.

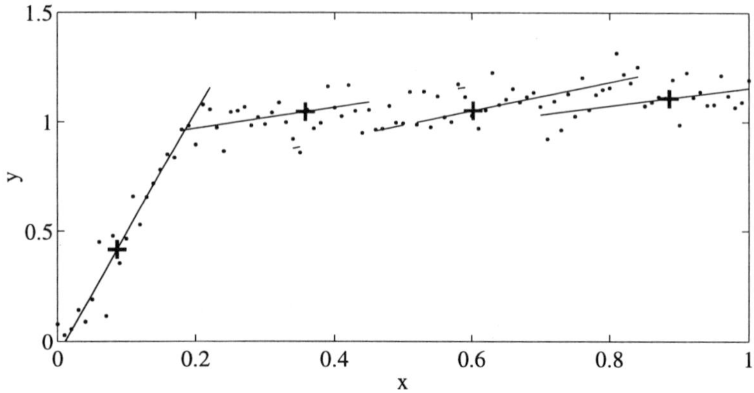

Figure 4.16. Initial partition with four local models.

Below is shown the result for $\gamma = 0.4$:

$$\mathbf{S}^{\gamma} = \begin{bmatrix} 1 & 0 & 0 & 0 \\ 0 & 1 & 1 & 0 \\ 0 & 1 & 1 & 0 \\ 0 & 0 & 0 & 1 \end{bmatrix} \xrightarrow{\text{Reduced row echelon}} \begin{bmatrix} 1 & 0 & 0 & 0 \\ 0 & 1 & 1 & 0 \\ 0 & 0 & 0 & 1 \\ 0 & 0 & 0 & 0 \end{bmatrix}.$$

The new number of cluster is $c' = \text{rank}(\mathbf{S}^{\gamma}) = 3$. Since the heuristic is not applied (only two clusters are merged), the new partition matrix can be computed by $\mathbf{U}' = \mathbf{S}^{\gamma}\mathbf{U}$. The clustering algorithm is now initialized with $\mathbf{U}'$ and the process repeats. □

Example 4.8 Consider the data set generated by equation $y = (x_1 - 1)^5 + (x_2 - 1)^5$, which was used in Example 4.5. A plot of the function surface is given in Figure 4.10a. The initial number of clusters c_{MAX} is set to 15. A number of values for γ are investigated in this example. Figure 4.17a shows the final number of clusters the CCM algorithm arrived at for different values of γ. For this smooth function, a more or less monotonic relation between γ and the number of cluster in the final model is observed. Figure 4.17b depicts the number of clusters during the iterative merging process for several values of γ. In most cases, the algorithm arrived at the final solution in two iterations, for some settings of γ, three iterations were needed. This means that the data only needed to be clustered two or three times, which is a significant saving in the computation effort, compared to the approach based on validity measures (Section 4.5.1). For γ from 0.15 to 0.25, the final number of clusters was four, the same result as obtained with the validity measures. As shown in Figure 4.17b, for $\gamma = 0.2$, the merging sequence of clusters was 15–6–4. □

Example 4.9 The purpose of this example is to demonstrate that clustering with the CCM algorithm can discover small clusters, which would be missed, if only clustering with random initialization were used (Kaymak and Babuška, 1995). A slight modification of the sinusoidal function from Example 4.2 is used to generate 100 identification data points. Note that there are several data points for $x \in [95, 100]$,

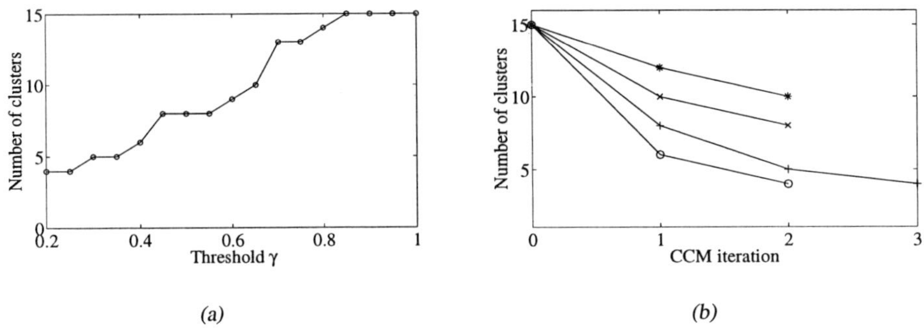

Figure 4.17. Graph (a) shows the final number of clusters found by CCM for different values of γ. In graph (b), the merging process is shown for $\gamma = 0.2$ ('o'), $\gamma = 0.25$ ('+'), $\gamma = 0.45$ ('$\times$'), and $\gamma = 0.65$ ('*').

which can be approximated by a separate local model. The CCM algorithm was initialized with nine clusters and $\gamma = 0.5$. The merging proceeded in the sequence 9–7–6 clusters. The final partition approximation of the function by six local linear models is shown in Figure 4.18. One can see that a very accurate description of the function is obtained, including the small cluster on the right side. Such a result cannot be obtained when starting from a random initial partition with six clusters, since the chance of finding the small cluster is very small. □

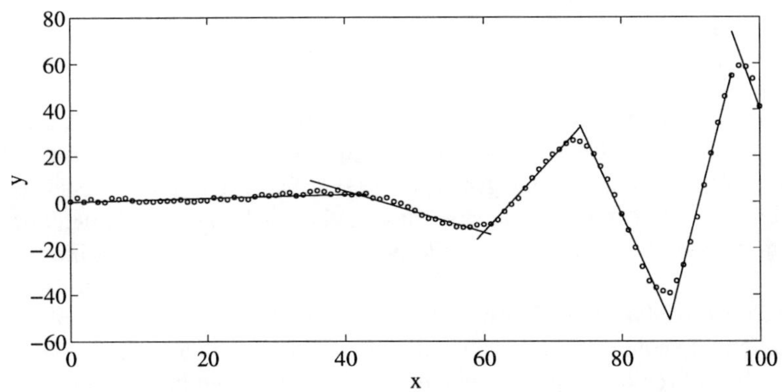

Figure 4.18. Approximation of a sinusoidal function.

The CCM algorithm can be modified and extended in several ways. Two are briefly mentioned here: automatic determination of the threshold, and adjustment of the cluster volumes.

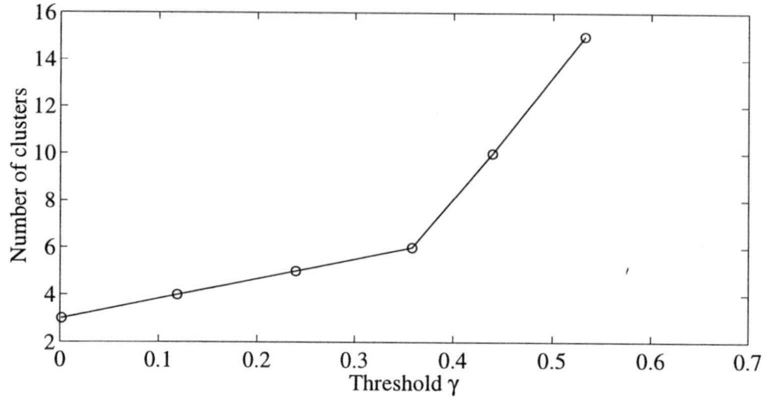

Figure 4.19. Number of clusters vers. γ with automatic threshold selection.

Automatic Threshold Determination. In order to get an initial idea on a suitable values of γ, the first run of the CCM algorithm can be done with merging clusters at each iteration, until some lower limit on c is reached. This can be achieved by merging two clusters at a time, simply by setting γ at each iteration to $\max_{i \neq j} s_{ij}$. A more efficient run can be obtained by setting γ to a lower value. This value can be determined in several ways, for instance, as the mean of the off-diagonal elements in $\mathbf{S}$:

$$\gamma = \frac{1}{c(c-1)} \sum_{i=1}^{c} \sum_{\substack{j=1 \\ j \neq i}}^{c} s_{ij} \,. \tag{4.42}$$

The elements in the upper triangle of $\mathbf{S}$ can also be clustered in two groups, using the fuzzy c-means algorithm. Thus, a group of compatible and incompatible clusters is obtained, and the compatible clusters, or a subset of compatible clusters, are merged. Figure 4.19 shows a CCM run with automatic threshold determination for the data from Example 4.8. A knee in the graph at $\gamma \approx 0.35$ indicates that above this threshold the clusters are highly compatible, and thus can be easily merged. When aiming at a simple model with few clusters, γ should be chosen at or below this point.

Adjustment of Cluster Volumes. The GK algorithm tends to prefer clusters of approximately equal volumes. This may give some problems in modeling systems where the size of regions approximated by linear models significantly differ, as shown in Example 4.3. This drawback can be avoided by using a modification of the CCM algorithm, where the volume constraints in GK clustering ρ_i are determined from the volumes of the merged clusters. The procedure is as follows: from the updated partition matrix $\mathbf{U}'$ (4.40), the cluster centers $\mathbf{v}_i$ are first computed using (3.9b). Then the cluster covariance matrices $\mathbf{F}_i$ are calculated using (3.18). The determinant of $\mathbf{F}_i$ is the volume of the cluster in the updated partition matrix. The volume constraints ρ_i in (3.17) are then set in the proportion of the cluster volumes. It must be noted that this procedure works well with problems where the size of the clusters differ, such as the one in Example 4.3. In other problems, where the clusters are approximately equal, the adaptation of cluster volumes may give unexpected results, in the sense that some clusters may grow in size, depending on the merging sequence.

The main advantage of successive cluster merging is that it helps the clustering algorithm to find better partitions than are found with a random initialization, and small clusters are not missed. During the CCM iterations, the clustering algorithm converges relatively fast, because of the good initialization from the previous iteration. The modified merging criteria are tailored to modeling and approximation problems where significant levels of noise on the data may occur. In this sense, they outperform the original method by Krishnapuram and Freg (1992), which is too conservative for such applications. The main disadvantage of the CCM algorithm is that it requires a threshold parameter which is problem dependent, and hence some tuning of this parameter for particular data may be needed.

4.6 Summary and Concluding Remarks

By means of product-space fuzzy clustering, a data set generated by a nonlinear system can be partitioned into fuzzy subsets of data that are locally described by linear submodels. Prior to clustering, the regression structure of the model must be selected, in order to properly represent the system's dynamics. The choice of the regressors has been discussed in Section 4.2. Problems where little prior knowledge is available are usually represented in an input–output form, using the NARX structure. When the structure of the system is (partly) known and when the state variables can be measured or reconstructed from other variables, state-space or hybrid semi-mechanistic modeling approaches should be preferred to black-box input–output models.

After the structure is determined, clustering in the product space of the regressors and of the regressand can be applied to partition the data, as described in Section 4.3. Since each cluster serves as a local linear model of the system, the clustering algorithm must be capable of detecting clusters which lie in linear subspaces. It was found that algorithms based on an adaptive distance measure perform generally better than the algorithms based on linear varieties such as the FCE algorithm, which has been used in the literature Yoshinari, et al. (1993). A significant advantage of clustering with adaptive distance measure is that the results are not affected by normalization of the data set. The FMLE algorithm proves to be useful for problems where the volumes of the different clusters differ considerably. This algorithm is, however, more sensitive to proper initialization.

The choice of the number of clusters was addressed in Section 4.5. Cluster validity measures and a compatible cluster merging (CCM) algorithm are there presented. The use of validity measures is computationally rather involved, and moreover, the performance of the different measures may vary from problem to problem. The average within-cluster distance and the fuzzy hypervolume were found to give consistently good results. A new measure, average cluster flatness which is based on the condition number of the cluster covariance matrix has been introduced. Also this measure performed well on most of the tested problems.

The CCM algorithm presented in this chapter is a modification of the original method proposed by Krishnapuram and Freg (1992) for straight line detection. An advantage of the CCM technique is that the number of clusters can be optimized without testing all the possibilities as is the case with the cluster validity measures. This may significantly reduce the computational effort. Moreover, by decreasing the number of clusters gradually a better solution can be found, since small regions with few data points can be captured by using initially a larger number of clusters. The final number of clusters found by CCM is, however, dependent on a user-defined threshold parameter. In practice, some tuning of this parameter may be needed.

5 CONSTRUCTING FUZZY MODELS FROM PARTITIONS

As shown in the previous chapter, a class of fuzzy clustering algorithms can be used to approximate a set of data by local linear models. Each of these models is represented by a fuzzy subset in the data set available for identification. In order to obtain a model useful for prediction or controller design, an additional step must be applied to generate a model independent of the identification data. Such a model can be represented either as a rule base or as a fuzzy relation. This chapter presents methods and algorithms for constructing fuzzy rule-based and relational models from the fuzzy partitions obtained by product-space clustering.

Section 5.1 addresses the construction of Takagi–Sugeno models. Different methods for generating the antecedent membership functions and estimating the consequent parameters are presented, as well as techniques for the simplification and reduction of fuzzy model obtained from data. Two simulation examples of dynamic systems are given. In Section 5.2, algorithms are given to generate linguistic and relational models from fuzzy partitions. Section 5.3 presents an approach to constructing a "low-level" relation, which describes the identified system as a fuzzy relation in the product space of the variables in the model.

5.1 Takagi–Sugeno Fuzzy Models

The idea of constructing TS fuzzy models by product-space fuzzy clustering is not entirely new. Yoshinari, et al. (1993) applied the fuzzy c-elliptotypes algorithm (see Section 3.5.2) to derive a TS fuzzy model to approximate static functions. Babuška and Verbruggen (1994a) and Zhao, et al. (1994) used the GK algorithm (Section 3.4) to identify nonlinear dynamic systems. In this section, different aspects in extraction of TS models from fuzzy partitions are addressed.

Each cluster obtained by product-space clustering of the identification data set can be regarded as a local linear approximation of the regression hypersurface. The global model can be conveniently represented as a set of affine Takagi–Sugeno (TS) rules:

$$R_i\colon\ \textbf{If } \mathbf{x} \text{ is } A_i \textbf{ then } y_i = \mathbf{a}_i^T\mathbf{x} + b_i, \quad i = 1, 2, \ldots, K\,. \tag{5.1}$$

The antecedent fuzzy sets A_i can be computed analytically in the antecedent product space, or can be extracted from the fuzzy partition matrix by projections. The consequent parameters $\mathbf{a}_i$ and b_i are estimated from the data using least-squares methods, or they can be extracted from the eigenstructure of the cluster covariance matrices. These methods are explained in the following sections.

5.1.1 Generating Antecedent Membership Functions

The antecedent membership functions can be obtained by projecting the fuzzy partition onto the antecedent variables, or by computing the membership degrees directly in the product space of the antecedent variables. These two methods are described in the following sections.

Generating Antecedent Membership Functions by Projection. The principle of this method is to project the multidimensional fuzzy sets defined point-wise in the rows of the partition matrix $\mathbf{U}$ onto the individual antecedent variables of the rules. These variables can be the original regression variables, in which case the projection is an orthogonal projection of the data. New, transformed antecedent variables can be obtained by means of eigenvector projection, using the p largest eigenvectors of the cluster covariance matrices. The eigenvector projection is useful for clusters which are opaque to the axis of the regression space, and cannot be represented by axis-orthogonal projection with a sufficient accuracy.

Axis-orthogonal Projection. This method projects the fuzzy partition matrix onto the axes of the antecedent variables x_j, $1 \le j \le p$. The TS rules are then expressed in the conjunctive form:

$$R_i\colon\ \textbf{If } x_1 \text{ is } A_{i1} \textbf{ and } \ldots \textbf{ and } x_p \text{ is } A_{ip} \textbf{ then } y_i = \mathbf{a}_i^T\mathbf{x} + b_i, \quad i = 1, \ldots, K\,. \tag{5.2}$$

In order to obtain membership functions for the antecedent fuzzy sets A_{ij}, the multidimensional fuzzy set defined point-wise in the ith row of the partition matrix is projected onto the regressors x_j by:

$$\mu_{A_{ij}}(x_{jk}) = \operatorname{proj}_j(\mu_{ik})\,. \tag{5.3}$$

A definition of the projection operator is given in Appendix A. When computing the degree of fulfillment $\beta_i(\mathbf{x})$ of the ith rule, the original cluster in the antecedent product space is reconstructed by applying the intersection operator in the Cartesian product space of the antecedent variables:

$$\beta_i(\mathbf{x}) = \mu_{A_{i1}}(x_1) \wedge \mu_{A_{i2}}(x_2) \wedge \ldots \wedge \mu_{A_{ip}}(x_p) . \tag{5.4}$$

Other t-norms, such as the product, can be used instead of the minimum operator. This reconstruction is generally not exact, and leads to a *decomposition error* in the case where the clusters are oblique to the axes, see Figure 5.1. This error can be partially compensated by global least-squares estimation of the consequent parameters, as described in Section 5.1.2. Despite the fact that through the axis-orthogonal projection some information may be lost, this method is useful, as it provides the possibility to interpret the model using the regressors directly.

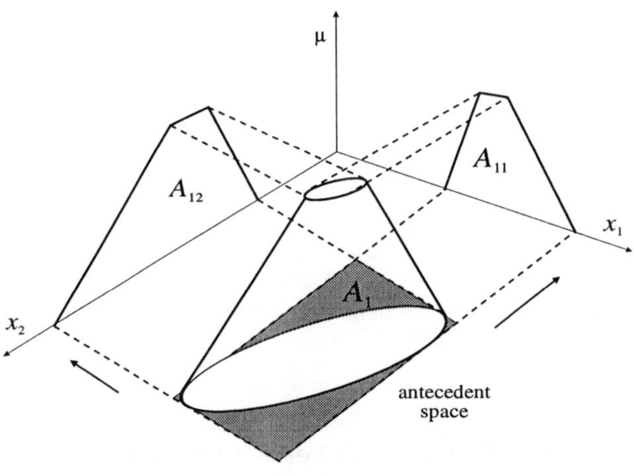

Figure 5.1. A decomposition error arises when a multidimensional fuzzy set is projected onto the antecedent variables and then composed back by applying fuzzy intersection in the antecedent space. The decomposition error is schematically shown in gray in the support of the fuzzy set.

Eigenvector Projection. The decomposition error can be reduced by means of eigenvector projection, using the information about the orientation of the clusters. Let $\mathbf{F}^x$ denote the partition of the cluster covariance matrix, corresponding to the regressor $\mathbf{x}$:

$$\mathbf{F}^x = [f_{ij}], \quad 1 \leq i, j \leq p, \tag{5.5}$$

where p is the dimension of the regressor $\mathbf{x}$. The unitary eigenvectors of $\mathbf{F}^x$ constitute an orthonormal basis oriented in the directions of the cluster ellipsoid axes. The regression matrix $\mathbf{X}$ is projected onto this new basis by:

$$\mathbf{X}_i' = \mathbf{X}\mathbf{H}_i, \tag{5.6}$$

where $\mathbf{H}_i \in \mathbb{R}^{p \times p}$ is a projection matrix having the unitary eigenvectors of $\mathbf{F}_i^x$ in its columns. In general, each cluster will have its own projection matrix. Consequently, each rule will have different antecedent variables, computed as linear combinations of the regressors:

$$R_i\text{: } \textbf{If } x'_{i1} \text{ is } A_{i1} \textbf{ and } \dots \textbf{ and } x'_{ip} \text{ is } A_{ip} \textbf{ then } y_i = \mathbf{a}_i^T \mathbf{x}'_i + b_i, \quad i = 1, \dots, K\,. \tag{5.7}$$

This transformation is similar to principal component analysis, and can be regarded as a kind of data pre-processing for each rule, as illustrated in Figure 5.2. The reduction of the decomposition error by means of eigenvector projection is demonstrated in Example 5.1.

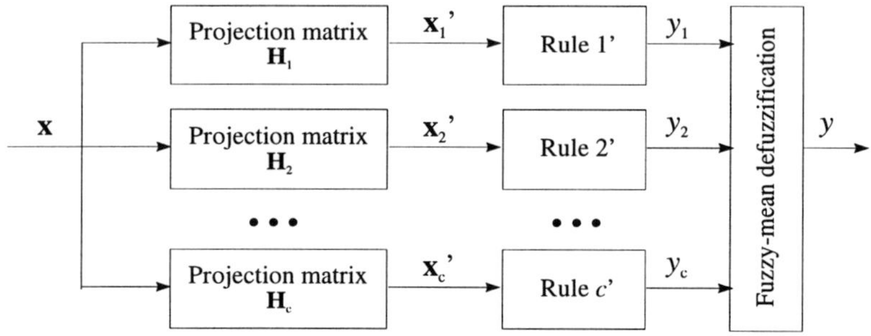

Figure 5.2. For each rule, a projection matrix $\mathbf{H}_i$ transforms the original antecedent vector $\mathbf{x}$ into a new vector $\mathbf{x}'_i = \mathbf{x}\mathbf{H}_i$.

Parameterization of Membership Functions. By projecting the ith row $\boldsymbol{\mu}_i$ of the fuzzy partition matrix $\mathbf{U}$ onto the antecedent variable x_j, a point-wise definition of the fuzzy set A_{ij} is obtained. In order to obtain a prediction model or a model suitable for control purposes, the antecedent membership functions must be expressed in a form that allows computation of the membership degrees, also for input data not contained in the data set $\mathbf{Z}$. This is achieved by approximating the point-wise defined membership function by some suitable parametric function, as depicted in Figure 5.3.

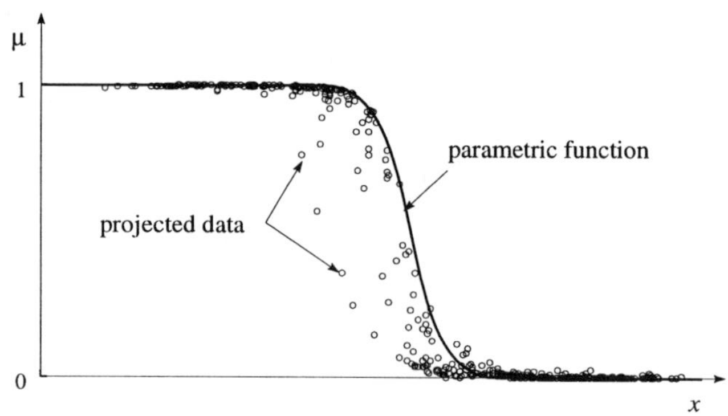

Figure 5.3. Approximation of the projected data by a parametric membership function.

The piece-wise exponential membership functions (A.5) proved to be suitable for the accurate representation of the actual cluster shape. This function is fitted to the envelope of the projected data by numerically optimizing its parameters. An advantage of this method over the multidimensional membership functions, described in the following section, is that the projected membership functions can always be approximated such that convex fuzzy sets are obtained. Moreover, asymmetric membership functions can be used to reflect the actual partition of the considered nonlinear regression problem.

Example 5.1 The GK algorithm was applied to a data set generated by the Rosenbrock Valley function: $y = 100(x_2 - x_1^2)^2 - (1 - x_1)^2$, shown in Figure 5.4a. The data set $\mathbf{Z}$ consisted of 1681 data triplets (x_1, x_2, y). The clustering parameters were $c = 6$, $m = 2$ and $\epsilon = 0.01$. Figure 5.4b presents a contour plot of the obtained partition projected to the antecedent space $X_1 \times X_2$. Shown are also the projections of the cluster centers onto the antecedent space and the directions of the two eigenvectors ϕ_{1i} and ϕ_{2i}, corresponding to the two largest eigenvalues of $\mathbf{F}_i$ for each cluster i.

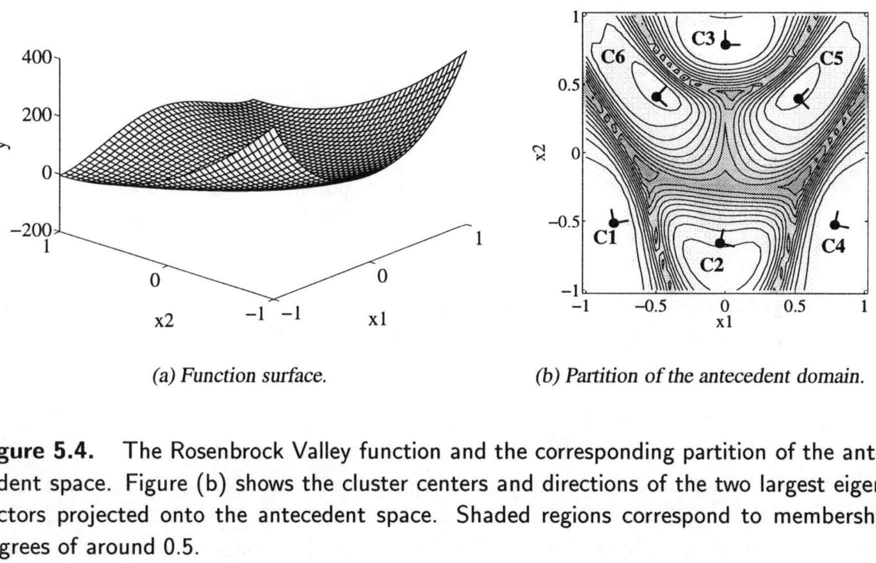

(a) Function surface.

(b) Partition of the antecedent domain.

Figure 5.4. The Rosenbrock Valley function and the corresponding partition of the antecedent space. Figure (b) shows the cluster centers and directions of the two largest eigenvectors projected onto the antecedent space. Shaded regions correspond to membership degrees of around 0.5.

One can see that the orientation of the clusters varies, clusters C_1 to C_4 are nearly parallel to the axes x_1, x_2, while clusters C_5 and C_6 are rotated by about ± 45 degrees. Projection of these rotated clusters results in inaccurate approximation of the clusters by conjunctive rules. Note that although cluster C_5 is quite elongated, due to its rotated position, the projections on X_1 and X_2 are rather broad, as shown in Figure 5.5 (a) and (b). The intersection of these membership functions results in a fuzzy set shown in Figure 5.6a. By comparing Figure 5.6a with the original partition in Figure 5.4b, one can see that a relatively large decomposition error arises.

Membership functions in Figure 5.5 (c) and (d) are obtained by projecting cluster C_5 onto the rotated axis frame x_1', x_2'. The projection matrix $\mathbf{H}_5$ has the eigenvectors

$\phi_{1,5}$ and $\phi_{2,5}$ in its columns:

$$H_5 = \begin{bmatrix} 0.71 & 0.70 \\ -0.70 & 0.71 \end{bmatrix}. \tag{5.8}$$

The new, transformed antecedent variables are given by $[x'_1, x'_2] = [x_1, x_2]H_5$. From Figure 5.5c, one can see that the projection of cluster 5 onto the new variable x'_1 results in a much narrower membership function than in Figure 5.5a.

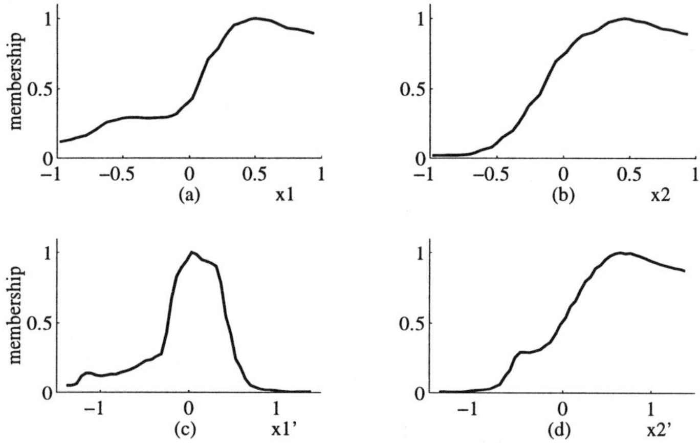

Figure 5.5. Membership functions (a) and (b) are obtained by orthogonal projection of cluster C_5. Membership functions (c) and (d) are obtained by eigenvector projection of the same cluster.

Figure 5.6b shows the intersection of the membership functions from Figure 5.5 (c) and (d) in the space of the original variables $X_1 \times X_2$. This intersection now much more accurately approximates cluster C_5. □

A drawback of eigenvector projection is that the interpretation of the rule antecedents may be hampered by the transformation of the original regressors to the new variables x'_i. However, eigenvector projection can lead to a reduction of the antecedent dimension, and to gaining information about the nature of the system's nonlinearity in the cases when all or some of the eigenvectors are equal to one another. In such a case, a global projection matrix $\mathbf{H}$ is obtained, which suggests a new model structure with the transformed antecedent variables, as demonstrated in the following example.

Example 5.2 As an example of an identification problem that may lead to a common projection matrix, consider a nonlinear system $y(k+1) = \big(y(k) - u(k)\big)^2$. Figure 5.7a shows the regression surface of this system, and Figure 5.7b presents level curves of the antecedent partition obtained by fuzzy clustering. The corresponding eigenvectors of the different clusters are practically identical. Consequently, a common projection matrix is obtained for all the five clusters:

$$H = \begin{bmatrix} 0.7071 & 0.7071 \\ -0.7071 & 0.7071 \end{bmatrix}, \tag{5.9}$$

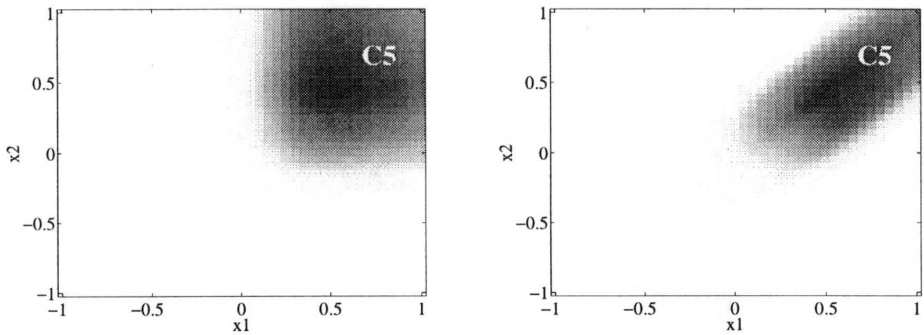

(a) (b)

Figure 5.6. Degree of fulfillment for the rule derived from cluster C_5, (a) is obtained by orthogonal projection, (b) by eigenvector projection. Dark shading represents large degree of fulfillment.

which defines a new axis frame rotated by 45 degrees with respect to the regressors $y(k)$, $u(k)$. This means that the transformed antecedent variables

$$x_1(k) = u(k) - y(k), \quad x_2(k) = u(k) + y(k) \tag{5.10}$$

are more relevant to the description of the system's nonlinearity than the original regressors $y(k)$ and $u(k)$. Moreover, projections onto x_1 and x_2 reveal that only x_1 is needed in the antecedent to distinguish the validity regions of the individual rules. This decreases the dimension of the antecedent and reveals the fact that the system has a linear dynamic part followed by a static nonlinearity. □

Multidimensional Antecedent Membership Functions. In this approach, the antecedent membership functions are represented analytically by computing an inverse of the distance from the cluster prototype. The membership degree is computed directly for the entire input vector (without the decomposition). The antecedents of the TS rules are simple propositions with multidimensional fuzzy sets given by eq. (5.1), and $\beta_i(\mathbf{x}) = \mu_{A_i}(\mathbf{x})$.

Recall that $\mathbf{F}^x = [f_{ij}]$, $1 \le i, j \le p$ includes all but the last column of the cluster covariance matrix. The corresponding norm-inducing matrix is given by:

$$\mathbf{A}_i^x = [\rho_i \det(\mathbf{F}_i^x)]^{1/p} (\mathbf{F}_i^x)^{-1} . \tag{5.11}$$

Let $\mathbf{v}_i^x = [v_{1i}, \ldots, v_{pi}]^T$ denote the projection of the cluster center onto X. The inner-product norm

$$D_{\mathbf{A}_i^x}(\mathbf{x}, \mathbf{v}_i^x) = (\mathbf{x} - \mathbf{v}_i^x)^T \mathbf{A}_i^x (\mathbf{x} - \mathbf{v}_i^x) . \tag{5.12}$$

measures the distance of the antecedent vector $\mathbf{x}$ from the projection of the cluster center $\mathbf{v}_i^x$. To transform the distance (a measure of dissimilarity) into the membership

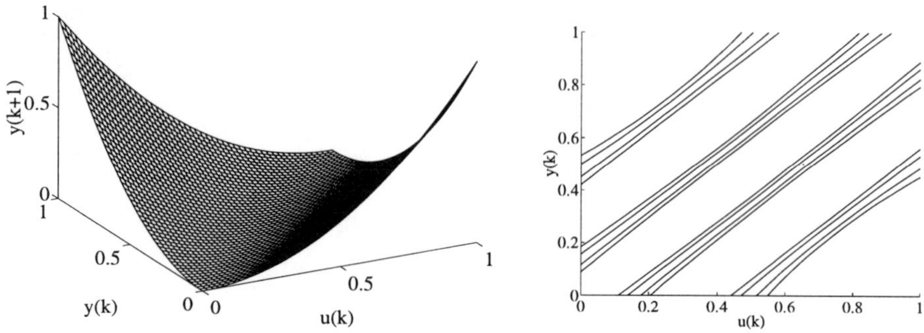

(a) System's surface.

(b) Partition of the antecedent domain.

Figure 5.7. Nonlinear system $y(k+1) = \big(y(k) - u(k)\big)^2$ and the corresponding partition of the antecedent space obtained with the GK algorithm. The contour lines represent the boundaries between the clusters.

degree (a measure of similarity), some kind of inversion must be applied. Two methods are described below.

Probabilistic Method. One possibility is to use the formula employed by the clustering algorithms:

$$\beta_i(\mathbf{x}) = \frac{1}{\sum_{j=1}^{c} \left(D_{\mathbf{A}_i^x}(\mathbf{x}, \mathbf{v}_i^x) / D_{\mathbf{A}_i^x}(\mathbf{x}, \mathbf{v}_j^x) \right)^{2/(m-1)}} \,. \tag{5.13}$$

This expression computes the degree of fulfillment of one rule relative to the other rules and the sum of the membership degrees of all the rules equals one, as in fuzzy clustering. Because of this constraint we call the method probabilistic.

Possibilistic Method. The degree of fulfillment of a rule can also be computed independently of the remaining rules, by applying the "possibilistic" transformation of the distance into the membership degree:

$$\beta_i(\mathbf{x}) = \frac{1}{1 + D_{\mathbf{A}_i^x}(\mathbf{x}, \mathbf{v}_i^x)} \,. \tag{5.14}$$

In both cases, the distance is computed in relation to the cluster center, which implies that the obtained multidimensional membership functions (5.13) and (5.14) have the same position (center) in the antecedent space. However, the shape of the membership functions may differ significantly. The probabilistic method generates fuzzy sets which are similar in shape to those obtained by clustering. These sets can be asymmetrical, which is an advantage, as they can accurately represent the clusters. At at the same time, however, these sets are often nonconvex, and relatively high membership degrees can be assigned to data points that are "outside" the clustered data set. The reason for this is that the normalization constraint forces the sum of the membership degrees to equal one. Consequently, the method gives reliable results only when the identification data set covers most of the antecedent domain.

Membership functions generated by the possibilistic method are always convex, as the distance function has a global minimum at $\mathbf{v}_i^x$ and increases monotonically in all directions from this point. They are also symmetrical around the cluster centers. Data points outside the identification data set are assigned low membership degrees in all clusters, since they are far from all cluster centers and the normalization of equation (5.13) is not present.

Example 5.3 Figure 5.8 illustrates the difference in the form of membership functions computed by projection and by equations (5.13) and (5.14) for the fuzzy partition obtained in Example 4.3 by the FMLE algorithm. One can see that the membership

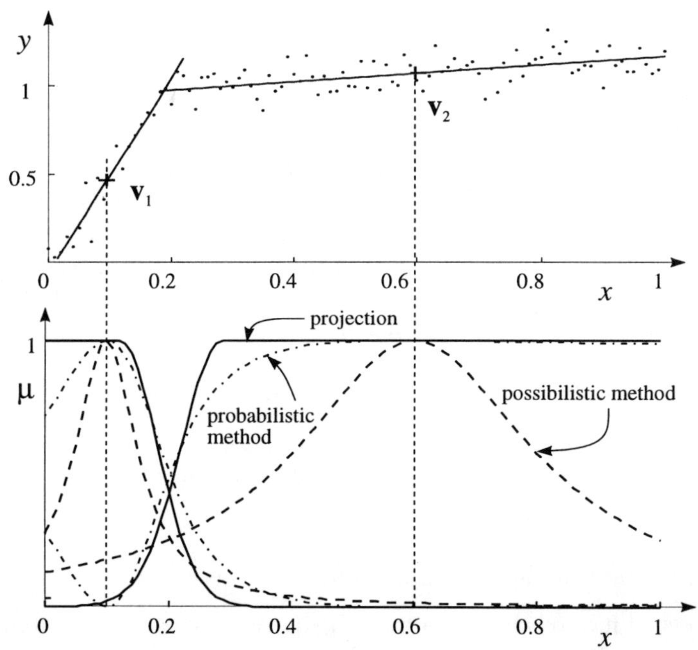

Figure 5.8. Comparison of membership functions generated by projection (solid line), and computed by the possibilistic method (5.14) (dashed line), and by the probabilistic method (5.13) (dashed-dotted line).

functions obtained by the possibilistic method (5.14) are symmetrical, and in this example do not reflect the validity of the local linear models correctly. The membership functions obtained by equation (5.13) represents better the partition, but due to the probabilistic constraint they are nonconvex (on the left-hand side). Projection gives the most plausible result in this case. □

On-line Adaptation of Membership Functions. An advantage of multidimensional membership functions is that they can be easily adapted on-line, as new data become available. This adaptation is achieved simply by recomputing the cluster centers and the covariance matrices. A data set $\mathbf{Z}$ of a constant size is kept in memory. Denote $\mathbf{z}_0^T = (\mathbf{x}_0^T, y_0)$ the new data vector. The partition is updated such that $\mathbf{z}_0$ replaces the vector $\mathbf{z}_{k^*}$ in Z whose regressor part has the smallest distance from $\mathbf{x}_0$:

$$k^* = \arg\min_k \, (\mathbf{x}_0 - \mathbf{x}_k)^T (\mathbf{x}_0 - \mathbf{x}_k) \,. \tag{5.15}$$

The cluster centers, covariance matrices and the partition matrix are updated by performing one or more iterations of the clustering algorithm. In this case, it is easier to compute the degrees of fulfillment by multidimensional membership functions, rather than by the projection. The latter technique requires the fitting of parametric membership functions to data, which is computationally more involved than the use of eq. (5.13) or (5.14).

Inversion of the TS Model. Finally, a remark should be made about the identification of an inverse TS model. A general TS model, consisting of rules (5.1), cannot be analytically inverted, because the consequents are functions of the antecedent variables. It is, however, possible to obtain an inverse model from the fuzzy partition. Recall that each cluster is an object which represents an implicit local linear relationship of the input and output data. Instead of projecting the clusters onto the space of regressors, one can generate membership functions for a subset of regressors, say $\{x_1, \ldots, x_{j-1}, x_{j+1}, \ldots, x_p\}$, and for the regressand y. Then a fuzzy model is obtained which describes an inverse relationship:

$$x_j = f^{-1}(y, x_1, \ldots, x_{j-1}, x_{j+1}, \ldots, x_p),$$

assuming, of course, that f is invertible. However, better results are usually obtained by identifying a fuzzy model of the process and by inverting that model, rather than identifying the inverse of the process.

5.1.2 Estimating Consequent Parameters

There are several methods to obtain the consequent parameters. Based on the geometrical interpretation of the TS model, the consequent parameters can be directly computed from the cluster prototypical points and the smallest eigenvectors of the cluster covariance matrices. This method assumes that errors are present in both the regressors and the regressand, and corresponds to the total least-squares solution[1] of the local linearization around the cluster center, as shown in this section.

A set of optimal parameters with respect to the model output can also be estimated from the identification data set by ordinary least-squares methods. This approach can be formulated as minimization of the total prediction error using the TS defuzzification

[1] Total least-squares estimation assumes that errors are present in both the regressor and regressand (van Huffel and Vandewalle, 1991).

formula (5.40), or as minimization of the prediction errors of the individual local models, solved as a set of c independent, weighted least-squares problems. If the model should serve as a numerical predictor, the global least-squares approach is usually preferred, as it gives the least prediction error. However, more accurate local models can be obtained by weighted ordinary or total least-squares methods.

Computing Consequent Parameters by Total Least-squares Method. The consequent parameters $\mathbf{a}_i$ and b_i of the affine TS model (5.1) can be derived from the geometrical structure of the clusters. Assume that the collection of c clusters approximates the regression surface. These clusters can be approximately regarded as p-dimensional linear subspaces of the regression space. The eigenvector ϕ_{in}, corresponding to the smallest eigenvalue λ_{in}, determines the normal vector to the hyperplane spanned by the remaining eigenvectors of that cluster (see Figure 5.9).

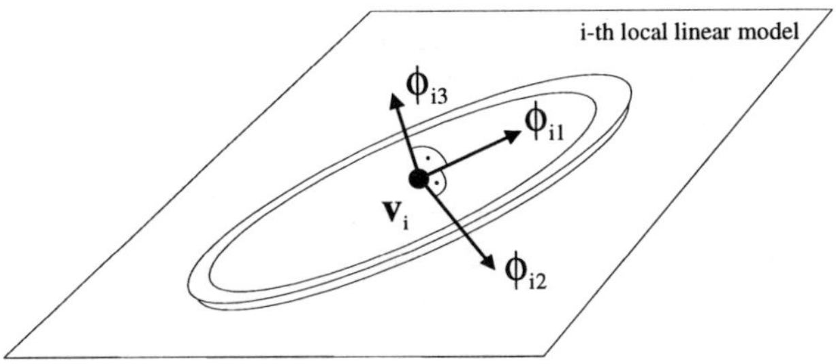

Figure 5.9. The position and orientation of the ith local linear model obtained by clustering is determined by the cluster center $\mathbf{v}_i$ and by the eigenvectors ϕ_{ij}, for $j = 1, 2, \ldots, n$.

To simplify the notation, the smallest eigenvector of the ith cluster will be denoted ϕ_i, omitting the subscript n. Recall that $\mathbf{z}^T = [\mathbf{x}^T, y]$ is the data vector and $\mathbf{v}_i$ is the cluster's prototype. The implicit normal form of the consequent's hyperplane is given by:

$$\phi_i^T (\mathbf{z} - \mathbf{v}_i) = 0\,. \tag{5.16}$$

This expression states that the inner product of the normal vector ϕ_i with any vector belonging to the hyperplane equals zero. For the following discussion, it is convenient to partition the prototype $\mathbf{v}_i$ into a vector $\mathbf{v}_i^x$ corresponding to the regressor $\mathbf{x}$, and a scalar v_i^y corresponding to the regressand y: $\mathbf{v}_i^T = \left[(\mathbf{v}_i^x)^T, v_i^y\right]$. The smallest eigenvector is partitioned in the same way as the cluster center, i.e.: $\phi_i^T = \left[(\phi_i^x)^T, \phi_i^y\right]$. Equation (5.16) can now be written as:

$$[(\phi_i^x)^T, \phi_i^y] \left([\mathbf{x}^T, y]^T - [(\mathbf{v}_i^x)^T, v_i^y]^T\right) = 0\,. \tag{5.17}$$

Carrying out the inner product leads to the following equality:

$$(\phi_i^x)^T (\mathbf{x} - \mathbf{v}_i^x) + \phi_i^y (y - v_i^y) = 0, \tag{5.18}$$

from which, by a simple algebraic manipulation, an explicit equation for the hyperplane is obtained:

$$y = \underbrace{\frac{-1}{\phi_i^y}(\phi_i^x)^T}_{\mathbf{a}_i^T} \mathbf{x} + \underbrace{\frac{1}{\phi_i^y}\phi_i^T \mathbf{v}_i}_{b_i} . \tag{5.19}$$

By comparing the above expression with the affine consequent of the TS rule (5.1), equations for $\mathbf{a}_i$ and b_i directly follow:

$$\mathbf{a}_i = \frac{-1}{\phi_i^y}\phi_i^x = \frac{-1}{\phi_{ip+1}}\left[\phi_{i1}, \phi_{i2}, \ldots, \phi_{ip}\right]^T , \tag{5.20a}$$

$$b_i = \frac{1}{\phi_i^y}\phi_i^T \mathbf{v}_i . \tag{5.20b}$$

Although these equations have been derived from the geometrical interpretation of the clusters in an intuitive way, it can be shown that $\mathbf{a}_i$ given by eq. (5.20a) is obtained as a solution of a weighted total least-squares (TLS) problem defined locally around the cluster center $\mathbf{v}_i$. The weights are the membership degrees contained in the ith row of the fuzzy partition matrix. Hence $\mathbf{v}_i$ is seen as a local operating point for the model. To obtain the affine linear form used in the TS rules (5.1), the offset parameters b_i are calculated using the estimates $\mathbf{a}_i$ and the cluster center $\mathbf{v}_i$. It is shown in the following that an expression equivalent to (5.20b) is obtained.

First, for each cluster, we transform the data by subtracting the cluster's center:

$$\Delta \mathbf{x}_k^i = \mathbf{x}_k - \mathbf{v}_i^x, \quad \Delta y_k^i = y_k - v_i^y . \tag{5.21}$$

The cluster's center $\mathbf{v}_i$ now becomes the origin and the local model describing the ith cluster can be given in a linear (homogeneous) form $\Delta y^i = \mathbf{a}_i^T \Delta \mathbf{x}^i$, instead of the affine form (5.1). To express the varying relevance of the different data samples to the ith local linear model, the data is weighted by $w_{ik} = \sqrt{(\mu_{ik})^m}$:

$$\tilde{\mathbf{x}}_k^i = w_{ik}\Delta \mathbf{x}_k^i, \quad \tilde{y}_k^i = w_{ik}\Delta y_k^i . \tag{5.22}$$

In order to formulate the problem in a matrix from, define the matrix $\tilde{\mathbf{X}}^i \in \mathbb{R}^{N \times p}$ and the column vector $\tilde{\mathbf{y}}^i \in \mathbb{R}^N$ as follows:

$$\tilde{\mathbf{X}}^i = \begin{bmatrix} (\tilde{\mathbf{x}}_1^i)^T \\ (\tilde{\mathbf{x}}_2^i)^T \\ \vdots \\ (\tilde{\mathbf{x}}_N^i)^T \end{bmatrix}, \quad \tilde{\mathbf{y}}^i = \begin{bmatrix} \tilde{y}_1^i \\ \tilde{y}_2^i \\ \vdots \\ \tilde{y}_N^i \end{bmatrix} . \tag{5.23}$$

Lemma 5.1 (Total least-squares solution) *Let the ith cluster be approximated by a local linear model in the transformed coordinates: $\Delta y^i = \mathbf{a}_i^T \Delta \mathbf{x}^i$. Let ϕ_i be the smallest eigenvector of the ith cluster's covariance matrix $\mathbf{F}_i$. The parameter vector $\mathbf{a}_i$, given by eq. (5.20a), is the unique solution of the linear system*

$$\tilde{\mathbf{y}}^i \approx \tilde{\mathbf{X}}^i \mathbf{a}_i \tag{5.24}$$

in the total least-squares sense.

Proof: The proof is straightforward and follows from Theorem 2.6 of van Huffel and Vandewalle (1991), stating that the standard TLS solution of (5.24) is given by:

$$\mathbf{a}_i = \frac{-1}{q^i_{p+1,p+1}}[q^i_{1p+1}, q^i_{2,p+1}, \ldots, q^i_{p,p+1}]^T, \tag{5.25}$$

where $\mathbf{q}^i_{p+1}$ is the last column of the matrix $\mathbf{Q}^i_2$ obtained by the singular value decomposition (SVD) of $[\tilde{\mathbf{X}}^i, \tilde{\mathbf{y}}^i]$:

$$[\tilde{\mathbf{X}}^i, \tilde{\mathbf{y}}^i] = \mathbf{Q}^i_1 \boldsymbol{\Sigma}^i (\mathbf{Q}^i_2)^T. \tag{5.26}$$

Here $\Sigma^i = \text{diag}\,(\sigma^i_1, \ldots, \sigma^i_{p+1})$ and the singular values are ordered such that $\sigma^i_1 \geq \sigma^i_2 \geq \ldots \geq \sigma^i_{p+1} > 0$. For any matrix $\mathbf{A}$ decomposed by SVD into $\mathbf{A} = \mathbf{Q}_1 \boldsymbol{\Sigma} \mathbf{Q}_2{}^T$, the columns of $\mathbf{Q}_2$ are the unit eigenvectors of $\mathbf{A}^T\mathbf{A}$ (Strang, 1976). In our case, $\mathbf{A} = [\tilde{\mathbf{X}}^i, \tilde{\mathbf{y}}^i]$ and

$$[\tilde{\mathbf{X}}^i, \tilde{\mathbf{y}}^i]^T[\tilde{\mathbf{X}}^i, \tilde{\mathbf{y}}^i] = \sum_{k=1}^{N} (\mu_{ik})^m (\mathbf{z}_k - \mathbf{v}_i)(\mathbf{z}_k - \mathbf{v}_i)^T = \mathbf{F}_i \sum_{k=1}^{N} (\mu_{ik})^m \tag{5.27}$$

where $\mathbf{F}_i$ is the cluster covariance matrix. The vector $\mathbf{q}^i_{p+1}$ is thus the smallest eigenvector of $\mathbf{F}_i$. □

Corollary 5.1 *For* $c = 1$, $\mathbf{a}$ *is the solution of the standard TLS problem* $\Delta\mathbf{y} \approx \mathbf{a}^T\Delta\mathbf{x}$, *since a single cluster results in the degenerated partition* $\mathbf{U} = \mathbf{1}^T$, $\mathbf{v}$ *is the sample mean and* $\mathbf{F}$ *is the sample covariance matrix.*

Now the offset parameter b_i of the affine consequent (5.1) is derived. In the incremental coordinates (5.22), the ith local linear model is given by:

$$\begin{aligned} \Delta y^i &= \mathbf{a}_i^T \Delta\mathbf{x} \\ (y - v^y_i) &= \mathbf{a}_i^T (\mathbf{x} - \mathbf{v}^x_i) \\ y &= \mathbf{a}_i^T \mathbf{x} + v^y_i - \mathbf{a}_i^T \mathbf{v}^x_i \end{aligned} \tag{5.28}$$

from which $b_i = v^y_i - \mathbf{a}_i^T \mathbf{v}^x_i$. By substituting for $\mathbf{a}_i^T$ from (5.20a), the following expression is obtained:

$$\begin{aligned} b_i &= v^y_i + \frac{1}{\phi^y_i}(\boldsymbol{\phi}^x_i)^T \mathbf{v}^x_i \\ &= \frac{1}{\phi^y_i}(\phi^y_i v^y_i + (\boldsymbol{\phi}^x_i)^T \mathbf{v}^x_i) \\ &= \frac{1}{\phi^y_i} \boldsymbol{\phi}_i^T \mathbf{v}_i, \end{aligned} \tag{5.29}$$

which is equivalent to (5.20b).

The adaptive distance (GK) clustering algorithm can be regarded as a method that decomposes a global nonlinear regression problem into a set of local linear TLS problems. This decomposition is based on minimization of the variance of the data from the cluster prototypes. The variance is measured by an inner product norm adapted iteratively to the shape of the cluster.

Instead of applying the weighted TLS, only the data samples that belong to the ith cluster to a degree greater than a specified threshold (α-level) can be used to obtain the estimate for the TS consequent parameters. This is achieved by computing the cluster means and covariance matrices from the α-cut of the partition $\mathbf{U}$ instead of the partition itself. The estimate of the parameters $\mathbf{a}_i$ and b_i is again given by (5.20a).

The close relationship between linear total least-squares estimation and fuzzy clustering with an adaptive distance allows the transfer or extension of many results known for linear TLS (van Huffel and Vandewalle, 1991) to the identification of nonlinear systems by fuzzy clustering. One of such extensions concerns the estimation of parameters in the case of different variance of the errors in the variables, which can be done in the linear case by using algorithms based on QR decomposition.

Example 5.4 This simulation example illustrates the parameter estimation properties of the adaptive distance clustering. Consider two linear functions:

$$y_1^0 = a_1 x_1^0 + b_1, \quad x_1^0 \in [0, 1], \tag{5.30a}$$

$$y_2^0 = a_2 x_2^0 + b_2, \quad x_2^0 \in [1, 2], \tag{5.30b}$$

with the parameters: $a_1 = 1$, $b_1 = 0$, $a_2 = -3$ and $b_2 = 4$. Noisy observations of both the regressor and the regressand are available:

$$x_1 = x_1^0 + \epsilon, \quad y_1 = y_1^0 + \epsilon, \quad x_2 = x_2^0 + \epsilon, \quad y_2 = y_2^0 + \epsilon, \tag{5.31}$$

where $\epsilon \sim N(0, \sigma^2)$. Two sets of experiments are considered, one with "small" standard deviation $\sigma = 0.2$ and one with "large" standard deviation $\sigma = 0.4$. For each standard deviation, two data sets are generated, one with a large number of samples and one with a small number of samples. Hence, in total, there are four different data sets. Denote N_1 and N_2 the number of samples in the data sets $\mathbf{Z}_1^T = [\mathbf{x}_1, \mathbf{y}_1]$ and $\mathbf{Z}_2^T = [\mathbf{x}_2, \mathbf{y}_2]$ respectively. Table 5.1 summarizes the specifications of the four generated data sets, denoted by "Case 1" to "Case 4".

Table 5.1. Specifications of the generated data sets.

	Std. deviation σ	No. of samples N_1, N_2	Total no. of samples N
Case 1	0.2	1000	2000
Case 2	0.2	20	40
Case 3	0.4	1000	2000
Case 4	0.4	50	100

For clustering, the data sets $\mathbf{Z}_1$ and $\mathbf{Z}_2$ are concatenated in one set: $\mathbf{Z} = [\mathbf{Z}_1, \mathbf{Z}_2]$ for each case. The total number of samples is $N = N_1 + N_2$. Figure 5.10 shows

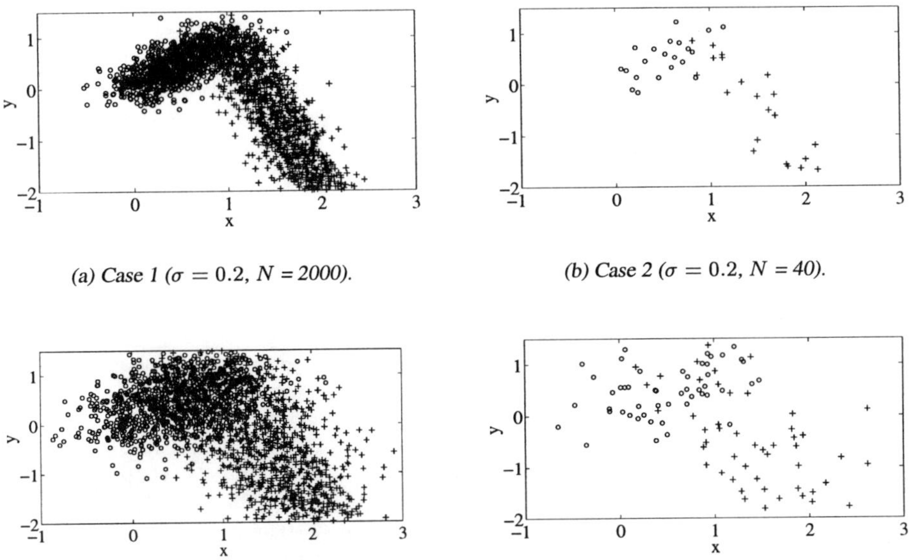

(a) Case 1 ($\sigma = 0.2$, $N = 2000$).

(b) Case 2 ($\sigma = 0.2$, $N = 40$).

(c) Case 3 ($\sigma = 0.4$, $N = 2000$).

(d) Case 4 ($\sigma = 0.4$, $N = 100$).

Figure 5.10. The generated data sets. Samples of $\mathbf{Z}_1$ are denoted by 'o', samples of $\mathbf{Z}_2$ by '+'.

plots of the four cases. In each case, a comparison is made of the parameter estimates obtained by:

1. Global TLS applied to $\mathbf{Z}_1$ and $\mathbf{Z}_2$ separately. This is simply the ideal case and it should give the best (reference) estimate. This approach is denoted as TLS 1. In practice, if the data set $\mathbf{Z}$ represents a nonlinear system, the separate sets $\mathbf{Z}_1$ and $\mathbf{Z}_2$, of course, cannot be obtained.

2. TLS applied to $\mathbf{Z}$ split at $x = 1$. Worse results can be expected in this situation because of wrong assignment of some data points along the hard boundary, due to the noise. This approach is denoted by TLS 2.

3. Gustafson–Kessel (GK) clustering with the two cluster centers initialized randomly. The fuzziness coefficient was set to $m = 1.2$ and the termination tolerance to 0.01. The setting of the fuzziness coefficient appeared to be quite crucial in this example, and it turns out that lower values are preferable to higher ones.

4. The fuzzy maximum likelihood estimates clustering (FMLE) initialized with the partition obtained by GK. The termination tolerance is again 0.01.

5. Ordinary least-squares method (OLS) applied to $\mathbf{Z}_1$ and $\mathbf{Z}_2$ separately. It is well known that OLS gives biased estimates for the *errors in variables* data such as (5.31). The results should confirm this fact.

To be able to compare the TLS and the clustering methods, only the slope parameter $\hat{a}$ is estimated from the data by TLS (the mean is subtracted from each data set). The offset parameter $\hat{b}$ is then computed from $\hat{a}$ and the means. For a meaningful comparison, 50 data sets are generated in each case and the average results are compared in terms of the estimated parameter values $\hat{a}_i$, $\hat{b}_i$, their squared biases $(a_i - \hat{a}_i)^2$, $(b_i - \hat{b}_i)^2$, and the estimated standard deviations $s(\hat{a}_i)$.

Table 5.2. Comparison of the parameter estimates for Case 1.

method	TLS 1	TLS 2	GK	FMLE	OLS
$\hat{a}_1$	1.0077	1.1497	0.8515	0.9112	0.6774
$(a_1 - \hat{a}_1)^2$	0.0001	0.0224	0.0220	0.0079	0.1041
$s(\hat{a}_1)$	0.0313	0.0418	0.0220	0.0204	0.0235
$\hat{a}_2$	-3.0008	-3.4054	-3.0913	-3.0617	-2.0319
$(a_2 - \hat{a}_2)^2$	0.0000	0.1643	0.0083	0.0038	0.9373
$s(\hat{a}_2)$	0.0691	0.0788	0.0606	0.0526	0.0478
$\hat{b}_1$	-0.0053	-0.0738	0.0349	0.0202	0.1602
$(b_1 - \hat{b}_1)^2$	0.0000	0.0054	0.0012	0.0004	0.0257
$\hat{b}_2$	3.9987	4.6942	4.1649	4.1065	2.5457
$(b_2 - \hat{b}_2)^2$	0.0000	0.4820	0.0272	0.0113	2.1149
iterations	—	—	9.3200	10.5800	—

Table 5.2 gives the results for Case 1. Note that the clustering methods applied to the global data set $\mathbf{Z}$ give estimates of the local parameters comparable to TLS applied to separate data sets (TLS 1), and better results than TLS applied to the split data set (TLS 2). The FMLE algorithm is slightly more accurate than GK, both in terms of the bias and the variance of the estimated parameters. Both algorithms needed in average about 10 iterations to converge. The OLS estimates are biased, as expected.

Table 5.3 summarizes the results for Case 2 (small number of samples). The outcome is comparable with Table 5.2, thus it can be concluded that, for this data, the clustering methods perform equally well for both small and large numbers of samples. Moreover, note that clustering gives lower estimated variance $s(\hat{a}_i)$ than TLS. This can be explained by the fact that some data points are shared for estimating both local models, and hence more data is effectively available than when using the data sets separately. The number of iterations needed remains unchanged, which suggests that for a well-determined partition, the number of data samples has little influence on the number of iterations. Finally, note that the GK and FMLE perform comparably, in fact, GK gives the best results for the second local submodel (also better than TLS).

Table 5.4 presents the results for Case 3, where the number of samples is the same as in Case 1, but the standard deviation is twice as high. The performance of TLS on

Table 5.3. Comparison of the parameter estimates for Case 2.

method	TLS 1	TLS 2	GK	FMLE	OLS
$\hat{a}_1$	0.9937	1.1744	0.8569	0.8987	0.7073
$(a_1 - \hat{a}_1)^2$	0.0000	0.0304	0.0205	0.0103	0.0857
$s(\hat{a}_1)$	0.2009	0.2927	0.1478	0.1322	0.1502
$\hat{a}_2$	-3.0050	-3.4034	-3.0011	-2.9627	-2.1700
$(a_2 - \hat{a}_2)^2$	0.0000	0.1627	0.0000	0.0014	0.6889
$s(\hat{a}_2)$	0.4405	0.5030	0.3554	0.3112	0.3163
$\hat{b}_1$	-0.0112	-0.0914	0.0421	0.0271	0.1353
$(b_1 - \hat{b}_1)^2$	0.0001	0.0084	0.0018	0.0007	0.0183
$\hat{b}_2$	3.9962	4.6902	3.9989	3.9242	2.7473
$(b_2 - \hat{b}_2)^2$	0.0000	0.4763	0.0000	0.0057	1.5693
iterations	—	—	10.0400	10.6000	—

the split data set (TLS 2) drops remarkably, while the clustering methods give results similar to Case 1. The average number of iterations has slightly increased.

Table 5.4. Comparison of the parameter estimates for Case 3.

method	TLS 1	TLS 2	GK	FMLE	OLS
$\hat{a}_1$	1.0227	4.7857	0.8223	0.8349	0.3406
$(a_1 - \hat{a}_1)^2$	0.0005	14.3314	0.0316	0.0273	0.4348
$s(\hat{a}_1)$	0.0642	0.8259	0.0418	0.0400	0.0300
$\hat{a}_2$	-3.0046	-4.3166	-3.2343	-3.3860	-1.0316
$(a_2 - \hat{a}_2)^2$	0.0000	1.7335	0.0549	0.1490	3.8748
$s(\hat{a}_2)$	0.1388	0.2103	0.1402	0.1441	0.0518
$\hat{b}_1$	-0.0142	-1.6033	0.0472	0.0198	0.3283
$(b_1 - \hat{b}_1)^2$	0.0002	2.5705	0.0022	0.0004	0.1078
$\hat{b}_2$	4.0020	6.3928	4.4408	4.6720	1.0442
$(b_2 - \hat{b}_2)^2$	0.0000	5.7255	0.1943	0.4516	8.7370
iterations	—	—	13.5400	10.9400	—

Table 5.5 presents the results for Case 4. TLS on the split data set (TLS 2) now gives useless results, while the GK algorithm is still comparable to the TLS 1 method. Note, however, that the estimates obtained by FMLE are much more biased in this case, which may suggest that the method is somewhat more sensitive to high levels of noise and a small number of samples. □

Table 5.5. Comparison of the parameter estimates for Case 4.

method	TLS 1	TLS 2	GK	FMLE	OLS
$\hat{a}_1$	1.2220	-7.0978	1.0645	0.1178	0.3684
$(a_1 - \hat{a}_1)^2$	0.0493	65.5743	0.0042	0.7783	0.3990
$s(\hat{a}_1)$	0.4302	836.1273	0.6796	2.1980	0.1349
$\hat{a}_2$	-3.2880	-4.7882	-3.6801	-7.5759	-1.0387
$(a_2 - \hat{a}_2)^2$	0.0829	3.1977	0.4626	20.9391	3.8467
$s(\hat{a}_2)$	0.7581	1.2387	0.9420	57.9089	0.2387
$\hat{b}_1$	-0.0975	3.8807	-0.0765	1.2657	0.3138
$(b_1 - \hat{b}_1)^2$	0.0095	15.0602	0.0058	1.6019	0.0985
$\hat{b}_2$	4.4236	7.1493	5.1674	11.7665	1.0533
$(b_2 - \hat{b}_2)^2$	0.1794	9.9179	1.3629	60.3185	8.6830
iterations	—	—	13.8600	20.3000	—

Estimating Consequent Parameters by Ordinary Least-squares Method. There are two approaches to estimate the consequent parameters $\mathbf{a}_i$, b_i by the ordinary least-squares technique:

1. Solving c independent weighted least-squares problems, one for each cluster. This approach is independent of the aggregation/defuzzification method used, as it estimates the parameters for each model separately (like the TLS method).

2. Solving a global least-squares problem following from the weighted mean defuzzification formula (5.40). This approach gives an estimate with the minimal prediction error.

The identification data and the membership degrees of the fuzzy partition are arranged in the following matrices:

$$\mathbf{X} = \begin{bmatrix} \mathbf{x}_1^T \\ \mathbf{x}_2^T \\ \vdots \\ \mathbf{x}_N^T \end{bmatrix}, \quad \mathbf{y} = \begin{bmatrix} y_1 \\ y_2 \\ \vdots \\ y_N \end{bmatrix}, \quad \mathbf{W}_i = \begin{bmatrix} \mu_{i1} & 0 & \cdots & 0 \\ 0 & \mu_{i2} & \cdots & 0 \\ \vdots & \vdots & \ddots & \vdots \\ 0 & 0 & \cdots & \mu_{iN} \end{bmatrix}. \tag{5.32}$$

The consequent parameters of the rule belonging to the ith cluster, $\mathbf{a}_i$ and b_i are concatenated into a single parameter vector: $\boldsymbol{\theta}_i$:

$$\boldsymbol{\theta}_i = \left[\mathbf{a}_i^T, b_i\right]^T . \tag{5.33}$$

Appending a unitary column to $\mathbf{X}$ gives the extended regressor matrix $\mathbf{X}_e$:

$$\mathbf{X}_e = [\mathbf{X}, \mathbf{1}] . \tag{5.34}$$

A Set of Weighted Least-squares Problems. Assuming that each cluster represents a local linear model of the system, the consequent parameter vectors θ_i, $i = 1, 2, \ldots, c$, can be estimated independently by the weighted least-squares method. The membership degrees μ_{ik} of the fuzzy partition serve as the weights expressing the relevance of the data pair $(\mathbf{x}_k, y_k)$ to that local model. If the columns of $\mathbf{X}_e$ are linearly independent and $\mu_{ik} > 0$ for $1 \leq k \leq N$, then

$$\boldsymbol{\theta}_i = \left[\mathbf{X}_e^T \mathbf{W}_i \mathbf{X}_e\right]^{-1} \mathbf{X}_e^T \mathbf{W}_i \mathbf{y} \tag{5.35}$$

is the least-squares solution of $\mathbf{y} = \mathbf{X}_e \theta + \epsilon$ where the kth data pair $(\mathbf{x}_k, y_k)$ is weighted by μ_{ik}. The parameters $\mathbf{a}_i$ and b_i are given by:

$$\mathbf{a}_i = [\theta_1, \theta_2, \ldots, \theta_p], \quad b_i = \theta_{p+1} . \tag{5.36}$$

If the columns of $\mathbf{X}_e$ are linearly dependent, techniques based on the orthogonal factorization of $\mathbf{X}_e$ should be used. It should be remarked that for a computer implementation, rather than using (5.35), it is more efficient to first multiply each row of $\mathbf{X}_e$ and $\mathbf{y}$ by $\sqrt{\mu_{ik}}$:

$$\tilde{\mathbf{X}}_i = \begin{bmatrix} \sqrt{\mu_{i1}}\mathbf{x}_{e1}^T \\ \sqrt{\mu_{i2}}\mathbf{x}_{e2}^T \\ \vdots \\ \sqrt{\mu_{iN}}\mathbf{x}_{eN}^T \end{bmatrix}, \quad \tilde{\mathbf{y}}_i = \begin{bmatrix} \sqrt{\mu_{i1}}y_1 \\ \sqrt{\mu_{i2}}y_2 \\ \vdots \\ \sqrt{\mu_{iN}}y_N \end{bmatrix}, \tag{5.37}$$

and to compute $\boldsymbol{\theta}_i$ by:

$$\boldsymbol{\theta}_i = \left[\tilde{\mathbf{X}}_i^T \tilde{\mathbf{X}}_i\right]^{-1} \tilde{\mathbf{X}}_i^T \tilde{\mathbf{y}}_i . \tag{5.38}$$

Also here, instead of applying the weighted least-squares approach, only the data samples that belong to the ith cluster to a degree greater than a specified threshold α can be used to obtain the estimate for the consequent parameters of the ith rule. In this case, instead of applying (5.37), $\tilde{\mathbf{X}}_i$ and $\tilde{\mathbf{y}}$ are constructed as:

$$\tilde{\mathbf{X}}_i = \{\mathbf{x}_{ek} \mid \mathbf{x}_{ek} \in \mathbf{X}_e, \ \mu_{ik} > \alpha\} \tag{5.39a}$$

$$\tilde{\mathbf{y}}_i = \{y_k \mid y_k \in \mathbf{y}, \ \mu_{ik} > \alpha\}, \tag{5.39b}$$

for some $\alpha \in [0, 1]$. The least-squares estimate of the parameters is given by eq. (5.38). If a sufficient number of data samples are available in each cluster, the α-cut approach gives a smaller bias, as only the data belonging in a high degree to the cluster are considered. However, thc variance of the estimate increases, as fewer data samples are available for each of the local estimation problems.

Global Least-squares Method. The weighted least-squares approach gives an optimal estimate of the parameters of the local models, but it does not provide an optimal TS model in terms of a minimal prediction error. In order to obtain an optimal global predictor, the aggregation of the rules has to be taken into account. When using the fuzzy-mean formula:

$$y = \frac{\sum_{i=1}^{K} \beta_i(\mathbf{x})[\mathbf{a}_i^T \mathbf{x} + b_i]}{\sum_{i=1}^{K} \beta_i(\mathbf{x})}, \tag{5.40}$$

the consequent parameter estimates can be obtained by solving a linear least-squares problem. The degree of fulfillment of the ith rule β_{ik} can be obtained from the fuzzy partition matrix $\mathbf{U}$, but the situation is slightly more complicated than in the case of the local weighted least-squares approach. Recall that each row of $\mathbf{U}$ contains a point-wise definition of the membership function for the data in the product space $X \times Y$. In order to obtain a point-wise definition of the membership function A_i in the antecedent space X, the ith row of $\mathbf{U}$ has to be projected onto the antecedent space:

$$\beta_{ik} = \text{proj}_{\mathbb{N}_p}(\mu_{ik}), \quad k = 1, \ldots, N, \tag{5.41}$$

where "proj" denotes the projection operator (Definition A.18). The result of this projection step is that a set of repeated vectors $\mathbf{x}_k$ in the data are assigned the maximum membership degree from this set. Another possibility is to compute the degrees of fulfillment after generating the antecedent membership functions, which can partially compensate for the decomposition error, see Section 5.1.1.

In order to write (5.40) in a matrix form for all the data $(\mathbf{x}_k, y_k)$, $1 \le k \le N$, denote $\mathbf{\Gamma}_i$ a diagonal matrix in $\mathbb{R}^{N\times N}$ having the normalized membership degree $\gamma_{ik} = \beta_{ik} / \sum_{j=1}^{c} \beta_{jk}$ as its kth diagonal element:

$$\mathbf{\Gamma}_i = \begin{bmatrix} \gamma_{i1} & 0 & \cdots & 0 \\ 0 & \gamma_{i2} & \cdots & 0 \\ \vdots & \vdots & \ddots & \vdots \\ 0 & 0 & \cdots & \gamma_{iN} \end{bmatrix}. \tag{5.42}$$

Denote $\mathbf{X}'$ the matrix in $\mathbb{R}^{N\times cN}$ composed from matrices $\mathbf{\Gamma}_i$ and $\mathbf{X}_e$

$$\mathbf{X}' = [\mathbf{\Gamma}_1 \mathbf{X}_e, \mathbf{\Gamma}_2 \mathbf{X}_e, \ldots, \mathbf{\Gamma}_c \mathbf{X}_e], \tag{5.43}$$

where $\mathbf{X}_e$ is given by eq. (5.34). Denote $\boldsymbol{\theta}'$ the vector in $\mathbb{R}^{c(p+1)}$ given by

$$\boldsymbol{\theta}' = \left[\boldsymbol{\theta}_1^T, \boldsymbol{\theta}_2^T, \ldots, \boldsymbol{\theta}_c^T\right]^T, \tag{5.44}$$

where $\boldsymbol{\theta}_i = [\mathbf{a}_i^T, b_i]^T$ for $1 \le i \le c$. The resulting least-squares problem $\mathbf{y} = \mathbf{X}'\boldsymbol{\theta}' + \epsilon$ has the solution:

$$\boldsymbol{\theta}' = \left[(\mathbf{X}')^T \mathbf{X}'\right]^{-1} (\mathbf{X}')^T \mathbf{y}. \tag{5.45}$$

From eq. (5.44) the parameters $\mathbf{a}_i$ and b_i are obtained by:

$$\mathbf{a}_i = [\theta'_{q+1}, \theta'_{q+2}, \ldots, \theta'_{q+p}]^T, \quad b_i = [\theta_{q+p+1}], \quad \text{with } q = (i-1)(p+1). \tag{5.46}$$

The global least-squares method gives a minimal prediction error, and thus it is suitable for deriving prediction models. At the same time, however, it biases the estimate of the local model parameters as demonstrated in Section 2.3.2. For the purpose of local interpretation and analysis of the TS model, the weighted least-squares approach is preferable.

5.1.3 Rule Base Simplification

One of the aspects that distinguish fuzzy modeling from other black-box approaches like neural networks, radial basis functions, or CMAC, is that fuzzy models are, to a certain degree, *transparent* to interpretation and analysis. However, for fuzzy models acquired from numerical data, the transparency is not achieved automatically. The same system can be described with a few rules using distinct and interpretable fuzzy sets, and also with a large number of highly overlapping fuzzy sets that can hardly be interpreted. It is the practice in current literature that both models are called fuzzy models, and both are labeled as transparent and *physically* interpretable (Wang, 1994).

When generating fuzzy models from data, some degree of redundancy, and thus also unnecessary complexity, cannot be avoided. The redundancy typically occurs in terms of *similar* fuzzy sets (membership functions). Similar fuzzy sets are highly overlapping fuzzy sets that describe almost the same region in the domain of a model variable. In such a case, the fuzzy model uses more fuzzy sets than necessary, since they represent more or less the same linguistic concept. Fuzzy sets estimated from data also can be similar to a universal set, adding no information to the model. Although simplification of a redundant fuzzy model can be left to the user, it is more convenient to automate this process. This section describes an algorithm developed by Setnes (1995) that uses a similarity measure to identify similar fuzzy sets in an initial fuzzy model, and replaces them by a common fuzzy set that represents the original ones. This common fuzzy set is substituted for the original ones in the rule base. If the redundancy in the model is high, merging similar fuzzy sets may result in equal rules that can be subsequently deleted from the rule base. This leads to the reduction of the number of rules. Compatible cluster merging, presented in Section 4.5.2, can also be seen as a simplification procedure. The main difference between CCM and the method described in this section is that CCM is applied during the clustering stage, in the product space of the model regressors and the regressand, while the similarity-based simplification is applied to the model obtained from the fuzzy partition. Consequently, the aim of CCM is to reduce the number of clusters (rules), while the method presented below merges similar membership functions.

Redundancy in Fuzzy Models. Acquisition of fuzzy models by product-space clustering may lead to redundancy when projecting the clusters onto the antecedent variables, as illustrated in Figure 5.11. The projection of clusters C_1, C_2 and C_3 onto antecedent variables x_1 and x_2 results in similarity between A_1 and A_2, and, further, B_3 is similar to the universal set, as it covers the entire domain of x_2.

Simplification of the rule base can be achieved by merging similar fuzzy sets, and by removing fuzzy sets similar to the universal set (these fuzzy sets do not contribute to the rule base). To quantify the similarity, a large number of similarity measures known

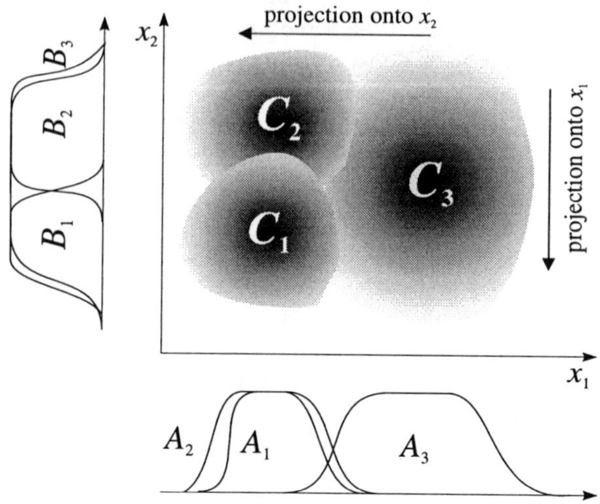

Figure 5.11. Redundancy in terms of similar membership functions obtained by cluster projection. Reproduced from (Babuška, et al., 1996), ©1996 IEEE.

from the literature can be applied. Based on an extensive research study (Setnes, 1995), the following similarity measure was selected for rule base simplification:

$$S(A,B) = \frac{\mid A \cap B \mid}{\mid A \cup B \mid} = \frac{\sum_{q=1}^{n_d} (\mu_A(x_q) \wedge \mu_B(x_q))}{\sum_{q=1}^{n_d} (\mu_A(x_q) \vee \mu_B(x_q))}, \tag{5.47}$$

where $\mid . \mid$ denotes the cardinality of a fuzzy set. Membership functions μ_A and μ_B are defined in the discrete domain $X = \{x_q \mid q = 1, 2, ..., n_d\}$.

Rule Base Simplification Algorithm. The simplification algorithm merges similar fuzzy sets iteratively using two thresholds: $\eta \in (0,1)$ for merging fuzzy sets that are similar to one another, and $\eta_r \in (0,1)$ for removing fuzzy sets similar to the universal set. In each iteration, the similarity between all fuzzy sets for each antecedent variable is considered. The pair of membership functions with the highest similarity $s \geq \eta$ are merged. The rule base is updated by substituting the new fuzzy set for the ones merged. The algorithm repeatedly evaluates the similarities in the updated rule base, until there are no more fuzzy sets for which $s \geq \eta$. Finally, fuzzy sets similar to the universal set are removed from the antecedents of the rules in which they occur. The algorithm only merges one pair of fuzzy sets per iteration. Merging two fuzzy sets A_{lq} and A_{mq} is accomplished by taking the support of the new fuzzy set A as the support of $A_{lq} \cup A_{mq}$. This guarantees preservation of the coverage of the antecedent space. The kernel of A is given by averaging the kernels of A_{lq} and A_{mq}, making a trade-off between the two rules l and m. The procedure is summarized in Algorithm 5.1.

Example 5.5 To demonstrate the rule base simplification algorithm, a TS fuzzy model is simulated to generate a set of input–output data. From this data, a fuzzy model of

Algorithm 5.1 (Rule base simplification algorithm)

Given a fuzzy rule base obtained by clustering, select the thresholds $\eta_r, \eta \in (0, 1)$.

Repeat:

Step 1: **Select the two most similar fuzzy sets in the rule base.** Calculate $s_{ijk} = S(A_{ij}, A_{kj})$, $j = 1, 2, \ldots, p$, $i, k = 1, 2, \ldots, K$. Select A_{lq} and A_{mq}, such that $s_{lmq} = \max\limits_{i,j,k,\; i \neq k} \{s_{ijk}\}$.

Step 2: **Merge the two most similar fuzzy sets and update the rule base.** If $S(A_{lq}, A_{mq}) \geq \eta$ merge A_{lq} and A_{mq} to create a new fuzzy set A and replace $A_{lq} = A$ and $A_{mq} = A$.

Until: no more fuzzy sets have similarity $s_{ijk} \geq \eta$.

Step 3: **Remove fuzzy sets similar to the universal set.** For each fuzzy set A_{ij} calculate $S(A_{ij}, U)$, where $\mu_U = 1$, $\forall x_j$. If $S(A_{ij}, U) \geq \eta_r$ remove A_{ij} from the antecedent of the ith rule.

the same TS structure is constructed by product-space fuzzy clustering. We will notice that the initial set of membership functions obtained by projecting the clusters onto the antecedent variables is larger than the original set of membership functions. Some of the membership functions are similar to each other and to the universal fuzzy set. By applying the simplification algorithm, the similar membership functions are replaced by a common one, and the antecedent propositions whose fuzzy sets are similar to the universal set are removed from the antecedents. Figure 5.12 shows the fuzzy systems to be identified from data (Sugeno and Kang, 1988).

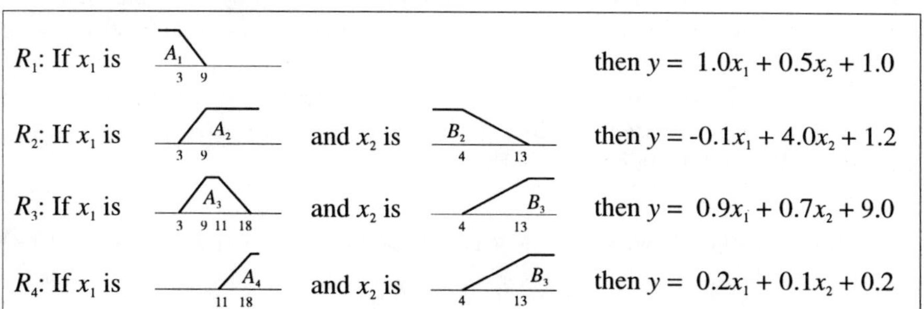

Figure 5.12. Rule base and membership functions of the fuzzy system. Reproduced from (Babuška, et al., 1996), ©1996 IEEE.

The rule base consists of four TS rules with two antecedent variables, $x_1, x_2 \in \mathbb{R}$, and one consequent variable, $y \in \mathbb{R}$. Note that the antecedent of rule R_1 contains only variable x_1, and the antecedents of R_3 and R_4 contain the same fuzzy set B_2. It is interesting to verify whether a rule base with the same structure and membership functions can be extracted from the data generated by this fuzzy system.

The output of the model is calculated by the fuzzy-mean method (5.40), where the degrees of fulfillment of the antecedents, β_i, are given by: $\beta_1 = \mu_{A_1}$; $\beta_2 = \mu_{A_2}\mu_{B_1}$; $\beta_3 = \mu_{A_3}\mu_{B_2}$; $\beta_4 = \mu_{A_4}\mu_{B_2}$. A set X of 500 input data points is generated to cover uniformly the antecedent space: $X = \{(x_{1j}, x_{2k})^T | j = 0, 1, \ldots, 25, k = 0, 1, \ldots, 20\}$. Figure 5.13 depicts the input–output mapping of the fuzzy systems.

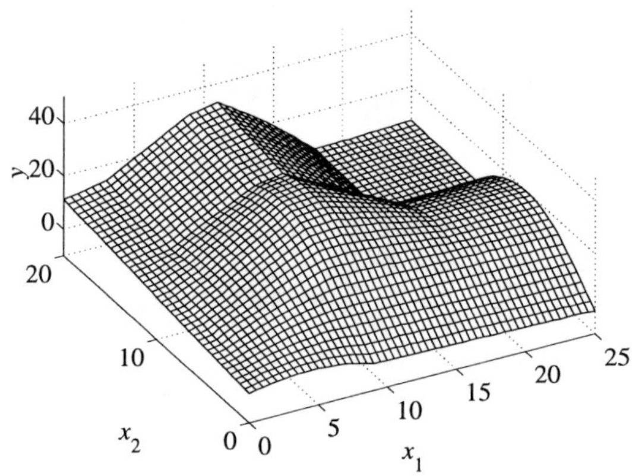

Figure 5.13. Input–output mapping of the system in Figure 5.12. Reproduced from (Babuška, et al., 1996), ©1996 IEEE.

The data set $\mathbf{Z}^T = [\mathbf{X}, \mathbf{y}]$ is clustered by the GK algorithm. From Figure 5.14, one can see that the original and the obtained antecedent partitions are quite similar even though the form, width, and location of the cluster boundaries (shaded areas) slightly differ.

Figure 5.15 shows the membership functions obtained by projecting the clusters on the antecedent variables and fitting them by the parametric exponential functions given by eq. (A.5).

Four membership functions are obtained for each antecedent variable, as there are four clusters in the antecedent product space and each is projected onto x_1 and x_2. The resulting TS model can be written by:

$$\textbf{If } x_1 \text{ is } A_i \textbf{ and } x_2 \text{ is } B_i \textbf{ then } y = a_{i1}x_1 + a_{i2}x_2 + b_i; \quad i = 1, \ldots, 4. \tag{5.48}$$

Membership functions B_3 and B_4 are quite similar to each other, and A_2 and B_1 cover the entire domains to a high degree, and can thus possibly be left out of the antecedents. These assumptions are now verified by applying the simplification algorithm. The threshold parameters are set to $\eta = 0.8$ for similarity among the membership functions,

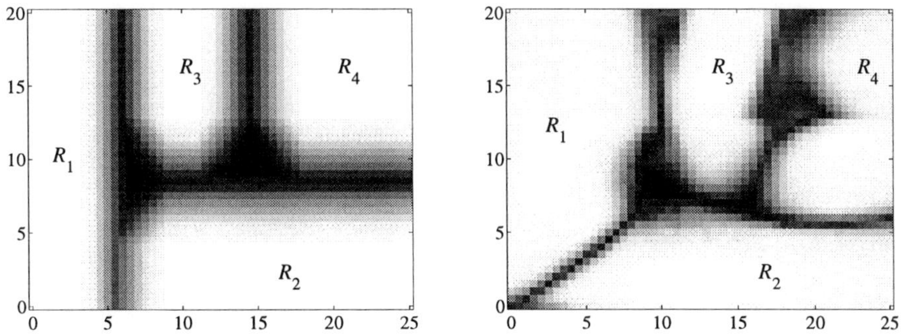

(a) Original partition.

(b) Partition obtained by clustering.

Figure 5.14. Fuzzy partition of the antecedent space. The dark areas correspond to the overlaps of the membership functions. Reproduced from (Babuška, et al., 1996), ©1996 IEEE.

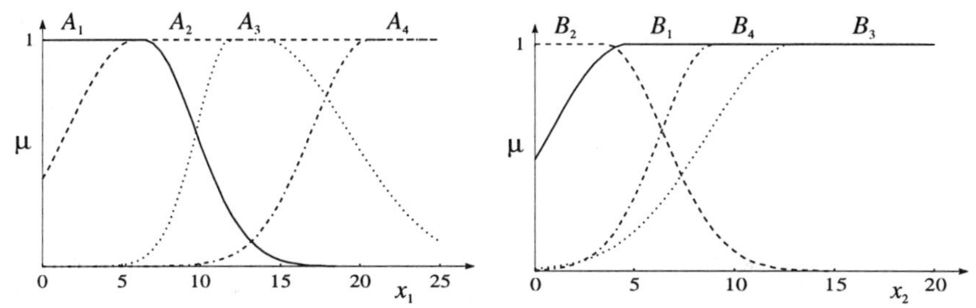

Figure 5.15. Membership functions A_i and B_i obtained by projecting clusters R_i onto x_1 and x_2, respectively. Reproduced from (Babuška, et al., 1996), ©1996 IEEE.

and $\eta_r = 0.95$ for similarity to the universal set. The result is the following rule base:

$$\begin{array}{ll} \textbf{If } x_1 \text{ is } A_1 & \textbf{then } \ldots \\ \textbf{If } x_1 \text{ is } A_2 \textbf{ and } x_2 \text{ is } B_2 & \textbf{then } \ldots \\ \textbf{If } x_1 \text{ is } A_3 \textbf{ and } x_2 \text{ is } B_3 & \textbf{then } \ldots \\ \textbf{If } x_1 \text{ is } A_4 \textbf{ and } x_2 \text{ is } B_3 & \textbf{then } \ldots \end{array} \tag{5.49}$$

The antecedent structure is now equivalent to the structure of the system shown in Figure 5.12. Also the obtained membership functions are in good agreement with the original definitions, see Figure 5.16. By lowering the threshold λ_r, also the fuzzy set A_2 would be classified as similar to the universal set and would be removed. The obtained rule base would be simpler than the original one, but the change of the input–output mapping would be negligible. □

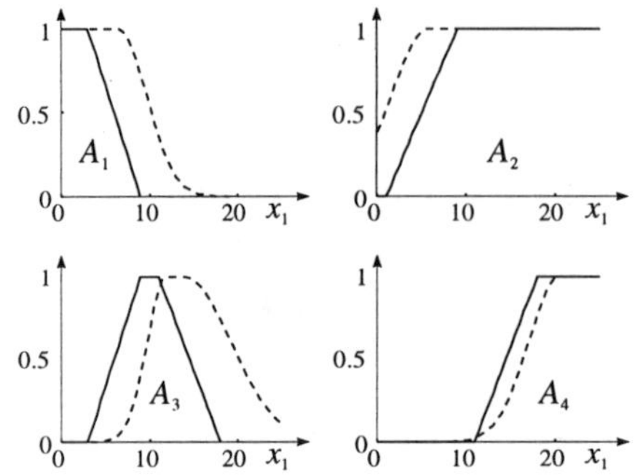

(a) Antecedent variable x_1.

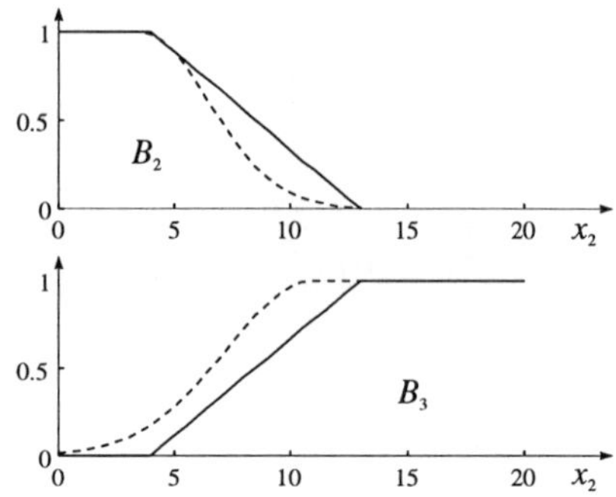

(b) Antecedent variable x_2.

Figure 5.16. Original membership functions (solid line) compared with the membership functions obtained by simplification (dashed line). Reproduced from (Babuška, et al., 1996), ©1996 IEEE.

5.1.4 Linguistic Approximation

Linguistic interpretation of the membership functions of a fuzzy rule base acquired from data can be obtained by means of *linguistic approximation*. Linguistic approximation is a method to assign an appropriate linguistic term to a given fuzzy set (Esragh and Mamdani, 1979; Sugeno and Yasukawa, 1993). In this section, a similarity measure is used to compare the fuzzy sets obtained from data to some reference fuzzy sets and their modifications by hedges and set-theoretic operations. The similarity of the antecedent fuzzy sets A_{ij} to the reference fuzzy sets and their linguistic modifications is evaluated for each antecedent variable x_i. The term with the highest degree of similarity is then selected to represent the membership function A_{ij} originally obtained from data.

Example 5.6 Three reference fuzzy sets Small, Medium and Big are defined for the antecedent variables x_1, x_2 from Example 5.5. The following hedges and set-theoretic operators were used to modify the meaning of the reference fuzzy sets:

linguistic hedge	operation	linguistic hedge	operation
very C_i	$\mu^2_{C_i}$	$C_i \cup C_j, \forall i,j,\ i \neq j$	$\min(\mu_{C_i} + \mu_{C_j}, 1)$
not very C_i	$1 - \mu^2_{C_i}$	rather C_i	$\text{int}\,(\mu_{C_i})$
more or less C_i	$\sqrt{\mu_{C_i}}$	fuzzy set "Any"	$\mu_{\text{Any}}(x) = 1,\ \forall x$

Here C_i stands for the reference fuzzy sets and "int" denotes the contrast intensification operator given by:

$$\text{int}\,(\mu_{C_i}) = \begin{cases} 2\mu^2_{C_i}, & \mu_{C_i} \leq 0.5 \\ 1 - 2(1 - \mu_{C_i})^2 & \text{otherwise.} \end{cases}$$

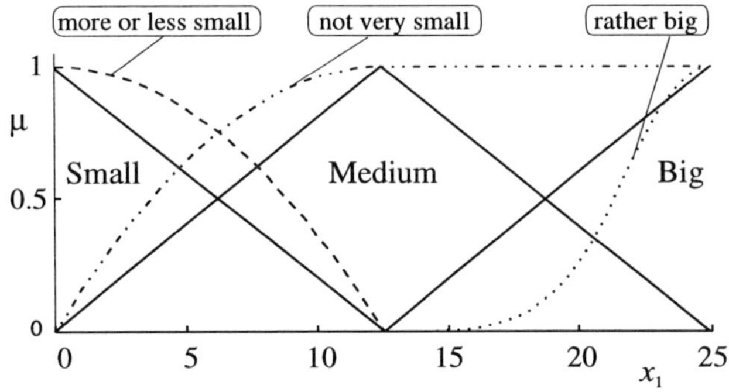

Figure 5.17. Reference fuzzy sets and their modifications by some linguistic hedges. Reproduced from (Babuška, et al., 1996), ©1996 IEEE.

The similarity of the antecedent fuzzy sets $A_1, \ldots, A_4$ and B_2, B_3 to the reference fuzzy sets and their modifications by the above-listed hedges is evaluated. As a result, a matrix Σ^1 is computed for the antecedent variable x_1. The ijth element of this matrix gives the similarity of the fuzzy set A_i to the corresponding reference fuzzy sets and their modifications. Analogously, the matrix Σ^2 is computed for the second antecedent variable x_2. The resulting linguistic approximation is then the one corresponding to the highest degree of similarity in each row of Σ^1 and Σ^2. The final rule base obtained is:

If x_1 is more or less Small	**and** x_2 is Any	**then** ...
If x_1 is Any	**and** x_2 is more or less Small	**then** ...
If x_1 is more or less Medium	**and** x_2 is Medium **or** Big	**then** ...
If x_1 is more or less Big	**and** x_2 is Medium **or** Big	**then** ...

□

Linguistic approximation can also be applied directly to the initial rule base obtained from the data. Thus, simplification of the rule base is also achieved, with similar results as obtained with Algorithm 5.1. An advantage of the simplification algorithm is that the user does not need to define the reference membership functions and to select the linguistic hedges. On the other hand, linguistic approximation does not require the user-defined thresholds.

5.1.5 Examples

This section presents two simulation examples to illustrate the individual steps of the identification technique for building TS fuzzy models from data. The first system is a highly nonlinear pH neutralization process, well known for posing problems to most black-box identification methods. It is shown that by fuzzy clustering, a TS fuzzy model is obtained which not only accurately represents the process, but also allows for local interpretation and analysis. The second example demonstrates that fuzzy models are also suitable for representing dynamic systems with hard (nonsmooth) nonlinearities.

pH Neutralization. A neutralization tank with three influent streams (acid, buffer and base) and one effluent stream is considered. The identification and validation data sets are obtained by simulating the model by Hall and Seborg (1989) for random changes of the influent base stream flow rate Q. The influent buffer stream and the influent acid stream are kept constant. The output is the pH in the tank. The identification data set, containing $N = 480$ samples with sampling time 15 s, is shown in Figure 5.18. This data set was obtained from (Johansen, 1994).

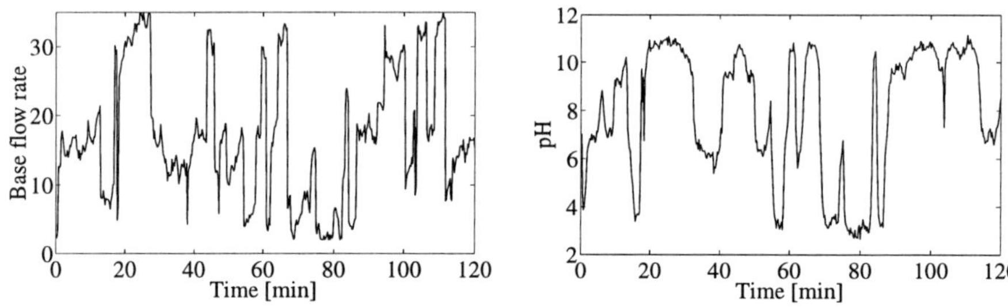

Figure 5.18. Identification data.

The process was approximated as a first-order discrete-time NARX model:

$$\mathrm{pH}(k+1) = f\big(\mathrm{pH}(k), Q(k)\big),$$

where k denotes the sampling instant, and f is an unknown relationship approximated by the TS fuzzy model. Based on prior knowledge, this structure was considered appropriate for the approximation of the pH–neutralization dynamics. The data matrix $\mathbf{Z}$ was constructed from the identification data set as:

$$\mathbf{Z} = \begin{bmatrix} \mathrm{pH}(1) & \mathrm{pH}(2) & \dots & \mathrm{pH}(N-1) \\ Q(1) & Q(2) & \dots & Q(N-1) \\ \mathrm{pH}(2) & \mathrm{pH}(3) & \dots & \mathrm{pH}(N) \end{bmatrix}$$

and it was clustered by the GK algorithm. To determine the number of clusters, the validity measures from Section 4.5.1 were applied. The upper limit on the number of clusters was set to $c = 10$, assuming that the nonlinearity of the process can be sufficiently well approximated by 10 local linear models. Figure 5.19 shows the result obtained with validity measures for $2 \le c \le 10$. One can see that all the considered validity measures attain their first extreme at $c = 4$.

The consequent parameters were obtained from the cluster covariance matrices and the cluster means, using equations (5.20). The resulting rule base is given by:

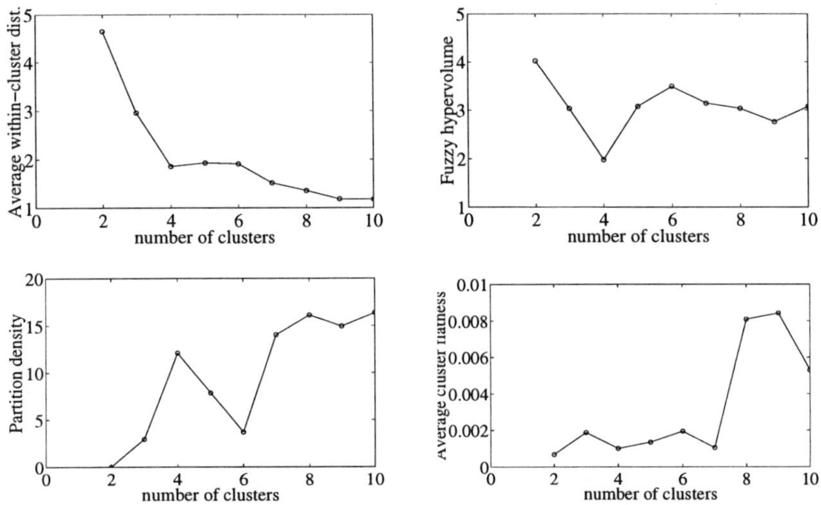

Figure 5.19. Validity measures.

1) **If** pH(k) **is** LOW **and** $Q(k)$ **is** LOW
then $\text{pH}(k+1) = 0.56\text{pH}(k) + 0.07Q(k) + 1.09$,

2) **If** pH(k) **is** MEDIUM **and** $Q(k)$ **is** BELOW MEDIUM
then $\text{pH}(k+1) = 0.60\text{pH}(k) + 0.10Q(k) + 1.27$,

3) **If** pH(k) **is** HIGH **and** $Q(k)$ **is** ABOVE MEDIUM
then $\text{pH}(k+1) = 0.40\text{pH}(k) + 0.08Q(k) + 4.37$,

4) **If** pH(k) **is** VERY HIGH **and** $Q(k)$ **is** HIGH
then $\text{pH}(k+1) = 0.81\text{pH}(k) + 0.01Q(k) + 1.77$.

The membership functions for the antecedent variables, obtained by fitting the cluster projections by piece-wise exponential functions (A.5), are shown in Figure 5.20.

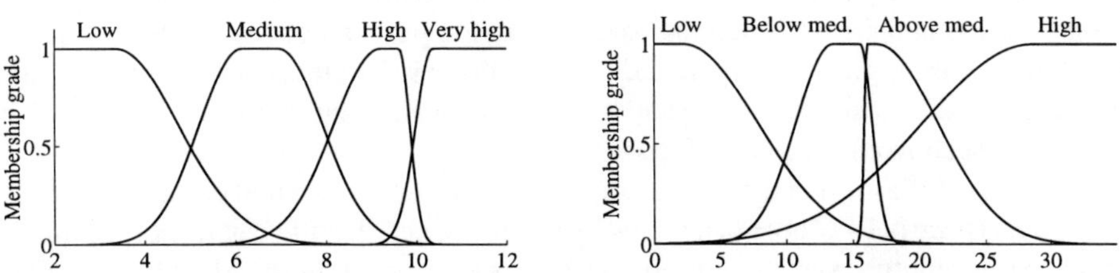

(a) pH. *(b) Base flow rate.*

Figure 5.20. Membership functions.

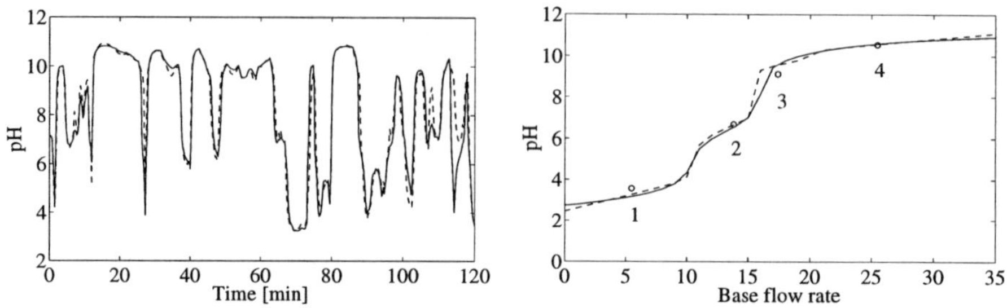

Figure 5.21. Left: model simulation on a validation data set. Right: steady-state responses (titration curves) of the true process and the model. The solid line is the process, the dashed line the model. The circles represent the centers of the obtained clusters.

Figure 5.21 shows the simulation of the obtained fuzzy model using a validation data set, and the steady-state responses (titration curves) of the "true" system and of the model. Both curves are in good agreement, and the results, obtained here from numerical data only, are comparable to those reported by Johansen (1994), who combined several different sources of information, e.g., smoothness assumptions, data ranges and simplified mass-balance equations.

In addition to accurate approximation of the system, the TS fuzzy model gives certain insights into the nonlinear dynamics of the system, as it is a set of locally valid linear ARX models. The validity regions for these models are defined by the antecedent membership functions. How the parameters of the local models (the gains and the time constants) vary from region to region can be observed. By converting the consequent models to the continuous-time domain, the following gains $\mathbf{K}$ and time constants τ are obtained:

$$\mathbf{K} = \{0.15,\ 0.25,\ 0.13,\ 0.06\}, \qquad \tau = \{25.68,\ 29.23,\ 16.49,\ 69.47\}\,.$$

From the step responses of the individual models shown in Figure 5.22, one can get an idea of the behavior of the model in each particular region.

It can be seen that the consequent model of rule 2, for instance, has a higher gain than the remaining models, which is in agreement with the region it represents, as given by the membership functions, and also by the position of the cluster center on the titration curve, given in Figure 5.21. Model 4, on the other hand, represents the saturated part of the titration curve and has a low gain and a large time constant, which can also be seen from the plot in Figure 5.18.

From the rule base and the membership functions, one can see that certain regions of the product space pH$\times Q$ are not covered by any rules, i.e., the rule base is not complete. This is because either very few or no data points are present in the identification data set for these regions. This is seen as an advantage of the clustering-based identification technique, as it gives the user information about the reliability of the obtained model in different regions of the model variables. This information can be used to design new identification experiments. The completeness of the rule base can be checked

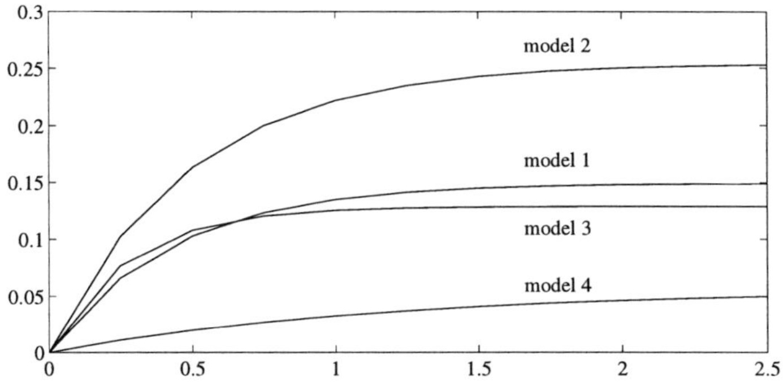

Figure 5.22. Step responses of the consequent models.

automatically by evaluating the degrees of fulfillment of the rules for different values of the antecedent variables.

Additional rules can be created by combining the unused membership functions in the antecedents. The consequent part can be defined by the user, based, for instance, on an approximate mechanistic model (Johansen, 1994), or can be computed by interpolating the neighboring rules. For the pH model, an additional rule can be created, for instance, to cover the domain of MEDIUM pH and LOW Q:

5) **If** pH(k) **is** MEDIUM **and** $Q(k)$ **is** LOW
then pH$(k+1) = 0.48$pH$(k) + 0.07Q(k) + 2.73$.

The consequent parameters in this rule were obtained by averaging two adjacent rules (i.e., rules with the two largest values of the fulfillment degree) in the original rule base.

Dead-zone Nonlinearity. In this example, an affine TS fuzzy model is used to identify a Hammerstein system, which consists of a series connection of a dead-zone/saturation nonlinearity and linear dynamics. Hard nonlinearities such as backlash, dead-zone or hysteresis pose problems to many modeling approaches that rely on assumptions about smoothness of the approximated function (neural networks, radial basis functions). From this point of view, fuzzy models are more flexible, as membership functions can represent gradual as well as abrupt transitions. Smooth, overlapping membership functions can be used to approximate smooth nonlinearities, while steep ones can capture discontinuities, hard nonlinearities and jump phenomena. Both types of membership functions can be naturally combined in one fuzzy model, providing a versatile framework for modeling nonlinear systems. The position, shape, and overlap of membership functions obtained from data give a good indication of the nature of the nonlinearity. At the same time, the consequent parameters provide estimates for the parameters of local linear models of the system.

A block diagram of the considered system, given in Figure 5.23a, is a simple example of a Hammerstein system often encountered in practice.

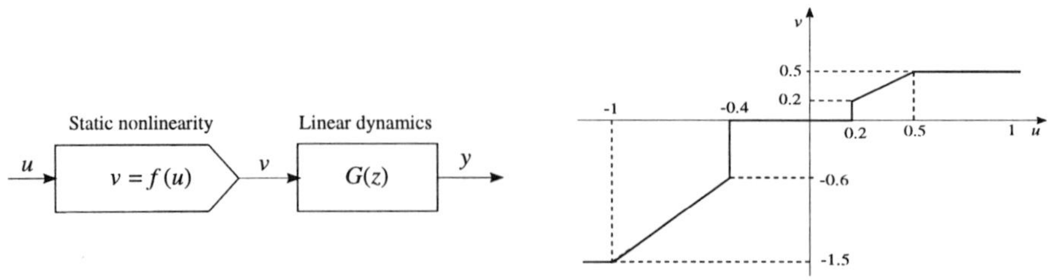

(a) Hammerstein system.

(b) Static nonlinearity.

Figure 5.23. A series combination of a static nonlinearity $v = f(u)$ and a linear dynamic system (a). The form and the parameters of the nonlinearity are given in graph (b).

The linear system is represented by a discrete-time first-order difference equation:

$$y(k+1) = ay(k) + bv(k) + \epsilon(k), \tag{5.50}$$

where $a = 0.8$, $b = 1$ and $\epsilon(k)$ is an independent random variable $\epsilon \sim N(0, 0.01)$. The nonlinearity is chosen as a complex nonlinear function $v = f(u)$ given by:

$$v = \begin{cases} \max(-1.5, 1.5u), & u \leq -0.4 \\ 0, & -0.4 \leq u \leq 0.2 \\ \min(u, 0.5), & u \geq 0.2\,. \end{cases} \tag{5.51}$$

A graph of this function is shown in Figure 5.23b. Note that the function is not symmetrical around zero, has different gains for positive and negative inputs, and the positive and negative saturation levels differ. It is interesting to investigate the degree to which the described identification method is capable of extracting this information from the input–output data of the complete system.

The system was simulated using a multi-sinusoidal input signal u shown in Figure 5.24a. The corresponding output y is given in Figure 5.24b. These two signals are the only observed variables of the system. The identification data set consists of 531 samples. For validation of the model, another data set was used.

A first-order NARX model $y(k+1) = f\big(y(k), u(k)\big)$ was chosen as a model of the system where the function f is represented by an affine TS fuzzy model. No further information was available to the identification algorithm. The number of clusters was found by applying the compatible cluster merging procedure (see Section 4.5.2), starting from nine clusters, with the threshold set to 0.4. Of the nine initial clusters, six were merged in one step and the procedure stopped at the final number of four clusters. Figure 5.25 shows membership functions for the antecedent variable $u(k)$, obtained by projecting the clusters and fitting the projected membership degrees by trapezoidal functions. These membership functions clearly reflect the nature of the nonlinearity at the input of the system, where the hard character of the dead zone is seen from the crisp transition between the adjacent membership functions. The softer transition at the positive part of the function results in more overlap between

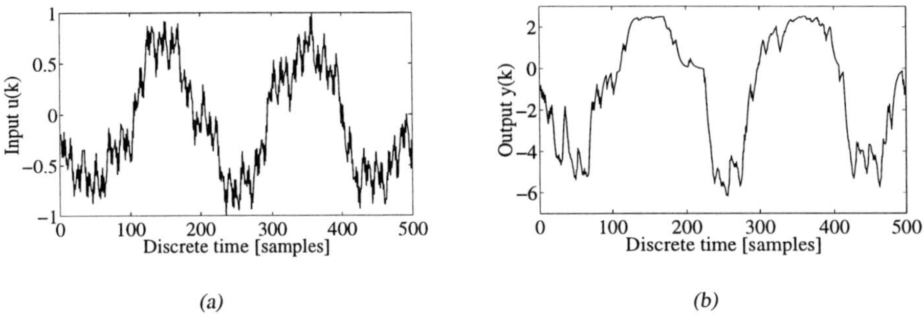

Figure 5.24. Identification data set.

the corresponding membership functions. The position of the overlaps between the membership functions indicate quite accurately the breakpoints of the nonlinearity.

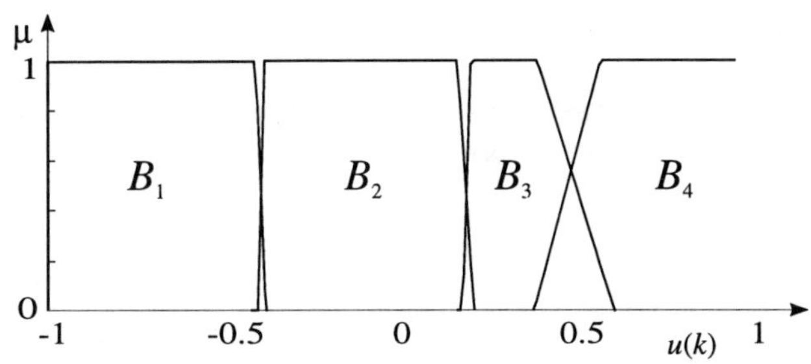

Figure 5.25. Parameterized trapezoidal membership functions for u.

Projection of the clusters onto the second regressor $y(k)$ results in fuzzy sets highly similar to the universal set, which indicates that this variable has no influence on the system's nonlinearity. The obtained rules are:

$$
\begin{array}{l}
\textbf{If } u(k) \text{ is } B_1 \textbf{ then } y(k+1) = 0.8007y(k) + 0.0007u(k) + 0.0001 \\
\textbf{If } u(k) \text{ is } B_2 \textbf{ then } y(k+1) = 0.8000y(k) + 1.5067u(k) + 0.0039 \\
\textbf{If } u(k) \text{ is } B_3 \textbf{ then } y(k+1) = 0.8050y(k) + 0.0061u(k) + 0.4932 \\
\textbf{If } u(k) \text{ is } B_4 \textbf{ then } y(k+1) = 0.7994y(k) + 1.0033u(k) - 0.0024
\end{array}
$$

The consequent parameters are computed by eq. (5.20). The parameter at $y(k)$, corresponding to the time constant of the system, is constant in all the rules, which correctly reflects the fact that the dynamics of the system are linear. The gains and offsets, however, vary in correspondence to the nonlinearity (5.51). Figure 5.26a shows the validation of the model on a validation data set. For comparison, Figure 5.26b shows also a global linear ARX model estimated from the same data as the fuzzy model, for which large errors occur due to the system's nonlinearity.

The structure of the model also allows reconstruction of the intermediate variable v. By comparing the form of the TS consequents, $y(k+1) = a_i y(k) + b_i u(k) + c_i$,

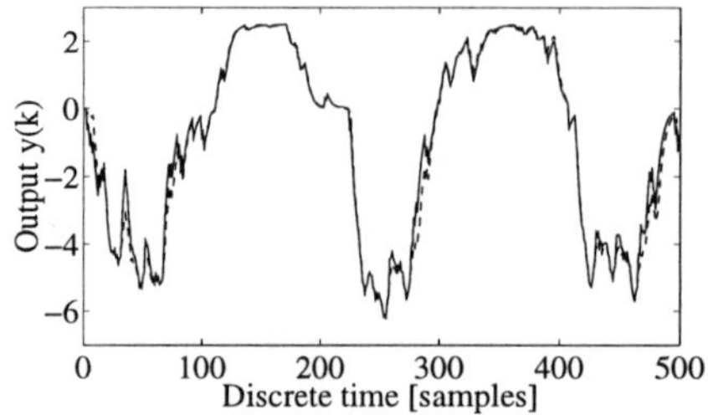

(a) Fuzzy model, RMS=0.0817.

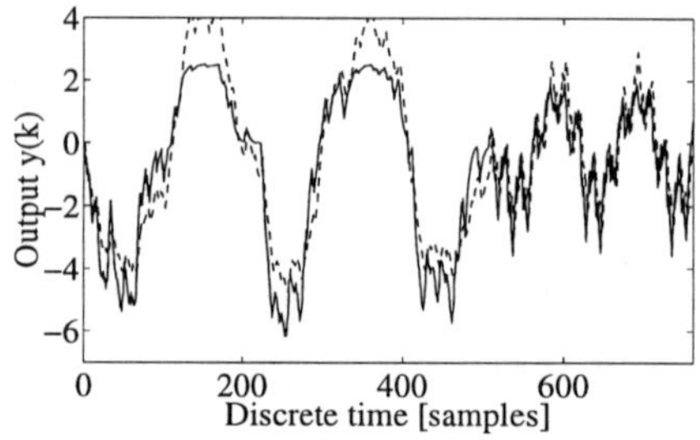

(b) Linear model, RMS=0.7897.

Figure 5.26. Validation of the fuzzy model and a linear ARX model.

with the linear part of the system given by eq. (5.50), one can see that $v(k)$ can be computed as a function of $u(k)$, using parts of the TS rules:

$$\textbf{If } u(k) \text{ is } B_i \textbf{ then } v(k) = b_i u(k) + c_i, \quad i = 1, \ldots, 4\,.$$

In this sense, the TS model can be seen as a (static) observer of the internal variable $v(k)$. Figure 5.27 shows that the form of the nonlinearity given in Figure 5.23b is accurately captured by the model.

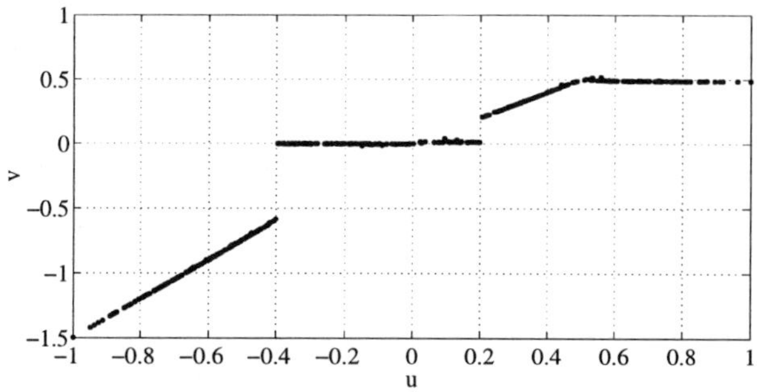

Figure 5.27. The shape of the nonlinearity reconstructed by the fuzzy model.

5.1.6 *Practical Considerations*

Methods for generating fuzzy models from data by product-space clustering have been described. This section summarizes the advantages and drawbacks of the presented approach, focusing mainly on its practical aspects.

Use of Prior Knowledge. Prior knowledge about the system is needed to postulate the structure of the (dynamic) model, as discussed in Section 4.2. Techniques based on validity measures and compatible cluster merging can then be applied to automatically determine the number of clusters in the data set. The number of clusters equals the number of rules in the initial fuzzy model generated from the partition. By evaluating the validity measures and prediction error criteria, a compromise can be found between the model complexity and the requirements on its performance.

A major advantage of the presented identification technique is that the obtained models are locally interpretable, and often correspond well with the local behavior of the identified process. The form and the overlap of the membership functions, extracted from the fuzzy partition, provide information about the character of the system's nonlinearity. By examining these results, one can gain insights into the behavior of the process, and also qualitatively validate the obtained model. Prior knowledge can also be incorporated in the obtained model, in the form of additional rules, based on approximate mechanistic models, step response identification, or heuristics and experience.

Model Validation. The validation of fuzzy models has several facets, namely the standard validation through numerical simulations and comparisons with the process data, analysis of the linear consequent models (stability, gains, time constants, nonminimum phase behavior, etc.) and analysis of the coverage of the input space by the rules. For an incomplete rule base, additional rules can be obtained from prior knowledge, or by local linearization of first-principle models. The antecedents of these rules can be created from unused combinations of membership functions in the initial model. The identification data usually cover only a fraction of the entire product space of the model variables. Therefore, the antecedents of the obtained rules include only those combinations of the linguistic terms which were identified from the data. However, in simulation or prediction, it is possible that regions not covered by any rules are entered. This situation can be detected by observing the degrees of fulfillment in the rule base. If no rule is activated above a specified threshold, an additional rule may be added to the rule base. The antecedent of this rule is given by the combination of linguistic terms which give the highest degree of fulfillment for the given data point, see 139 for an example. This is a major difference from many other purely black-box techniques, such as neural networks, CMAC, splines, etc., where the validation relies entirely on numerical simulation.

Model Complexity. Product-space clustering methods have the potential to generate less complex models than methods based on optimization of the antecedent membership functions, such as adaptive spline methods (Brown and Harris, 1994), Takagi–Sugeno's approach (Takagi and Sugeno, 1985; Sugeno and Tanaka, 1991), table look-up schemes (Wang, 1994), etc. The reason is that the rules are created directly in the product space of the model variables, and take into account the form of the system's nonlinearity and coverage of the space by identification data. More rules are obtained in regions characterized by complex nonlinear behavior, while (almost) linear regions are described by fewer rules. The information about the system's nonlinearity is reflected in the obtained membership functions. No rules are generated in the regions that do not contain any data. With regard to generalization capabilities, this can be seen as a drawback, as the model gives no output for these regions. However, one can see it as an advantage, since analysis of the rule base completeness helps in validating the model and designing new identification experiments. Additional rules also can be supplied by the expert, using prior knowledge.

Redundancy may occur in the obtained model in terms of similar membership functions, describing the same region in the domain of a particular variable. This is a result of cluster projection, as shown in Section 5.1.3. Techniques based on similarity measures are applied to simplify the initial fuzzy models acquired from data.

Robustness of the Identification Method. Clustering requires sufficiently informative data. Piece-wise constant or binary signals, often used for the identification of linear systems, are not suitable. The technique also cannot use data sets with linearly dependent variables, as the clustering algorithm runs into numerical problems. Practical experience shows, that in this sense, the technique is somewhat more sensitive to poor data than other approaches.

Large data sets in combination with high-dimensional regression problems may drastically increase the computational demands of the clustering algorithm. The technique proved to be practically useful for medium-complexity identification problems, i.e., data sets up to several thousand data points and low-order dynamic systems with few input variables. Large problems need to be decomposed (if possible), or solved by using hybrid approaches, such as the one described in Sections 4.2.4 and 7.3. The clustering approach can be applied to well-behaved, smooth nonlinear systems, as well as to systems whose behavior changes abruptly as a result of hard nonlinearities. The character of the process nonlinearity is reflected in the form and overlap of the obtained membership functions.

Other Applications of Product-space Fuzzy Clustering. Apart from the acquisition of fuzzy models, clustering can be used as a data analysis and preprocessing step for other techniques, such as training neural networks (te Braake, et al., 1997). Besides, the use of product-space fuzzy clustering is not only restricted to the identification of nonlinear systems. It can also be used to detect changes in the behavior of linear systems, and hence can be applied to monitoring and fault detection.

5.2 Linguistic and Relational Models

In the previous section we explained how TS fuzzy models can be derived from fuzzy partitions obtained by product-space fuzzy clustering. In this section we present methods to derive singleton models, fuzzy relational models and linguistic models. The main motivation is that linguistic and relational models can be more easily interpreted in terms of natural language than TS models. As discussed in Chapter 2, linguistic, relational, and singleton models have constant consequents, and thus also similar interpolation properties. Therefore, the method to construct the antecedent membership functions is identical for all of them and is described first. Then, different techniques to estimate the consequent parameters, or the fuzzy relation which defines the rule base are presented.

5.2.1 *Extraction of Antecedent Membership Functions*

The main idea is to construct the membership functions such that the linear submodel represented by one cluster is obtained by interpolation among linguistic fuzzy rules (Babuška and Verbruggen, 1995b). Piece-wise linear approximation can be achieved by using a singleton fuzzy model (Section 2.1.8), or a linguistic model where the consequent sets are defuzzified first and then the fuzzy-mean method is applied (Section 2.1.5). Fuzzy relational models with the sum–product composition also provide linear interpolation. In the following discussion, the term "linguistic model" stands for the three above-mentioned types of models.

In order to represent the piece-wise linear mapping of the TS model, the cores of the membership functions in the linguistic model are chosen such that they coincide with the intersection points of the adjacent membership functions of the affine TS model. This is because each TS rule by itself results in a locally linear input-output mapping, while, in the linguistic model, the linear relation is a consequence of the interpolation between the neighboring rules. Additional sets must be placed at the extreme points of the domain. Figure 5.28 illustrates this idea on an example of a TS model with three rules:

$$R_i : \textbf{If } x \text{ is } A_i \textbf{ then } y_i = a_i x + b_i, \quad i = 1, 2, 3\,.$$

The membership functions A_1, A_2, A_3 are shown in the middle part of Figure 5.28. The piece-wise linear input–output mapping shown in the upper part this figure can be approximately represented by the following four linguistic rules:

R_1: **If** x is A'_1 **then** y is B_1 R_3: **If** x is A'_3 **then** y is B_1
R_2: **If** x is A'_2 **then** y is B_2 R_4: **If** x is A'_4 **then** y is B_2

Formally, this transformation can be described as follows. Consider first a TS fuzzy model with a scalar input and scalar output:

$$R_i\text{: } \textbf{If } x \text{ is } A_i \textbf{ then } y = \mathbf{a}_i\mathbf{x} + b_i, \quad i = 1, \ldots, K\,. \tag{5.52}$$

Let the fuzzy sets A_i, $i = 1, 2, \ldots, K$ be ordered such that

$$\sup\ \text{core}(A_i) < \inf\ \text{core}(A_{i+1}), \quad i = 1, 2, \ldots, K-1\,. \tag{5.53}$$

This also ensures that the cores of the fuzzy sets A_j are disjunct. Further, it is assumed that the TS rules are of the V type, as defined in Section 2.3.2. Denote $A' = \{a'_j \mid j = 1, \ldots, K+1\}$ the set of the intersection points of the adjacent fuzzy sets A_j and the infimum and supremum of the domain X:

$$A' = \inf X \cup \{\text{core}(\text{norm}(A_j \cap A_{j+1})) \mid j = 1, \ldots, K-1\} \cup \sup X, \tag{5.54}$$

where norm$(\cdot)$ denotes normalization of a fuzzy set (see Appendix A). Now, triangular membership functions A'_j of the linguistic model can be constructed such that they form a partition and their cores are the points a'_j. These membership functions are given by:

$$\mu_{A'_1}(x) = \max\left(0, \min(1, \frac{a'_2 - x}{a'_2 - a'_1})\right), \tag{5.55a}$$

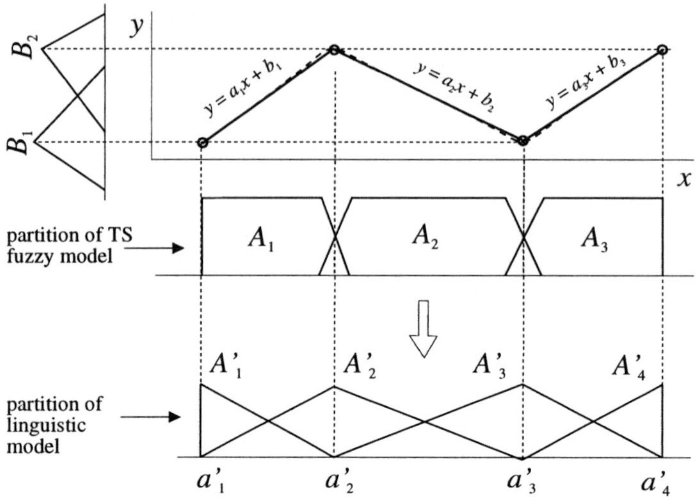

Figure 5.28. Cores of triangular membership functions in the linguistic model are chosen at the points where the adjacent membership functions of the TS model intersect and at the extreme points of the domain. Reproduced from (Babuška and Verbruggen, 1995b), ©1995b IEEE.

$$\mu_{A'_j}(x) = \max\left(0, \min(\frac{x - a'_{j-1}}{a'_j - a'_{j-1}}, \frac{a'_{j+1} - x}{a'_{j+1} - a'_j})\right), \quad 2 \le j \le K \quad (5.55b)$$

$$\mu_{A'_{K+1}}(x) = \max\left(0, \min(\frac{x - a'_K}{a'_{K+1} - a'_K}, 1)\right). \quad (5.55c)$$

In the case of a MISO singleton model, the conjunctive form of the antecedent is considered:

$$R_i\text{: } \textbf{If } x_1 \text{ is } A'_{i1} \textbf{ and } \ldots \textbf{ and } x_p \text{ is } A'_{ip} \textbf{ then } y = b_i\,. \quad (5.56)$$

The membership functions are derived per variable x_j, in the same way as for a SISO model. To obtain a complete linguistic model, it remains to identify the rule consequents for all combinations of the antecedent fuzzy sets. Depending on the type of the fuzzy model, there are several possibilities, as discussed in the following sections.

5.2.2 Estimation of Consequent Parameters

In the singleton model (5.56), the optimal consequent parameters b_i can be estimated by least-squares techniques. Let $\mathbf{x}_k = [x_{1k}, x_{2k}, \ldots, x_{pk}]^T$ denote the kth input vector, for which the degree of fulfillment $\beta_i(\mathbf{x}_k)$ of the ith rule is given by:

$$\beta_i(\mathbf{x}_k) = \mu_{A'_{i1}}(x_{1k}) \wedge \mu_{A'_{i2}}(x_{2k}) \wedge \cdots \wedge \mu_{A'_{in}}(x_{pk})\,.$$

The output y_k of the model corresponding to the input $\mathbf{x}_k$ is computed by the fuzzy-mean defuzzification method:

$$y_k = \frac{\sum_{i=1}^{K} \beta_i(\mathbf{x}_k) b_i}{\sum_{i=1}^{K} \beta_i(\mathbf{x}_k)} . \tag{5.57}$$

Let $\mathbf{\Gamma}$ denote the matrix in $\mathbb{R}^{N \times K}$ having the normalized degree of fulfillment $\gamma_{ki} = \beta_i(\mathbf{x}_k) / \sum_{j=1}^{K} \beta_i(\mathbf{x}_k)$ as its kith element, let $\mathbf{y}$ denote the vector in $\mathbb{R}^N$ having y_k as its kth component, and let $\mathbf{b} = [b_1, b_2, \ldots, b_K]^T$ denote the vector containing the consequent parameters. The least-squares problem given by eq. (5.57), written in a matrix form, $\mathbf{y} = \mathbf{\Gamma b} + \epsilon$, has the solution:

$$\mathbf{b} = \left[\mathbf{\Gamma}^T \mathbf{\Gamma}\right]^{-1} \mathbf{\Gamma}^T \mathbf{y} . \tag{5.58}$$

The singleton model may be more accurate than the corresponding TS model, as it reduces the approximation error of the TS model, see Sections 2.3.2 and 7.2. Note that, because triangular membership functions are used in the rule antecedent, this model can be implemented as a simple look-up table with the index vectors given by the cores of the antecedent fuzzy sets. In this sense, the described method can be regarded as a general algorithm for determining irregularly spaced index vectors for piece-wise multi-linear approximations.

Example 5.7 The purpose of this example is to demonstrate the properties of the different methods for consequent estimation. Consider the function $y = x^4 + \epsilon$, where $\epsilon \sim N(0, \sigma)$. First, a noise-free data set is generated ($\sigma = 0$), which consists of 100 points uniformly sampled in $[0, 1]$, as shown in Figure 5.29a. A second data set is generated for $\sigma = 0.1$. Moreover, samples for $x \in (0.6, 0.9)$ are missing, and an "outlier" data point is added at the bottom right-hand side, see Figure 5.29b.

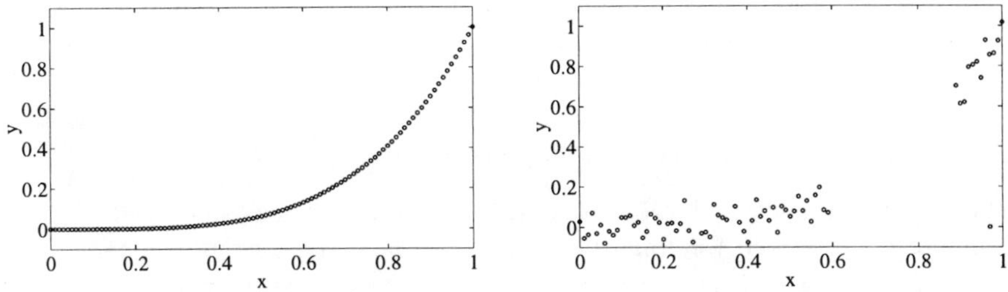

(a) Noise-free data. (b) Corrupted data.

Figure 5.29. Graph (a) shows input–output data generated by function $y = x^4$. Graph (b) shows a data set with missing samples for $x \in (0.6, 0.9)$ and additive noise from $N(0, 0.1)$.

The noise-free data is clustered in three clusters, using the GK algorithm. The effect of noise is considered for the estimation of the consequents only. Figure 5.30

shows the membership functions for x, which are obtained by the method described in Section 5.2.1.

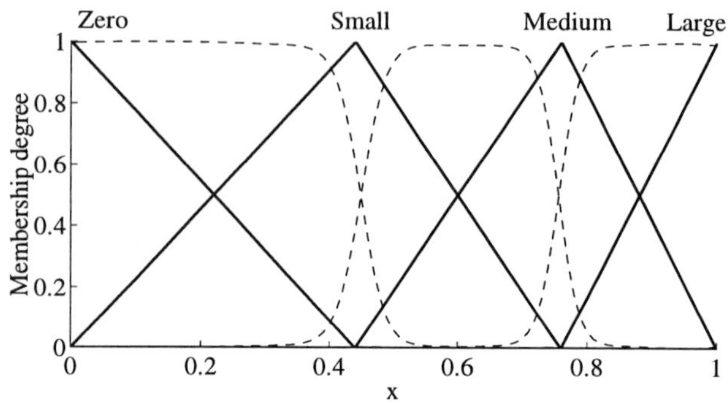

Figure 5.30. Antecedent membership functions.

The data is approximated by four fuzzy rules with singleton consequents:

$$\begin{array}{lll} R_1: & \textbf{If } x \text{ is Zero} & \textbf{then } y = b_{\text{Zero}} \\ R_2: & \textbf{If } x \text{ is Small} & \textbf{then } y = b_{\text{Small}} \\ R_3: & \textbf{If } x \text{ is Medium} & \textbf{then } y = b_{\text{Medium}} \\ R_4: & \textbf{If } x \text{ is Large} & \textbf{then } y = b_{\text{Large}} \end{array} \tag{5.59}$$

Using the least-squares method (5.58), the estimated vector of consequent parameters is $\mathbf{b}_1 = [-0.02,\ 0.06,\ 0.41,\ 0.97]$ for the noise-free data, and $\mathbf{b}_2 = [-0.05,\ 0.06,\ 0.52,\ 0.86]$ for the corrupted data. Figure 5.31 shows the fit for both data sets. The RMS errors are 0.0073 and 0.1130 for the noise-free and noisy data sets, respectively. One can see that in both cases the model captures the underlying function quite accurately. Note, however, that because of the least-squares fit, some of the consequent estimates may attain values outside the range of admissible outputs, such as $b_{11} < 0$. In Figure 5.31b, one can also observe that the outlier biases the estimate. □

5.2.3 *Convertion of Singleton Model into Relational Model*

In the singleton model, each rule has, generally, a different consequent, which makes the model less suitable for linguistic interpretation. The singleton model can be simplified by replacing the consequent parameters by a small number of fuzzy sets. There are several methods to obtain the consequent membership functions. One possibility is to cluster the singletons into a specified number of clusters, using standard algorithms, such as fuzzy c-means, see Section 3.3. In the rules, each singleton is then replaced by the prototype of the cluster in which it has the highest degree of membership. The shape of the consequent membership function is determined from the fuzzy partition matrix using the projection techniques from Section 5.1.1. Another possibility is to define triangular membership functions by using the method from Section 5.2.1. If some prior knowledge is available, the consequent membership functions may be specified by the user.

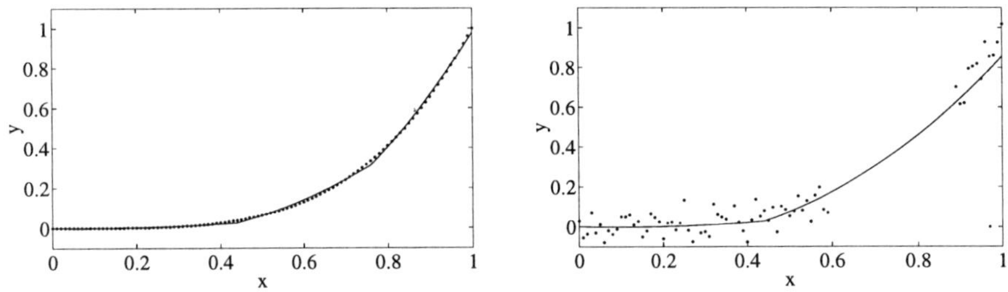

(a) Noise-free data.

(b) Corrupted data.

Figure 5.31. Model fit on the noise-free data (a) and corrupted data (b).

Once the consequent membership functions are available, the singleton model can be expressed in the form of a fuzzy relational model. The ijth element of the fuzzy relation is computed as a membership degree of the singleton b_i in the consequent fuzzy sets B_j, as described in Section 2.11. Clearly, if only two adjacent consequent fuzzy sets overlap, the relation has only two nonzero elements in each row. Such a model can be interpreted in the form of linguistic rules with two weighted consequents in each rule, e.g.:

$$\textbf{If } x \text{ is Small } \textbf{then } y \text{ is Small } (0.7),\ y \text{ is Medium } (0.2)\,. \tag{5.60}$$

An approximate linguistic description can be obtained by using only the consequent terms with the maximum membership degree.

Example 5.8 The singleton model from Example 5.7 can be expressed in a relational form. First, membership functions are defined for the consequent variable y. Here, triangular membership functions were chosen such that their cores are images of the cores of the antecedent sets A_i via the function $y = x^4$, and the functions form a partition. This is an ideal setting, since the intersections of the cores of the antecedent and consequent fuzzy sets belong to the relation $x = y^4$. In order to illustrate the weighting of the consequents in the fuzzy relation, the fuzzy set corresponding to A_2 is removed. This means that, even in the noise-free case, R_2 must contain two weighed consequents. Figure 5.32 shows the membership functions for y.

The relations R_1^s and R_2^s obtained from the singleton model for the noise-free and noisy data are respectively:

$$R_1^s = \begin{bmatrix} 1.00 & 0.00 & 0.00 \\ 0.83 & 0.17 & 0.00 \\ 0.00 & 0.89 & 0.11 \\ 0.00 & 0.03 & 0.97 \end{bmatrix}, \qquad R_2^s = \begin{bmatrix} 1.00 & 0.00 & 0.00 \\ 0.81 & 0.19 & 0.00 \\ 0.00 & 0.72 & 0.28 \\ 0.00 & 0.22 & 0.78 \end{bmatrix}. \tag{5.61}$$

Recall that the rows of each relation correspond to the individual rules (antecedent terms in this example) and the columns correspond to the consequent linguistic terms.

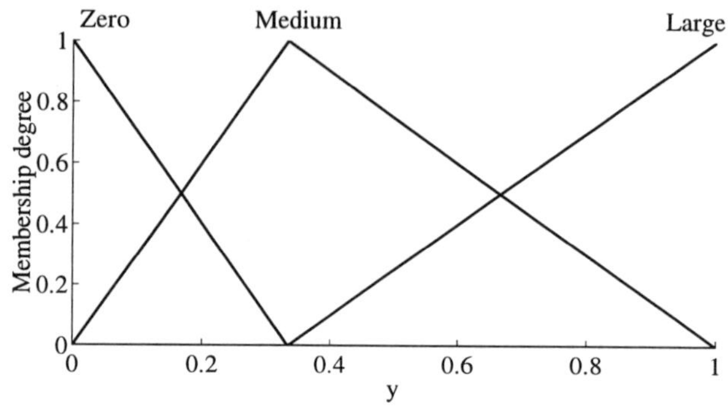

Figure 5.32. Consequent membership functions.

For both relations, the linguistic approximation reads:

If x is Zero **then** y is Zero
If x is Small **then** y is Zero
If x is Medium **then** y is Medium
If x is Large **then** y is Large

which gives a good description of the parabolic function. Numerical performance of the relational model is identical to the singleton model. □

5.2.4 *Estimation of Fuzzy Relations from Data*

The fuzzy relation R can also be directly estimated from the data. It is usually obtained as an approximate solution of a set of fuzzy relational equations, using the φ-composition (implication) operator, or a t-norm (Pedrycz, 1984; Pedrycz, 1993). The use of φ-composition is mathematically equivalent to constructing a fuzzy relation from a set of fuzzy rules, as described in Section 2.1.4, since the φ-composition operator is a residuated implication. A major difference is, however, that in the identification of fuzzy relational models, the relation applies to the linguistic terms and not to the domain elements (cf. Section 2.11). The relation R is computed by:

$$R = \bigcap_{k=1}^{N} (\boldsymbol{\beta}_k \varphi Y_k), \tag{5.62}$$

where $Y_k = [\mu_{B_1}(y_k), \mu_{B_2}(y_k), \ldots, \mu_{B_M}(y_k)]$ denotes the vector of membership degrees of the output y_k in the M consequent fuzzy sets B_j, N is the number of data points, and $\boldsymbol{\beta}_k$ denotes the vector of the degrees of fulfillment. For the Łukasiewicz implication, for instance, the partial relation $R_k = [(r_{ij})_k]$, obtained from the kth measurement, is given by:

$$(r_{ij})_k = \min(1, 1 - \beta_{ki} + \mu_{B_j}(y_k)) \,. \tag{5.63}$$

The overall relation R is the intersection of the partial relations R_k:

$$R = \bigcap_{k=1}^{N} R_k \,. \tag{5.64}$$

Instead of the φ-composition, Pedrycz (1984) applies a t-norm (cf. Section 2.1.4). For instance, for the minimum t-norm, the kth partial relation R_k is given by:

$$(r_{ij})_k = \min(\beta_{ki}, \mu_{B_j}(y_k)), \tag{5.65}$$

and the overall relation R is computed as the union of the partial relations R_k:

$$R = \bigcup_{k=1}^{N} R_k \,. \tag{5.66}$$

With respect to identification, the main difference between these two approaches is in the way in which they resolve conflicts in data. The term conflict should be understood in a very general sense. First, it refers to the fact that the inputs and outputs are quantized by means of overlapping membership functions, and hence conflicts necessarily occur for data at the boundary of adjacent membership functions. Secondly, conflicts occur in the presence of noise and outliers in the data. This aspect can be illustrated by continuing the example from the previous sections.

Example 5.9 Let us estimate the fuzzy relation for the data generated in Example 5.7, using the same consequent membership functions from Example 5.8. By applying the φ-composition, eq. (5.63) and (5.64), relations R_1^L and R_2^L are obtained for the noise-free and noisy data, respectively. Similarly, the t-norm (minimum) method, equations (5.65) and (5.66), gives relations R_1^m and R_2^m. Let us first examine the noise-free case:

$$R_1^L = \begin{bmatrix} 1.00 & 0.00 & 0.00 \\ 0.89 & 0.11 & 0.00 \\ 0.00 & 0.88 & 0.00 \\ 0.00 & 0.00 & 0.90 \end{bmatrix}, \qquad R_1^m = \begin{bmatrix} 1.00 & 0.08 & 0.00 \\ 0.92 & 0.44 & 0.00 \\ 0.56 & 1.00 & 0.44 \\ 0.00 & 0.54 & 1.00 \end{bmatrix}. \tag{5.67}$$

One can see that the relation R_1^L is quite similar to the one obtained in Example 5.8 from the singleton consequents. The main difference is that the partial conflict of the data due to the fuzzy quantization results in a lower value of the highest membership degrees in the rows of the relation (for instance the third and the fourth row). More precisely, there are samples in the data set that belong to the antecedent fuzzy set Large and, simultaneously, do not to a very high degree belong to the consequent fuzzy set Large. As an example, take the pair $(x_k, y_k) = (0.9, 0.66)$ with the corresponding membership degrees $\boldsymbol{\beta}_k = [0,\ 0,\ 0.42,\ 0.58]$ and $Y_k = [0,\ 0.51,\ 0.49]$. For this data pair, the partial relation R_k^L reads:

$$R_k^L = \begin{bmatrix} 1.00 & 1.00 & 1.00 \\ 1.00 & 1.00 & 1.00 \\ 0.58 & 1.00 & 1.00 \\ 0.41 & 0.92 & 0.91 \end{bmatrix}, \tag{5.68}$$

which implies that due to the intersection operator, the element $r_{4,3}^L$ in the overall relation R^L cannot have greater value than 0.91. Data pairs for which x is fully a member of an antecedent fuzzy set and y is never fully a member of any of the consequent fuzzy sets result in two or more nonzero elements in the corresponding row of the fuzzy relation. An example is the second row of R_1^L which has no exact consequent fuzzy set available (it was removed on purpose), and hence must result in interpolation between two neighboring consequents.

The relation R_1^m built by the t-norm operator has quite different properties. Conflicts in data do not result in a decrease of the membership degree of the most activated consequent fuzzy set, but in an increase of the remaining membership degrees. This is because the overall relation is built as a union, which simply "adds" each new piece of information, without taking care of the overall consistency. A typical example of this behavior is the third row, where all the three consequent terms have a relatively high membership degree, even though an exact consequent term Medium is available for the antecedent Medium. This complicates the linguistic interpretation and deteriorates the approximation (compare Figure 5.33a and b).

To demonstrate the effects of noise and missing data, let us compute the relations for the corrupted data set:

$$R_2^L = \begin{bmatrix} 0.86 & 0.02 & 0.00 \\ 0.64 & 0.09 & 0.00 \\ 0.54 & 0.74 & 0.53 \\ 0.00 & 0.00 & 0.13 \end{bmatrix}, \qquad R_2^m = \begin{bmatrix} 0.98 & 0.40 & 0.00 \\ 0.91 & 0.59 & 0.00 \\ 0.47 & 0.45 & 0.46 \\ 0.87 & 0.58 & 1.00 \end{bmatrix}. \tag{5.69}$$

Due to the increased level of conflict (noise), the relation R_2^L has lower maxima per row. An extreme situation can be observed in rule 4, where the presence of the outlier in the data causes the largest membership degree to drop significantly, indicating a conflict. Such a rule then has a lower credibility, but it does not bias the output, as shown in Figure 5.33c. In relation R_2^m, this conflict results in a nonconvex output fuzzy set (the last row), which after defuzzification severely biases the numerical output, see Figure 5.33d.

Finally, concentrating on the third rule, where the data are missing, one can see that in the case of R_2^L, all three consequent fuzzy sets have relatively high membership degrees. Note that this situation can be quite clearly distinguished from a conflict, which results in low membership degrees. This is, however, not the case with the relation R_2^m

One can clearly see, that the identification method based on a t-norm leads to a rather poor numerical performance (apart from a worse transparency and a more difficult linguistic interpretation of the model). The model obtained by the least-squares method gives the highest numerical accuracy, as can be see from the Table 5.6. □

In this section, methods to derive linguistic and relational models from the fuzzy partition obtained by product-space clustering have been presented. The main motivation is to give interpretation to the model, in terms of natural language. A consistent manner of deriving the antecedent membership functions is presented, which significantly differs from the gradient-descent learning schemes usually considered in the literature (Jang, 1993; Pedrycz, 1993; Pedrycz, 1995). For numerical reasons, the singleton model may be preferred to the TS model, since it does not suffer from large approximation errors. The singleton model can be directly inverted, because its consequents do not depend

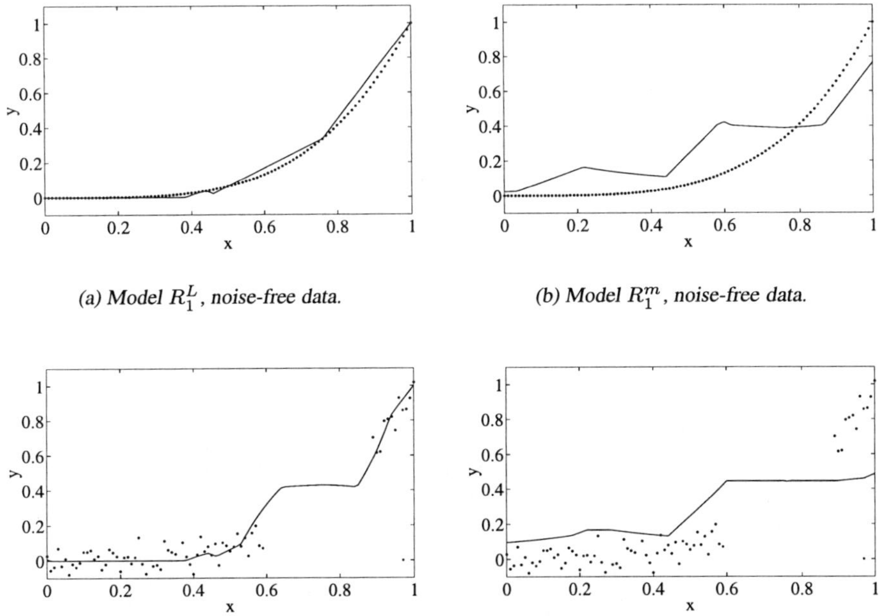

(a) Model R_1^L, noise-free data.

(b) Model R_1^m, noise-free data.

(c) Model R_1^L, corrupted data.

(d) Model R_1^m, corrupted data.

Figure 5.33. Numerical performance of relational models estimated from data by applying the φ-composition and the minimum t-norm.

Table 5.6. Comparison of relational models.

	noise-free data	corrupted data
least squares	0.0073	0.1130
φ-composition	0.0289	0.1267
t-norm	0.1520	0.2165

on the inputs. This property is used to design a nonlinear controller (Section 6.1). To estimate fuzzy relations from data, techniques based on fuzzy implications are preferable to those using t-norms, as they lead to semantically more consistent models.

5.3 Low-level Fuzzy Relational Models

Using the information obtained by fuzzy clustering, a fuzzy relation can be computed that allows the calculation of the model's output by means of relational composition:

$$B = A \circ R, \tag{5.70}$$

where A (B) is the input (output) fuzzy set and R is the fuzzy relation. Note that this relation is different from the relation employed into fuzzy relational models used in the previous section. We call this relation a "low-level" relation, since its elements describe the degree of association between the individual discrete elements of the input and output domains. In fuzzy relational models, on the other hand, the relation defines the interactions between the linguistic terms defined in the model input and output domains, i.e., it establishes a rule base.

It is assumed that the system identification problem has been transformed into the following static regression problem (see Section 4.2):

$$y \approx f(\mathbf{x}), \tag{5.71}$$

where $\mathbf{x} = [x_1, x_2, \ldots, x_p]^T$ is the regression vector, $\mathbf{x} \in (X_1 \times X_2 \times \ldots \times X_p) \subset \mathbb{R}^p$ and $y \in Y \subset \mathbb{R}$ is the regressand. The symbol $\approx$ represents the fact that y is not an exact function of $\mathbf{x}$. In the model, the functional relationship (5.71) can be represented by means of a fuzzy relation:

$$R: (X_1 \times X_2 \times \ldots \times X_p \times Y) \to [0, 1] \,. \tag{5.72}$$

The Euclidean space $(X_1 \times X_2 \times \ldots \times X_p \times Y)$ of the system variables is called the *context* of the system (Mesarovic and Takahara, 1989) and is denoted Z. The $[\mathbf{x}, y]$ pair is called an *event*. The relation R is a fuzzy restriction on the system's context and it can be seen as a possibility distribution of the events on Z. An α-cut R_α is then a set of events possible with a degree greater or equal to α.

By clustering the data set $S = \{(\mathbf{x}_k, y_k) | k = 1, 2, \ldots, N\}$, groups of locally "similar" points are obtained. Each cluster is determined by its prototype, and possibly also by other objects such as the cluster covariance matrix or the norm inducing matrix (see Chapter 3). The global system relation R is obtained as a union of the clusters. Since the set $\mathbf{Z}$ of the available data represents only a limited sample from the system's behavior, a generalization procedure is required to provide R as a model for all possible data. This procedure for obtaining such a model is described below.

The system's context Z is discretized using an orthogonal grid. The grid is obtained as the Cartesian product of discretizations in the individual components of Z denoted: $\Xi_j = \{\xi_{j1}, \xi_{j2}, \ldots, \xi_{jN_j}\}$, $j = 1, 2, \ldots, p$ for the regressors x_j and $\Upsilon = \{\upsilon_1, \upsilon_2, \ldots, \upsilon_{N_{p+1}}\}$ for the regressand y. N_j is the number of discretization levels in the jth component of Z. The discretized system's context is denoted by $D = (\Xi_1 \times \Xi_2 \times \ldots \times \Xi_p \times \Upsilon)$. Since the relation R is a $(p+1)$-dimensional array, it is convenient to represent D as a $(p+1)$-dimensional vector array: $D = [\boldsymbol{\delta}_\mathbf{q}]$, where $\boldsymbol{\delta}_\mathbf{q}$ is a $(p+1)$-dimensional vector and $\mathbf{q}$ denotes a vector of indices $\mathbf{q} = [q_1, q_2, \ldots, q_{p+1}]$, $q_j \in 1, 2, \ldots, N_j$.

Example 5.10 As an example of a system context, consider a two-dimensional Euclidean space $X \times Y$. The discretized context D is then:

$$D = \begin{bmatrix} (\xi_1, \upsilon_1) & (\xi_1, \upsilon_2) & \cdots & (\xi_1, \upsilon_{N_2}) \\ (\xi_2, \upsilon_1) & (\xi_2, \upsilon_2) & \cdots & (\xi_2, \upsilon_{N_2}) \\ \vdots & \vdots & \cdots & \\ (\xi_{N_1}, \upsilon_1) & (\xi_{N_1}, \upsilon_2) & \cdots & (\xi_{N_1}, \upsilon_{N_2}) \end{bmatrix}. \tag{5.73}$$

□

Once the system's context is discretized, a relation $R_i = [(r_{\mathbf{q}})_i]$ is first computed for each cluster:

$$(r_{\mathbf{q}})_i = \frac{1}{1 + D(\boldsymbol{\delta}_{\mathbf{q}}, \mathbf{v}_i)}, \qquad \forall q, \tag{5.74}$$

where $D(\cdot, \cdot)$ is the distance measure used in the clustering algorithm and $\mathbf{v}_i$ is the prototype of the ith cluster. Equation (5.74) transforms distance as a measure of dissimilarity of a data point to the cluster prototype into a membership degree. Also other transformations can be used in this step, such as $(r_{\mathbf{q}})_i = \exp(-D(\boldsymbol{\delta}_{\mathbf{q}}, \mathbf{v}_i))$. Note that the expression for computing membership degrees in the FCM-like clustering algorithms, see eq. (3.9a), is not suitable here, since it makes the sum of membership degrees for each point equal to one. Hence, also the events that do not belong to the relation would be assigned relatively high membership degrees. For more details refer to Section 3.6 on possibilistic clustering. The global system's relation is computed as the union of the individual relations R_i:

$$R = \bigcup_{i=1}^{c} R_i \,. \tag{5.75}$$

The use of the maximum union operator would result in dips in the regions where the clusters overlap. Instead, the union can be computed by a parametric t-conorm with a suitably chosen parameter, such that R is a convex fuzzy set. Figure 5.34 gives an example of fuzzy unions for several different values of the parameter α in the Yager t-conorm given by:

$$\mu_{A \cup B} = \min(1, [\mu_A^\alpha + \mu_B^\alpha]^{\frac{1}{\alpha}}), \quad \alpha > 0\,. \tag{5.76}$$

For $\alpha = 1$ the bold (Łukasiewicz) union is obtained. Algorithm 5.2 summarizes the procedure derived above.

Algorithm 5.2 (Low-level fuzzy relational model.)

Given the cluster prototypes:

Step 1: Discretize the domains of the system's variables.

Step 2: For each cluster compute the relation R_i by using (5.74).

Step 3: Compute R by using (5.75).

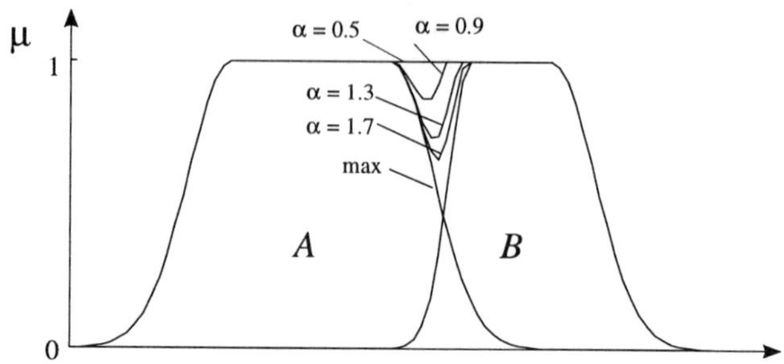

Figure 5.34. Fuzzy union obtained using the Yager t-conorm for different values of α.

Once R is obtained, the fuzzy set B of the response variable y can be derived from the fuzzy sets A_j of the regressors x_j, $j = 1, 2, \ldots, p$ by means of relational composition:

$$B = (A_1 \times A_2 \times \ldots \times A_p) \circ R\,. \tag{5.77}$$

The fuzzy sets A_j and B are defined point-wise in the discrete domains Ξ_j and Υ respectively. A crisp input value is represented as a singleton. In order to obtain a numerical output, the output fuzzy set can be defuzzified using the mean-of-maxima defuzzification (2.33). This method is used because the domain elements with the largest membership degrees are most representative for the clusters, and consequently also for the derived fuzzy relation R.

Example 5.11 The described method is illustrated using a simulation of a nonlinear autoregressive dynamic system (Ikoma and Hirota, 1993) given by:

$$x(k+1) = f\big(x(k)\big) + \epsilon(k), \qquad f(x) = \begin{cases} 2x - 2, & 0.5 \le x \\ -2x, & -0.5 < x < 0.5 \\ 2x + 2, & x \le -0.5 \end{cases} \tag{5.78}$$

where $\epsilon(k) \sim N(0, \sigma^2)$ with $\sigma = 0.3$. From the generated data $x(k)$, $0 \le k \le 200$, with initial condition $x(0) = 0.1$, the first 100 points are used for identification and the rest for model validation. Figure 5.35 shows the identification data set.

It is assumed that the only prior knowledge is that the data were generated by a nonlinear autoregressive system:

$$x(k+1) = f\big(x(k), x(k-1), \ldots, x(k-p+1)\big), \tag{5.79}$$

where p is the system order. Here $\varphi(k) = [x(k), x(k-1), \ldots, x(k-p+1)]^T$ is the regression vector and $x(k+1)$ is the regressand. The data matrix $\mathbf{Z}$ for clustering is constructed from the identification data:

$$\mathbf{Z} = \begin{bmatrix} x(p) & x(p+1) & \cdots & x(N-1) \\ \ldots & \ldots & \ldots & \ldots \\ x(1) & x(2) & \cdots & x(N-p) \\ x(p+1) & x(p+2) & \cdots & x(N) \end{bmatrix}. \tag{5.80}$$

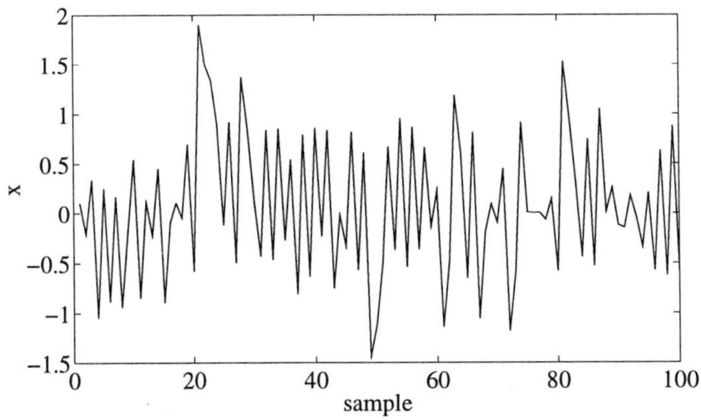

Figure 5.35. Identification data for the autoregressive system.

To identify the system, one first needs to find the order p. In this example, the average cluster flatness validity measure was calculated for a range of model orders $1 \leq p \leq 5$ and numbers of clusters $2 \leq c \leq 7$. The resulting matrix is shown in Figure 5.36b. The minimum value (printed in boldface) was obtained for $p = 1$ and $c = 3$. In Figure 5.36a, the average cluster flatness ν is plotted as a function of c for orders $p = 1, 2$. Note that this function may have several local minima, of which the first is usually chosen in order to obtain a simple model.

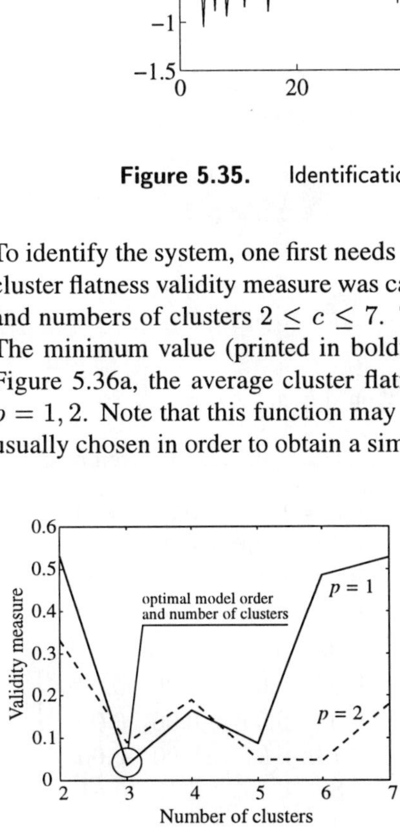

(a)

ν (number of clusters)	model order 1	2	3	4	5
2	0.53	0.33	0.50	4.64	1.27
3	**0.03**	0.08	0.07	0.21	2.45
4	0.16	0.19	5.62	0.36	1.60
5	0.08	0.04	0.06	0.18	0.27
6	0.48	0.04	0.43	0.22	0.51
7	0.52	0.18	2.07	0.12	0.13

(b)

Figure 5.36. Average cluster flatness for different model orders and number of clusters.

The corresponding first-order relational model is

$$A(k+1) = A(k) \circ R, \tag{5.81}$$

where $A(k)$ and $A(k+1)$ are fuzzy sets and R is the fuzzy relation: R: $(\Xi \times \Xi) \to [0, 1]$. Ξ is a set of discrete values in the domain of the system state $X \subset \mathbb{R}$. The limits of X

are determined from the available data, from Figure 5.35 one can see that $x \in [-2, 2]$. This interval is discretized with a step of 0.25: $\Xi = \{-2, -1.75, \ldots, 0, \ldots, 1.75, 2\}$. The discretized context D is thus given by $D = (\Xi \times \Xi)$. The fuzzy relation R is computed using (5.74) and (5.75) where the t-conorm due to Yager (5.76) with $\alpha = 0.5$ is used to compute the fuzzy union. The resulting fuzzy relation is depicted in Figure 5.37.

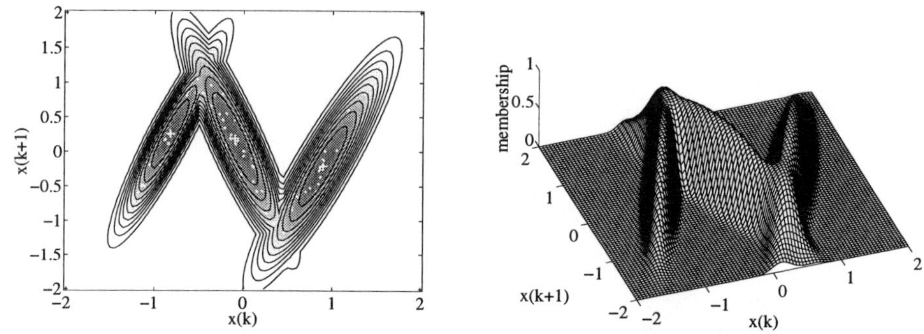

(a) Contour plot. (b) 3D plot.

Figure 5.37. The fuzzy relation obtained from the clusters.

A part of the relation for a subset of Ξ: $\{-1, -0.75, \ldots, 0.75, 1\}$ is given below:

$$R = \begin{bmatrix} 0.00 & 0.09 & 0.56 & 0.90 & 0.91 & 0.56 & 0.09 & 0.00 & 0.00 \\ 0.00 & 0.00 & 0.00 & 0.07 & 0.52 & 0.93 & 0.98 & 0.72 & 0.25 \\ 0.00 & 0.00 & 0.00 & 0.00 & 0.00 & 0.08 & 0.51 & 0.85 & 0.92 \\ 0.00 & 0.00 & 0.01 & 0.14 & 0.56 & 0.88 & 0.97 & 0.93 & 0.76 \\ 0.04 & 0.28 & 0.73 & 0.95 & 0.99 & 0.92 & 0.63 & 0.18 & 0.02 \\ 0.73 & 0.82 & 0.81 & 0.63 & 0.27 & 0.04 & 0.00 & 0.00 & 0.00 \\ 0.87 & 0.80 & 0.50 & 0.12 & 0.01 & 0.00 & 0.00 & 0.00 & 0.00 \\ 0.50 & 0.86 & 0.98 & 0.90 & 0.56 & 0.13 & 0.01 & 0.00 & 0.00 \\ 0.01 & 0.12 & 0.54 & 0.89 & 0.99 & 0.89 & 0.53 & 0.12 & 0.01 \end{bmatrix}$$

The maximal element in each row is emphasized in order to highlight the correspondence between R and Figure 5.37b. By using relational composition (5.81), a fuzzy set $A(k+1)$ for the predicted state $x(k+1)$ can be computed based on a fuzzy set $A(k)$ of the current state $x(k)$. As shown in Figure 5.38a, for $A(k)$ = "about 0" a prediction $A(k+1)$ = "about 0" is obtained. Similarly, $A(k)$ = "about 0.75" results in $A(k+1)$ = "about -0.5", see Figure 5.38b.

In order to obtain numerical output values, the fuzzy set $A(k+1)$ is defuzzified, using the mean-of-maxima method. In Figure 5.39a, the defuzzified values are plotted for the values in Ξ together with the original state transition map (5.78). One can see that the relational model gives a rather accurate approximation. It also outperforms the model of Ikoma and Hirota (1993). Figure 5.39b shows the defuzzified output along

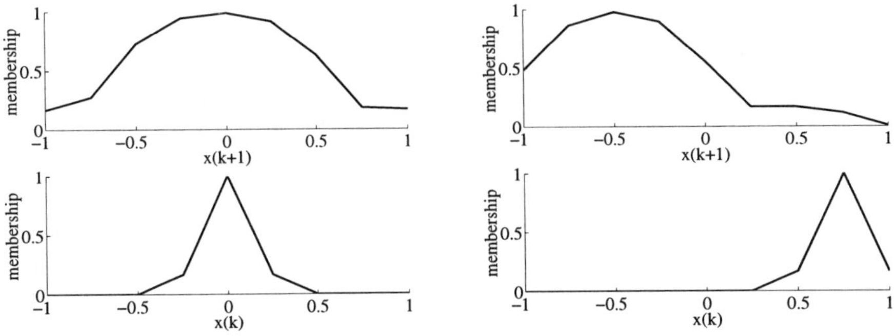

(a) (b)

Figure 5.38. Output obtained by relational composition.

with α-cut intervals of the output fuzzy set for $\alpha = 0.8$. These intervals determine a set of possible values $x(k+1)$ given $A(k)$ (with the degree of possibility 0.8 or higher).

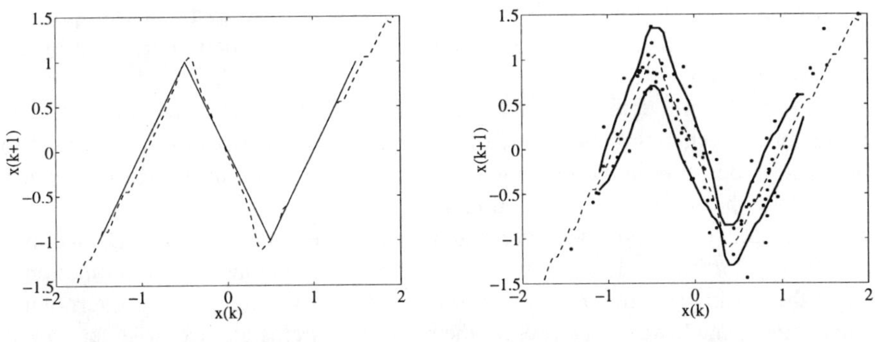

(a) Defuzzified output. (b) 0.8-cut.

Figure 5.39. Graph (a) shows the defuzzified output of the model (dashed line) compared with the original system (solid line). Graph (b) gives the bounds obtained as the 0.8-cut of the output fuzzy set superimposed on the identification data.

The relation R only conveys information about the regions covered by the identification data and little generalization occurs outside these regions. This can be seen as an advantage, since the model gives implicitly low confidence for regions where it has to extrapolate the identification data and the user is thus aware of the decreased reliability of the model output. Simultaneously, it can become a drawback when data is missing in some regions within the range of the identification data set. For such regions, the model may not provide any output. □

5.4 Summary and Concluding Remarks

Several identification methods have been presented in this chapter. Section 5.1 dealt with techniques for building Takagi–Sugeno fuzzy models. Two main tasks can be distinguished: the generation of the antecedent membership functions, and the estimation of the consequent parameters. The antecedent membership functions can be obtained by orthogonal projection of the fuzzy partition matrix onto the antecedent variables, or by projection onto the principal directions of the cluster, using the eigenvectors of the fuzzy partition matrix. Orthogonal projection is usually preferable, as it allows for better interpretation of the obtained model. Eigenvector projection, however, leads to a more accurate representation of the fuzzy cluster by reducing the decomposition error which occurs when orthogonal projection is applied to clusters oblique to the axes. An alternative method is to compute the antecedent membership functions directly in the product space, using an inverse of the distance measure of the particular clustering algorithm. This method is computationally simple and gives accurate results if the antecedent space is well covered by the identification data.

The consequent parameters can be computed from the cluster centers and covariance matrices, or can be estimated by least-squares methods. It was shown that the cluster covariance matrices provide (locally) optimal estimates of the consequents in the total least-squares sense. The consequent parameters can be estimated by weighted or global ordinary least-squares techniques, as well. The weighted least-squares approach estimates the parameters of each consequent model independently of the other ones, and is useful when accurate local models are sought. The global least-squares approach, on the other hand, aims at the minimization of the global prediction error, and yields an optimal predictor.

The initial fuzzy models obtained by fuzzy clustering are often redundant in terms of similar antecedent membership functions. This similarity arises from the projection of the clusters onto the individual antecedent variables. Similarity measures can be used to reduce the complexity in the model.

In Section 5.2, algorithms were presented to generate linguistic and relational models from fuzzy partitions. The construction of the antecedent membership functions is based on the assumption that the model provides (approximately) linear interpolation among its rules, and hence can represent the hyperplanar clusters. Least-squares methods can be used to estimate consequent parameters in a singleton model, or a fuzzy relation can be established among the antecedent and consequent linguistic terms. Methods using the φ-composition and t-norms have been presented and analyzed.

Relationships between the input and output variables can also be approximated by a low-level fuzzy relation, which describes the system as a fuzzy subset on the product space of the variables in the model. A method to compute this relation has been presented in Section 5.3. An advantage of this approach is that no projection of clusters is needed, since the relation is built directly in the product space of the regressands and the regressor. However, the domains of the system's variables must be discretized. For higher-order and multivariable systems, the computational requirements may become severe. The model is also purely numerical and does not allow for qualitative (linguistic) interpretation, as opposed to the other methods presented.

6 FUZZY MODELS IN NONLINEAR CONTROL

This chapter addresses the design of different nonlinear controllers based on a fuzzy model of the process. In the presence of strong nonlinearities, nonlinear controllers can offer significant advantages over linear control techniques, and sometimes may be the only solution to develop a reliable controller. The previous chapters dealt with methods to obtain fuzzy models of dynamic systems. Once a reasonably accurate fuzzy model of the considered process is available, it can be used off-line to design a nonlinear controller for that process, or it may become a part of a model-based control scheme. Most methods for off-line control pertain to the homogenous or affine Takagi–Sugeno structure. Standard methods from linear system theory can be applied to design local feedback controllers for the individual consequent models. Stability of the closed loop can be analyzed by the Lyapunov method (Tanaka and Sugeno, 1992). Furthermore, the control design problem can be formulated by using linear matrix inequalities (Tanaka, et al., 1996). As quite an extensive literature is available on this topic (Palm, et al., 1997), this chapter will be devoted to the latter approach, i.e., model-based fuzzy control. Two particular techniques are considered: model inverse control and model-based predictive control.

For stable systems whose inverted dynamics are stable, a nonlinear controller can be designed by inverting the fuzzy model. In Section 6.1, an algorithm is given to exactly invert a singleton fuzzy model. In the ideal situation with no modeling errors and disturbances, such a controller yields perfect control with zero steady-state errors. In practice, however, one has to deal with both modeling errors and disturbances, which can be accomplished by incorporating the inverse model controller in the well-known internal model control scheme.

A fuzzy model can also predict the process behavior, and can be used in model-based predictive control (MBPC). An optimal control strategy is obtained by minimizing a cost function which usually includes the difference between the desired and predicted outputs, and a penalty on the control effort. Constraints on the control inputs and other process variables or their rates can be included as a part of the optimization problem. In MBPC with nonlinear models, generally a nonconvex optimization problem must be solved at each control sample, which hampers practical applicability of the nonlinear MBPC. To circumvent the optimization problems, a scheme is described in Section 6.2, which is based on a discrete search technique (a branch-and-bound method). In MBPC, fuzzy sets can also be applied at a higher level to formulate the control objectives and constraints.

Simulation examples of a heat transfer systems and a pH neutralization process are presented in Sections 6.3 and 6.4, respectively, to demonstrate the application of inverse model and predictive control to semi-realistic problems.

6.1 Control by Inverting Fuzzy Models

An effective way to design a controller for a nonlinear process is to invert the process model. Inverse model control can be applied to a class of systems that are open-loop stable (or that are stabilizable by feedback) and whose inverse is stable as well, i.e., the system does not exhibit nonminimum phase behavior. In theory, the use of an inverse model possess the advantages of open-loop (feedforward) control, i.e., inherent stability and "perfect" control with zero error both in dynamic transients and in steady states (Economou, et al., 1986).

In the previous chapters, it has been demonstrated that fuzzy modeling can be used to accurately approximate nonlinear systems. This section shows that the class of fuzzy models with singleton consequents can be analytically (exactly) inverted. The nonlinear controller based on a fuzzy model inverse takes into account the process nonlinearities, including the inherent saturation (level) constraints of the control inputs and other process variables, and is capable of controlling a nonlinear system in the entire operating range. To deal with additional constraints, for instance rate constraints on the control inputs, the inverse model scheme is combined with a predictive control strategy, as described in Section 6.2.1.

6.1.1 Singleton Model

Consider a SISO singleton fuzzy model. In this section, the rule index is omitted, in order to simplify the notation. The considered fuzzy rule is then given by the following expression:

$$\begin{aligned}&\textbf{If } y(k) \text{ is } A_1 \textbf{ and } y(k-1) \text{ is } A_2 \textbf{ and } \ldots \textbf{ and } y(k-n_y+1) \text{ is } A_{n_y}\\ &\textbf{and } u(k-n_d+1) \text{ is } B_1 \textbf{ and } \ldots \textbf{ and } u(k-n_u-n_d+2) \text{ is } B_{n_u}\\ &\textbf{then } y(k+1) \text{ is } c,\end{aligned} \tag{6.1}$$

where $A_1, \ldots, A_{n_y}$ and $B_1, \ldots, B_{n_u}$ are fuzzy sets, n_d is a delay from the input to the output and c is a fuzzy singleton. Introduce a state vector $\mathbf{x}(k)$ containing the $n_u - 1$ past inputs, the $n_y - 1$ past outputs and the current output, i.e., all the antecedent variables in (6.1) except $u(k - n_d + 1)$:

$$\mathbf{x}(k) = [y(k), \ldots, y(k - n_y + 1), u(k - n_d), \ldots, u(k - n_u - n_d + 2)]^T . \quad (6.2)$$

The corresponding fuzzy sets are composed into one multidimensional state fuzzy set X, by applying a t-norm operator on the Cartesian product space of the state variables: $X = A_1 \times \ldots \times A_{n_y} \times B_2 \times \ldots \times B_{n_u}$. To simplify the notation, substitute B for B_1. Rule (6.1) now can be written by:

$$\textbf{If } \mathbf{x}(k) \text{ is } X \textbf{ and } u(k - n_d + 1) \text{ is } B \textbf{ then } y(k+1) \text{ is } c . \quad (6.3)$$

Note that the transformation of (6.1) into (6.3) is only a formal simplification of the rule base which does not change the order of the model dynamics, since $\mathbf{x}(k)$ is a vector and X is a multidimensional fuzzy set. Let M denote the number of fuzzy sets X_i defined for the state $\mathbf{x}(k)$ and N the number of fuzzy sets B_j defined for the input $u(k - n_d + 1)$. Assume that the rule base consists of all possible combination of sets X_i and B_j, the total number of rules is then $K = M \cdot N$. The entire rule base can be represented as a table:

	$u(k - n_d + 1)$			
$\mathbf{x}(k)$	B_1	B_2	$\ldots$	B_N
X_1	c_{11}	c_{12}	$\ldots$	c_{1N}
X_2	c_{21}	c_{22}	$\ldots$	c_{2N}
$\vdots$	$\vdots$	$\vdots$	$\vdots$	$\vdots$
X_M	c_{M1}	c_{M2}	$\ldots$	c_{MN}

(6.4)

By using the product t-norm operator, the degree of fulfillment of the rule antecedent $\beta_{ij}(k)$ is computed by:

$$\beta_{ij}(k) = \mu_{X_i}(\mathbf{x}(k)) \cdot \mu_{B_j}(u(k - n_d + 1)) \quad (6.5)$$

The output $y(k+1)$ of the model is computed as an average of the consequents c_{ij} weighted by the normalized degrees of fulfillment β_{ij}:

$$\begin{aligned} y(k+1) &= \frac{\sum_{i=1}^{M}\sum_{j=1}^{N} \beta_{ij}(k) \cdot c_{ij}}{\sum_{i=1}^{M}\sum_{j=1}^{N} \beta_{ij}(k)} = \\ &= \frac{\sum_{i=1}^{M}\sum_{j=1}^{N} \mu_{X_i}(\mathbf{x}(k)) \cdot \mu_{B_j}(u(k - n_d + 1)) \cdot c_{ij}}{\sum_{i=1}^{M}\sum_{j=1}^{N} \mu_{X_i}(\mathbf{x}(k)) \cdot \mu_{B_j}(u(k - n_d + 1))} . \end{aligned} \quad (6.6)$$

Example 6.1 Consider a fuzzy model of the form $y(k+1) = f(y(k), y(k-1), u(k))$ where two linguistic terms $\{low, high\}$ are used for $y(k)$ and $y(k-1)$ and three terms

$\{small, medium, large\}$ for $u(k)$. The complete rule base consists of $2 \times 2 \times 3 = 12$ rules:

If $y(k)$ is *low* and $y(k-1)$ is *low* and $u(k)$ is *small* **then** $y(k+1)$ is c_{11}
If $y(k)$ is *low* and $y(k-1)$ is *low* and $u(k)$ is *medium* **then** $y(k+1)$ is c_{12}

...

If $y(k)$ is *high* and $y(k-1)$ is *high* and $u(k)$ is *large* **then** $y(k+1)$ is c_{43}

In this example $\mathbf{x}(k) = [y(k), y(k-1)]$, $X_i \in \{(low \times low), (low \times high), (high \times low), (high \times high)\}$, $M = 4$ and $N = 3$. The rule base is represented by the following table:

$\mathbf{x}(k)$	$u(k)$ *small*	*medium*	*large*
X_1 (*low* × *low*)	c_{11}	c_{12}	c_{13}
X_2 (*low* × *high*)	c_{21}	c_{22}	c_{23}
X_3 (*high* × *low*)	c_{31}	c_{32}	c_{33}
X_4 (*high* × *high*)	c_{41}	c_{42}	c_{43}

(6.7)

□

6.1.2 Inversion of the Singleton Model

The inversion method is first explained for model (6.1) with a unit delay ($n_d = 1$). For systems with larger delays, the same method can be applied after performing $n_d - 1$ steps of prediction with the fuzzy model, as explained in Remark 3 on p. 168. When $n_d = 1$, the rule-based model (6.3) corresponds to the following regression model:

$$y(k+1) = f\big(\mathbf{x}(k), u(k)\big), \tag{6.8}$$

The inputs of the model are the current state $\mathbf{x}(k) = [y(k), \ldots, y(k-n_y+1), u(k-1), \ldots, u(k-n_u+1)]^T$ and the current input $u(k)$. The output is a prediction of the system's output at the next sample $y(k+1)$. The objective of the control algorithm is to compute the control input $u(k)$, such that the system output at the next sampling instant is equal to the desired (reference) output $r(k+1)$. This can be achieved by inverting the model of the process. Given the current state $\mathbf{x}(k)$ and the reference $r(k+1)$, the control input is given by:

$$u(k) = f^{-1}(\mathbf{x}(k), r(k+1)), \tag{6.9}$$

where the reference $r(k+1)$ was substituted for $y(k+1)$. Generally, it is difficult to find the inverse function f^{-1}. The method presented below makes use of the singleton model structure. For each particular state $\mathbf{x}(k)$, the multivariate mapping (6.8) can be reduced to a univariate mapping

$$y(k+1) = f_x(u(k)), \tag{6.10}$$

where the subscript x denotes that f_x is obtained for the particular state $\mathbf{x}$. From this mapping, the inverse mapping $u(k) = f_x^{-1}(r(k+1))$ can be easily found, provided the model is invertible. The concept of invertibility and the respective conditions for the fuzzy model are defined more exactly below.

Definition 6.1 (Model invertibility) *A fuzzy model f given by the rule base (6.3) and the defuzzification method (6.6) is called* invertible *(monotonic), if $\forall x \in X$ and $\forall y \in Y$, a unique u exists such that $y = f(x, u)$.*

Lemma 6.1 (Invertibility conditions) *Denote $b_j = \text{core}(B_j)$ the core of B_j. The singleton model, given by the rule base (6.3) and the defuzzification method (6.6), is invertible if and only if:*

1. *the core of each B_j is a single point, i.e., $|b_j| = 1$, $j = 1, 2, \ldots, N$, and*
2. $b_1 < b_2 < \ldots < b_N \longrightarrow c_{i1} < c_{i2} < \ldots < c_{iN}$, *or*
   $b_1 < b_2 < \ldots < b_N \longrightarrow c_{i1} > c_{i2} > \ldots > c_{iN}$, $i = 1, 2, \ldots, M$.

Proof: When $1 \leq j \leq N$ exists such that $|b_j| > 1$, $u_k \neq u_l$ can be found such that $\mu_{B_j}(u_k) = \mu_{B_j}(u_l)$ and hence an $\mathbf{x}$ exists for which $f(\mathbf{x}, u_k) = f(\mathbf{x}, u_l)$. The second condition reflects the monotonicity of the model output for all N-tuples of rules contained in the rows of the table in eq. (6.4). If the model output c_{ij} is monotonic in b_j for each $1 \leq i \leq M$, then also the entire rule base is monotonic, since eq. (6.6) results in linear interpolation. □

Example 6.2 Figure 6.1 presents an example where both conditions of Lemma 6.1 are violated. Fuzzy set B_2 does not meet the condition $|\text{core}(B_2)| > 1$ and further $\text{core}(B_3) > \text{core}(B_1) \rightarrow c_{i3}(k) < c_{i1}(k)$ while $\text{core}(B_4) > \text{core}(B_3) \rightarrow c_{i4}(k) > c_{i3}(k)$. □

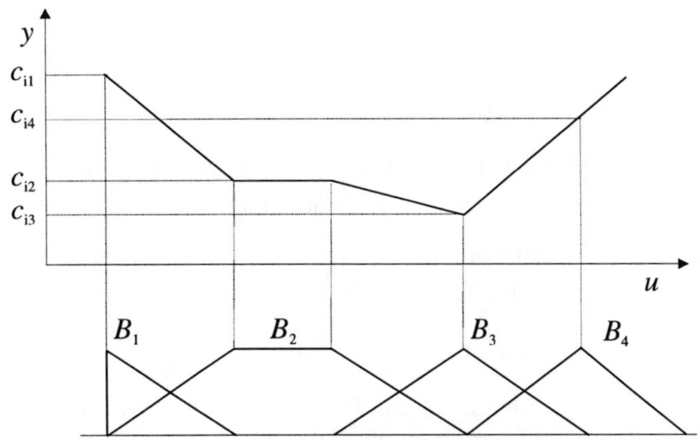

Figure 6.1. Example of a singleton model which does not satisfy the invertibility conditions.

The inverse of the fuzzy model is formulated in Lemma 6.2.

Lemma 6.2 (Inverse of singleton fuzzy model) *Let the process be represented by an invertible singleton fuzzy model (6.3) with the defuzzification method given by (6.6). Further, let the antecedent membership functions form a partition, i.e., let* $\sum_{i=1}^{M} \mu_{X_i}(\mathbf{x}) = 1, \ \forall \mathbf{x}$, *and* $\sum_{j=1}^{N} \mu_{B_j}(u) = 1, \ \forall u$. *For a given state* $\mathbf{x}(k)$, *the control law based on an inverse of this singleton model is given by the following rules:*

$$\textbf{\textit{If}}\ r(k+1)\ \textit{is}\ C_j(k)\ \textbf{\textit{then}}\ u(k)\ \textit{is}\ B_j, \quad j = 1, 2, \ldots, N, \tag{6.11}$$

where $C_j: Y \to [0,1]$ *are fuzzy sets with triangular membership functions given by:*

$$\mu_{C_1}(y) = \max\left(0, \min(1, \frac{c_2 - y}{c_2 - c_1})\right), \tag{6.12a}$$

$$\mu_{C_j}(y) = \max\left(0, \min(\frac{y - c_{j-1}}{c_j - c_{j-1}}, \frac{c_{j+1} - y}{c_{j+1} - c_j})\right), \quad 1 < j < N, \tag{6.12b}$$

$$\mu_{C_N}(y) = \max\left(0, \min(\frac{y - c_{N-1}}{c_N - c_{N-1}}, 1)\right), \tag{6.12c}$$

where the cores c_j *are given by:*

$$c_j = \sum_{i=1}^{M} \mu_{X_i}(\mathbf{x}(k)) c_{ij}, \quad j = 1, \ldots, N, \tag{6.13}$$

and are ordered such that $c_1 \le c_2 \le \cdots \le c_N$. *Fuzzy sets* B_j *are ordered accordingly. The inference of the rules (6.11) is accomplished by the fuzzy-mean method:*

$$u(k) = \sum_{j=1}^{N} \mu_{C_j}(r(k+1)) b_j, \tag{6.14}$$

where b_j *are the cores of* B_j. *The series connection of the controller and the inverse model, shown in Figure 6.2, gives an identity mapping (perfect control):*

$$y(k+1) = f_x(u(k)) = f_x(f_x^{-1}(r(k+1))) = r(k+1), \tag{6.15}$$

when $u(k)$ *exists such that* $r(k+1) = f\big(\mathbf{x}(k), u(k)\big)$. *When no such* $u(k)$ *exists, the difference* $|r(k+1) - f_x(f_x^{-1}(r(k+1)))|$ *is minimal.*

Proof: Using the product t-norm for the *and* connective in the rule antecedent, the rule (6.3) can be rewritten in the following form:

$$\textbf{If}\ \mathbf{x}(k)\ \text{is}\ X_i\ \textbf{then}\ (\textbf{if}\ u(k)\ \text{is}\ B_j\ \textbf{then}\ y(k+1)\ \text{is}\ c_{ij})\,. \tag{6.16}$$

For a given state $\mathbf{x}(k)$, the degree of fulfillment of the first antecedent proposition, "$\mathbf{x}(k)$ is X_i", can be calculated as $\mu_{X_i}(\mathbf{x}(k))$. Then the consequents of the rules containing a particular B_j, i.e., columns in (6.4), can be aggregated (there are M such rules for each B_j). The aggregated consequents $c_j(k)$ are given by eq. (6.13). A set of N rules for the input $u(k)$ is obtained:

$$\textbf{If}\ u(k)\ \text{is}\ B_j\ \textbf{then}\ y(k+1)\ \text{is}\ c_j(k), \qquad j = 1, \ldots, N\,. \tag{6.17}$$

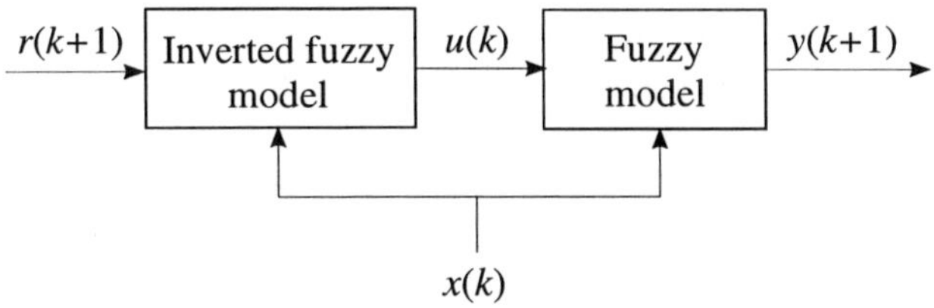

Figure 6.2. Series connection of the fuzzy model and the controller based on the inverse of this model.

Each of the above rules is inverted by exchanging the antecedent and the consequent, which yields the following rules:

$$\textbf{If } y(k+1) \text{ is } c_j(k) \textbf{ then } u(k) \text{ is } B_j \qquad j = 1, \ldots, N\,. \tag{6.18}$$

Since $c_j(k)$ are singletons, it is necessary to interpolate between the consequents $c_j(k)$ in order to obtain $u(k)$. This interpolation is accomplished by triangular fuzzy sets C_j defined by equations (6.12) and (6.13). To show that the series connection of the controller and of the inverse model gives an identity mapping, or a mapping with a minimal error, two situations must be distinguished:

1. $r(k+1) \in [c_1, c_N]$, i.e., the desired output is in the range of $f_x(u(k))$, i.e., can be reached from the current state within one sample.

2. $r(k+1) \notin [c_1, c_N]$, , i.e., the desired output cannot be reached from the current state in one sample.

First, consider case 1. For the sake of brevity, the time argument k is omitted. As the input fuzzy sets B_j form a partition, at most two adjacent control rules are active simultaneously. Denote these two rules by indices j and $j+1$. Using equations (6.12) and (6.14), the control input is given by:

$$u = \frac{r - c_j}{c_{j+1} - c_j} b_j + \frac{c_{j+1} - r}{c_{j+1} - c_j} b_{j+1} = \frac{r(b_j - b_{j-1}) + c_{j+1}b_{j+1} - c_j b_j}{c_{j+1} - c_j}\,. \tag{6.19}$$

Analogously, the output of the fuzzy model is given by:

$$y = \frac{u - b_j}{b_{j+1} - b_j} c_j + \frac{b_{j+1} - r}{b_{j+1} - b_j} c_{j+1}\,. \tag{6.20}$$

After substituting u from eq. (6.19) into eq. (6.20), and carrying out some elementary algebraic manipulations, the following expression is obtained:

$$y = \frac{r(b_j - b_{j+1}) + c_{j+1}b_{j+1} - c_{j+1}b_j}{(c_{j+1} - c_j)(b_{j+1} - b_j)} c_j + \frac{c_j b_j - c_j b_{j+1} - r(b_j - b_{j+1})}{(c_{j+1} - c_j)(b_{j+1} - b_j)} c_{j+1}\,. \tag{6.21}$$

Further simplification leads to:

$$y = \frac{c_{j+1} - r}{(c_{j+1} - c_j)} c_j + \frac{r - c_j}{(c_{j+1} - c_j)} c_{j+1} = \frac{c_{j+1}c_j - rc_j + rc_{j+1} - c_j c_{j+1}}{(c_{j+1} - c_j)} = r\,. \tag{6.22}$$

Now consider the second situation. In the case that $r(k+1) > c_N$, $\mu_{C_N}(r(k+1)) = 1$, which gives the control input $u(k) = b_N$. In the fuzzy model, $\mu_{B_N}(u(k)) = 1$, which yields the model output $y(k+1) = c_N$. As $c_N > c_j$, $1 \leq j \leq N-1$, the difference $|r(k+1) - y(k+1)|$ is the least possible. Analogically for $r < c_1$, one obtains $y(k+1) = c_1$ as a result of the control action $u(k) = b_1$, which is the best possible control command, since $c_1 < c_j$, $2 \leq j \leq N$. □

Remark 1: Note that the membership functions C_j in eq. (6.12) depend on the values c_j, which in turn depend on the process state, see eq. (6.13). The control law is hence dynamically varying according to the state.

Remark 2: For a noninvertible rule base (a rule base that does not satisfy the conditions in Lemma 6.1), a set of possible control commands can be found by splitting the rule base into two or more invertible parts. For each part, a control action is found by inversion, using Lemma 6.2. Among these control actions, only one has to be selected, which requires some additional criteria, such as minimal control effort (minimal $u(k)$ or $|u(k) - u(k-1)|$), etc. Note, however, that this solution is in general different from the solution obtained by predictive control where the control effort is included in the objective function, see Section 6.2.1. The invertibility of the fuzzy model can also be checked in run-time, by checking the monotonicity of the aggregated consequents c_j with respect to the cores of the input fuzzy sets b_j, see eq. (6.13). This is useful, since nonlinear models can be noninvertible only locally, resulting in a kind of exception in the inversion algorithm. Moreover, for models adapted on line, this check is necessary.

Remark 3: For fuzzy models with input delays $y(k+1) = f\big(\mathbf{x}(k), u(k-n_d+1)\big)$, the inversion cannot be applied directly, as it would give a control action $u(k-n_d+1)$, $n_d - 1$ steps delayed. In order to generate the appropriate value $u(k)$, the model must be inverted $n_d - 1$ samples ahead, i.e., $u(k) = f^{-1}(r(k+n_d), x(k+n_d-1))$, where

$$\begin{aligned} x(k+n_d-1) &= [y(k+n_d-1), \ldots, y(k+1), \ldots \\ &\quad y(k-n_y+n_d), u(k-1), \ldots, u(k-n_u+1)] \,. \end{aligned} \tag{6.23}$$

The unknown values, $y(k+1), \ldots, y(k+n_d-1)$, are predicted recursively using the fuzzy model:

$$\begin{aligned} y(k+i) &= f\big(x(k+i-1), u(k-n_d+i)\big), \\ x(k+i) &= [y(k+i), \ldots, y(k-n_y+1+i), u(k-n_d+i), \ldots \\ &\quad u(k-n_u+2-n_d+i)], \quad \text{for } i = 2, \ldots, n_d \,. \end{aligned} \tag{6.24}$$

The inversion procedure is summarized in Algorithm 6.1. Apart from the computation of the membership degrees, both the model and the controller can be implemented using standard matrix operations and linear interpolations, which makes the algorithm suitable for real-time implementation. The algorithm is illustrated step by step in Example 6.3.

Algorithm 6.1 (Inverse of singleton fuzzy model)

Step 1: **Measure or estimate the current system state.** For input–output models this means to measure the output $y(k)$ and update the state vector $\mathbf{x}(k)$, see eq. (6.2).

Step 2: **Compute the cores c_j, using eq. (6.13).**

Step 3: **Check invertibility.** If nonmonotonicity is detected, partition the rule base (6.11) into two or more rule bases containing only monotonic rules.

Step 4: **Construct membership functions, using eq. (6.12).** If several rule bases result from Step 3, separate membership functions are built for each rule base.

Step 5: **Compute the control input $u(k)$, using eq. (6.14).** If several rule bases result from Step 3, a control input is computed for each rule base, and selection is done on the basis of additional criteria.

Example 6.3 Consider the fuzzy model from Example 6.1, which is, for convenience, repeated below:

$\mathbf{x}(k)$	$u(k)$ *small*	*medium*	*large*
X_1(*low* × *low*)	c_{11}	c_{12}	c_{13}
X_2(*low* × *high*)	c_{21}	c_{22}	c_{23}
X_3(*high* × *low*)	c_{31}	c_{32}	c_{33}
X_4(*high* × *high*)	c_{41}	c_{42}	c_{43}

Given the state $\mathbf{x}(k) = [y(k), y(k-1)]$, the degree of fulfillment of the first antecedent proposition "$\mathbf{x}(k)$ is X_i", is calculated as $\mu_{X_i}(\mathbf{x}(k))$. For X_2, for instance, $\mu_{X_2}(\mathbf{x}(k)) = \mu_{\text{low}}(y(k)) \cdot \mu_{\text{high}}(y(k-1))$. Using (6.13), one obtains the cores $c_j(k)$:

$$c_j(k) = \sum_{i=1}^{4} \mu_{X_i}(\mathbf{x}(k)) c_{ij}, \quad j = 1, 2, 3. \tag{6.25}$$

An example of membership functions for fuzzy sets C_j, obtained by eq. (6.12), is shown in Figure 6.3:

Assuming that $b_1 < b_2 < b_3$, the model is (locally) invertible if $c_1 < c_2 < c_3$ or if $c_1 > c_2 > c_3$. In such a case, the following rules are obtained:

1) **If** $r(k+1)$ is $C_1(k)$ **then** $u(k)$ is B_1

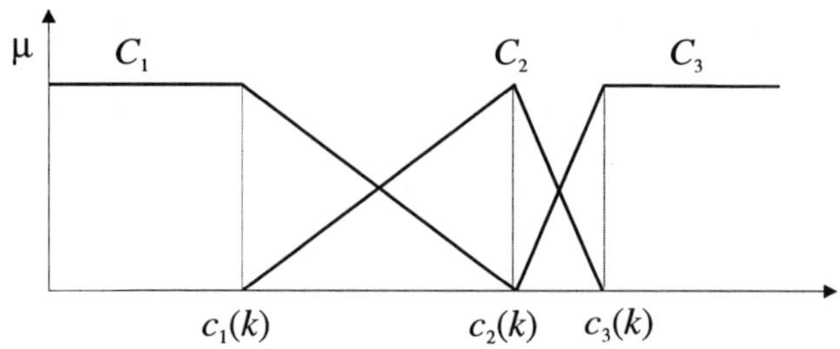

Figure 6.3. Fuzzy partition created from $c_1(k)$, $c_2(k)$ and $c_3(k)$.

2) **If** $r(k+1)$ is $C_2(k)$ **then** $u(k)$ is B_2
3) **If** $r(k+1)$ is $C_3(k)$ **then** $u(k)$ is B_3

Otherwise, if the model is not invertible, e.g., $c_1 > c_2 < c_3$, the above rule base must be split into two rule bases. The first one contains rules 1 and 2, and the second one rules 2 and 3. □

To control the process, the inverse model can be employed in an open-loop (feedforward) control scheme, or in an open-loop feedback scheme (Ku and Athans, 1973; de Vlieger, 1983). In the feedforward scheme, the state of the controller (6.2) is updated using the output of the model only. No feedback from the process output exists in this scheme, and hence stable control is guaranteed for stable, minimum-phase systems. However, a model-plant mismatch will cause a steady-state error at the process output. This error can be compensated by some kind of feedback, using, for instance, the IMC scheme presented in Section 6.1.3.

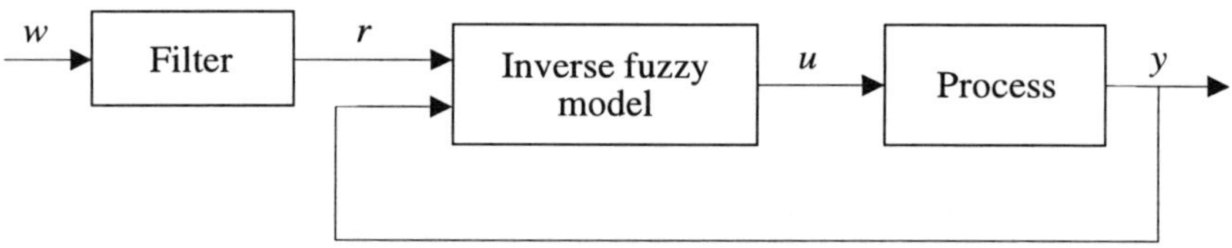

Figure 6.4. Control diagram of the controller based on fuzzy model inverse.

Figure 6.4 depicts the open-loop feedback control scheme with an inverse fuzzy model. The controller, in fact, operates in an open loop (does not use the error between the reference and the process output), but the current output (state) of the process, or an estimate of the state from an observer is used at each sample to update the internal state $\mathbf{x}$ of the controller. This can improve the prediction accuracy and eliminate offsets. At the same time, however, the direct updating of the model state may not be desirable in the presence of noise or a significant model–plant mismatch, in which cases it can cause oscillations or instability.

Both schemes involve a reference-shaping filter, usually a first-order or second-order reference model, whose task is to generate the desired dynamics and to avoid peaks in the control action for step-like references. When a continuous-time process is controlled, an additional function of the filter is to avoid undesired behavior between the sampling instants, the intersampling ripple. In the linear case, this is achieved by filtering the reference by the dynamics of the process (Isermann, 1977).

Inverse model control can only be applied to systems with a minimum phase behavior (systems whose inverted dynamics are stable). Nonminimum phase systems (also called inverse response systems) would be destabilized by the inverse model controller, as it would always attempt to drive the system directly to the reference. Nonminimum phase systems can be effectively dealt with by means of predictive control, see Section 6.2.

6.1.3 Compensation of Disturbances and Modeling Errors

Disturbances acting on the process, measurement noise and model-plant mismatch cause differences in the behavior of the process and of the model, which results in an error between the reference and the process output. Two mechanism to compensate this error are presented: the internal model control (IMC) scheme and on-line adaptation of the fuzzy model.

Internal Model Control. Figure 6.5 depicts the internal model control scheme, which consists of three blocks: the controller based on an inverse of the process fuzzy model, the fuzzy model, and a feedback filter. The control system has two inputs, the reference and the measurement of the process output, and one output, the control action.

The purpose of the fuzzy model working in parallel with the process is to subtract the effect of the control action from the process output. If the predicted and the measured process outputs are equal, the error e is zero and the controller works in an open-loop configuration. If a disturbance d acts on the process output, the feedback signal e is equal to the influence of the disturbance and is not affected by the effects of the control action. This signal is simply subtracted from the reference. With a perfect process model, the IMC scheme is hence able to cancel the effect of unmeasured output-additive disturbances.

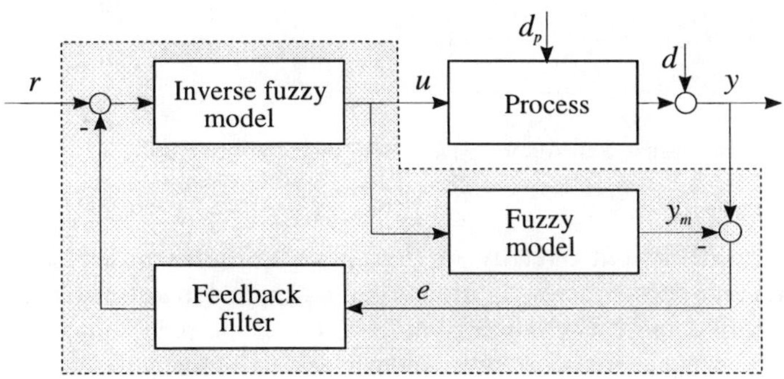

Figure 6.5. Internal model control scheme.

Two basic properties of the ideal IMC are inherent stability and perfect control. Inherent stability means that if the controller and the process are input–output stable and a perfect model of the process is available, the closed-loop system is input–output stable. If the system is not input–output stable, but it can be stabilized by feedback, IMC can still be applied. Perfect control means that if the controller is an exact inverse of the model, and the closed-loop system is stable, then the control is error-free, i.e., $y(k) = r(k), \forall k$.

However, in practice, the model is never an exact representation of the process. The feedback signal then contains both the effect of unmeasured disturbances and of the modeling errors, and it becomes a true feedback. For large modeling errors, it deteriorates the performance of the control system and may introduce stability problems. An important difference between linear and nonlinear systems is that for linear systems, disturbances can be assumed to act additively at the output, while for nonlinear systems, unmeasured disturbances d_p acting on the process generally cause differences between the model and the process. The feedback filter is introduced in order to filter out the measurement noise and to stabilize the loop by reducing the loop gain. With nonlinear systems and models, the filter must be designed empirically.

On-line Adaptation. On-line adaptation can be applied to cope with the mismatch between the process and the fuzzy model. In many cases, a mismatch occurs as a consequence of (temporary) changes of process parameters. To deal with these phenomena, especially if their effects vary in time, the fuzzy model can be adapted in the control loop. Since the control actions are derived by inverting the model on line, the controller is adjusted automatically. Figure 6.6 depicts the IMC scheme with on-line adaptation of the consequent parameters in the fuzzy model.

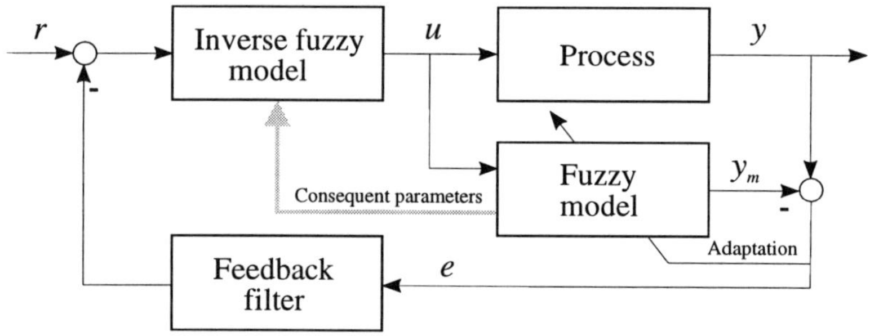

Figure 6.6. Adaptive model-based control scheme.

Since the model output given by eq. (6.6) is linear in the consequent parameters, standard recursive least-squares algorithms can be applied to estimate the consequent parameters from data. It is assumed that the rules of the fuzzy model are given by (6.1) and the consequent parameters are indexed sequentially by the rule number. The column vector of the consequents is denoted by $\mathbf{c}(k) = [c_1(k), c_2(k), \ldots, c_K(k)]^T$, where K is the number of rules. The normalized degrees of fulfillment denoted by

$\gamma_i(k)$ are computed by:

$$\gamma_i(k) = \frac{\beta_i(k)}{\sum_{j=1}^{K} \beta_j(k)}, \quad i = 1, 2, \ldots, K. \tag{6.26}$$

They are arranged in a column vector $\gamma(k) = [\gamma_1(k), \gamma_2(k), \ldots, \gamma_K(k)]^T$. The consequent vector $\mathbf{c}(k)$ is updated recursively by:

$$\mathbf{c}(k) = \mathbf{c}(k-1) + \frac{\mathbf{P}(k-1)\gamma(k)}{\lambda + \gamma^T(k)\mathbf{P}(k-1)\gamma(k)} \left[y(k) - \gamma^T(k)\mathbf{c}(k-1)\right], \tag{6.27}$$

where λ is a constant forgetting factor, which influences the tracking capabilities of the adaptation algorithm. The smaller the λ, the faster the consequent parameters adapt, but the algorithm is more sensitive to noise. Therefore, the choice of λ is problem dependent. The covariance matrix $\mathbf{P}(k)$ is updated as follows:

$$\mathbf{P}(k) = \frac{1}{\lambda} \left[\mathbf{P}(k-1) - \frac{\mathbf{P}(k-1)\gamma(k)\gamma^T(k)\mathbf{P}(k-1)}{\lambda + \gamma^T(k)\mathbf{P}(k-1)\beta(k)}\right]. \tag{6.28}$$

The initial covariance is usually set to $\mathbf{P}(0) = \alpha \cdot \mathbf{I}$, where $\mathbf{I}$ is a $K \times K$ identity matrix and α is a large positive constant.

The above algorithm adapts only the consequent parameters, while the antecedent membership functions remain unchanged. In many practical situations, it is indeed not necessary to adapt the antecedents, as the structure of the process nonlinearity does not change, and the antecedent partition derived through off-line identification remains valid. A typical example is compensation for constant offsets or changes in the process gain (scaling of the nonlinearity).

If a good initial model is not available, the membership function parameters can be adapted using some nonlinear optimization technique. Various approaches from the literature can be applied, such as error back-propagation (Jang, 1992), nonlinear programming (Jang, et al., 1997) or genetic algorithms (Klawonn, et al., 1994). In addition, fuzzy clustering can be repeated regularly, after gathering new data or on-line, see Section 5.8.

6.2 Predictive Control

Model-based predictive control (MBPC) is a general methodology for solving control problems in the time domain. It is based on three main concepts:

1. A model is used to predict the process output at future discrete time instants, over a *prediction horizon*.

2. A sequence of future control actions is computed over a *control horizon* by minimizing a given objective function.

3. Only the first control action in the sequence is applied, the horizons are moved towards the future and optimization is repeated. This is called the *receding horizon* principle.

Because of the optimization approach and the explicit use of the process model, MBPC can realize multivariable optimal control, deal with nonlinear processes, and can efficiently handle constraints. This section addresses the use of fuzzy sets in the various components of predictive control, namely at the modeling level, in optimization and in the specification of the control objectives.

6.2.1 Basic Concepts

Prediction and Control Horizons. The future process outputs are predicted over the *prediction horizon* H_p using a model of the process. The predicted output values, denoted $\hat{y}(k+i)$ for $i = 1, \ldots, H_p$, depend on the state of the process at the current time k and on the future control signals $u(k+i)$ for $i = 0, \ldots, H_c - 1$, where $H_c \leq H_p$ is the *control horizon*. The control signal is manipulated only within the control horizon and remains constant afterwards, i.e., $u(k+i) = u(k+H_c-1)$ for $i = H_c, \ldots, H_p - 1$, see Figure 6.7.

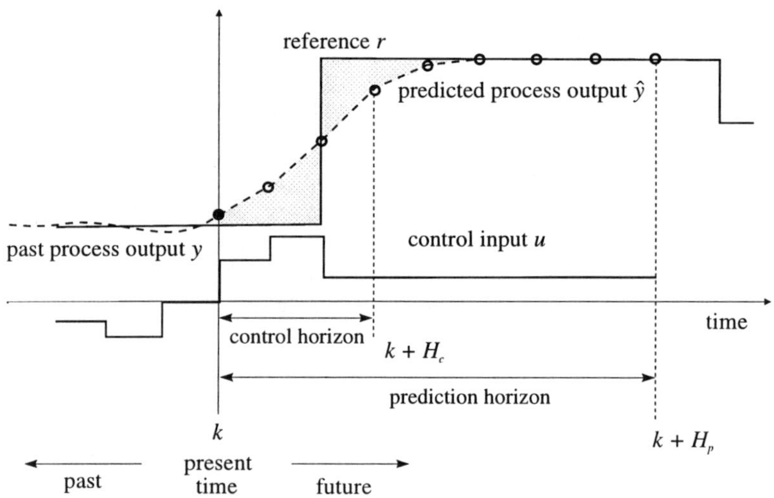

Figure 6.7. The basic principle of model-based predictive control.

Objective Function. The sequence of future control signals $u(k+i)$ for $i = 0, \ldots, H_c - 1$ is computed by optimizing a given objective (cost) function, in order to bring and keep the process output as close as possible to the given reference trajectory r, which can be the set-point itself or, more often, some filtered version of it. Most often, the objective functions used are modifications of the following quadratic function (Clarke, et al., 1987):

$$J = \sum_{i=1}^{H_p} \alpha_i \left[r(k+i) - \hat{y}(k+i)\right]^2 + \sum_{i=1}^{H_c} \beta_i \Delta u(k+i-1)^2 . \tag{6.29}$$

The first term accounts for minimizing the variance of the process output from the reference, while the second term represents a penalty on the control effort (related, for instance, to energy). The latter term can also be expressed by using u itself, or other filtered forms of u, depending on the problem (Soeterboek, 1992). The parameters α_i and β_i define the weighting of the output error and the control effort with respect to each other, and with respect to the prediction step.[1] For systems with a dead time of n_d samples, only outputs from time $k + n_d$ are considered in the objective function, because outputs before this time cannot be influenced by the control signal $u(k)$. Similar reasoning holds for nonminimum phase systems. Additional "soft" constraints can be included in the cost function. "Hard" constraints, e.g., level and rate constraints of the control input, or other process variables can be specified as a part of the optimization problem. Generally, any other suitable cost function can be used, but for a quadratic cost function, a linear, time-invariant model, and, in the absence of constraints, an explicit analytic solution of the above optimization problem can be obtained. Otherwise, numerical (usually iterative) optimization methods must be used.

Receding Horizon Principle. Only the control signal $u(k)$ is applied to the process. At the next sampling instant, the process output $y(k+1)$ is available and the optimization and prediction can be repeated with the updated values. This is called the receding horizon principle. The control action $u(k+1)$ computed at time step $k+1$ will be generally different from the one calculated at time step k, since more up-to-date information about the process is available. This concept is similar to the open-loop control strategy discussed in Section 6.1.2. Also here, the model can be used independently of the process, in a pure open-loop setting.

Predictive control can be regarded as a generalization of the inverse model control approach. Assume that $H_p = H_c = 1$ and $\boldsymbol{\beta} = \mathbf{0}$ in eq. (6.29), and that a control command $u(k)$ exists such that $y(k+1) = r(k+1)$ (i.e., the inverse exists), a global optimum of the functional (6.29) yields $J^* = 0$ and $u(k)^* = u(k)$. In other words, without constraints and without penalizing the control action, the one-step-ahead predictive control strategy is equivalent to the inverse model control strategy, where the inverse is computed numerically, by means of function minimization. Extending the prediction and control horizons, adding the control signal to the objective function, and including constraints can be regarded as generalization of model inverse control.

Advantages and Drawbacks of Predictive Control. MBPC has been widely applied in industry and has been accepted both by control engineers and plant operators (Richalet, 1993). It is easy to understand for practitioners with a limited knowledge of control, and the tuning of the controller is quite intuitive. MBPC is also applicable to a wide range of processes including multi-variable nonlinear systems with dead times, nonminimum phase, or unstable systems, and can deal with constraints in a systematic way. It can also make efficient use of known future references for path following, set-point tracking, batch process control, etc. Finally, it is a methodology open to extensions and new approaches in process modeling and optimization (Camacho and Bordons, 1995).

[1] Tuning rules were suggested for these parameters (Soeterboek, 1992).

The performance of MBPC depends, for a large part, on the predictive accuracy of the model, and as the accuracy decreases, so does also the performance of the controller. Consequently, the major part of the MBPC design effort and cost is related to modeling and identification (Richalet, 1993). With linear models, the rest of the control design boils down to the use of some MBPC package and to the tuning of the various parameters (horizons, weighting constants).

The chosen model must be capable of predicting the future process outputs, must be simple to implement, cost-effective to simulate, and easy to understand. Complex, nonlinear and partially known systems are still not amenable to the application of conventional predictive control, since reliable models for these systems cannot be obtained by using conventional approaches based on physical modeling or linear system identification. Moreover, in the presence of nonlinearities and constraints, a nonconvex optimization problem must be solved at each sampling period. This hampers the application of MBPC to fast systems where iterative optimization techniques cannot be properly used, due to short sampling times and extensive computation times.

The following sections address the above problems in the framework of fuzzy modeling and fuzzy decision-making. An optimization technique is developed, based on a combination of the branch-and-bound optimization scheme with the inverse model control algorithm presented in Section 6.1. To illustrate the methodology, an example of application to a laboratory heat transfer process is given, where predictive control is compared to inverse model control. The presented approach is also compared to an approach based on neural network modeling and input–output linearization, using the pH titration process from Section 5.1.5.

6.2.2 Fuzzy Models in MBPC

Fuzzy models exhibit features of both purely numerical and qualitative models. The defuzzification of the output allows the use of fuzzy models for numerical prediction of the process behavior, in the same way as one can use other types of nonlinear models, e.g., neural networks, radial basis functions, etc. The main difference is that the rule-based structure of fuzzy models can give a certain insight into the prediction, and allow for interpretation and analysis of the process behavior, for instance, for monitoring purposes. This advantage of fuzzy models can be employed, among other, in the development of an intelligent operator interface (Ali and Heuer, 1995). Another advantage of fuzzy models is that imprecise prior knowledge expressed in terms of fuzzy rules can be included in the initial model, and can be further refined using process data, see Section 6.5. From the implementational point of view, fuzzy models with singleton consequents, and models of the TS type, are simple and computationally efficient.

A fuzzy model acting as a numerical predictor of the process' output can be directly integrated in the MBPC scheme shown in Figure 6.8. The IMC scheme is usually employed to compensate for the disturbances and modeling errors, see Section 6.1.3.

The most straightforward way of minimizing (6.29) is to use some numerical optimization technique, such as the Nelder-Mead method (Walsh, 1975), or sequential quadratic programming (Gill, et al., 1981). These algorithms, however, require a significant computing power which may be a serious obstacle for the real-time im-

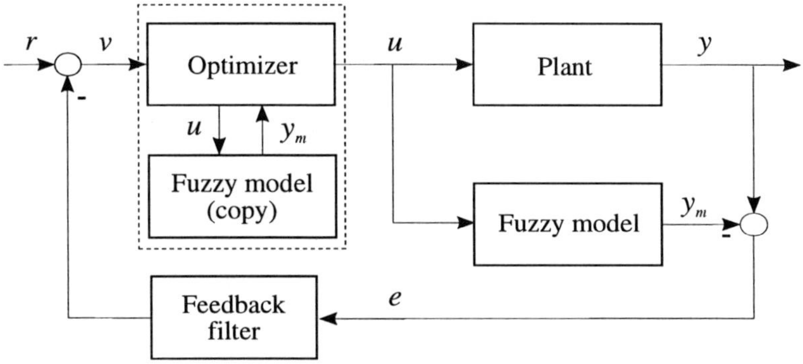

Figure 6.8. Fuzzy model in the MBPC scheme with an internal model and a feedback to compensate for disturbances and modeling errors.

plementation. Moreover, the algorithm can be trapped in a local minimum, which may result in undesired control actions and poor performance. Also, the number of iterations needed to reach a solution can drastically differ from sample to sample. Premature termination of the iteration process is often necessary to fulfill the real-time constraints, which results in inferior solutions, and hence a decreased control performance. Alternative optimization schemes are sought which can solve the optimization problem satisfactorily, i.e., find the global minimum under constraints and fulfill the real-time requirements (Henson and Kurtz, 1994; Nevistić and Morari, 1995; Botto, et al., 1998).

One possibility is to look for nonlinear models of a specific structure, e.g., nonlinear affine models. For instance, an input–output TS model which does not contain the inputs in the antecedent part is an affine nonlinear form:

$$\begin{aligned} R_i: \quad & \textbf{If } y(k) \text{ is } A_{i1} \textbf{ and } \ldots \textbf{ and } y(k-n_y+1) \text{ is } A_{in_y} \textbf{ and} \\ & u(k-1) \text{ is } B_{i2} \textbf{ and } \ldots \textbf{ and } u(k-n_u+1) \text{ is } B_{in_u} \textbf{ then} \\ & y(k+1) = \sum_{j=1}^{n_y} a_{ij} y(k-j+1) + \sum_{j=1}^{n_u} b_{ij} u(k-j+1) + c_i, \\ & i = 1, 2, \ldots, K. \end{aligned} \tag{6.30}$$

Using the TS defuzzification formula (2.54), the output $y(k+1)$ can be written as:

$$\begin{aligned} y(k+1) &= \sum_{j=1}^{n_y} \left(\sum_{i=1}^{K} \lambda_i(y(k), \ldots, u(k-n_u+1)) a_{ij} \right) y(k-j+1) + \\ &\quad + \sum_{j=1}^{n_u} \left(\sum_{i=1}^{K} \lambda_i(y(k), \ldots, u(k-n_u+1)) b_{ij} \right) u(k-j+1) + \\ &\quad + \sum_{i=1}^{K} \lambda_i(y(k), \ldots, u(k-n_u+1)) c_i, \end{aligned} \tag{6.31}$$

The above expression is a special case of a nonlinear affine form:

$$\begin{aligned} y(k+1) &= g\left(y(k), \ldots, u(k-1), \ldots, u(k-n_u+1)\right) + \\ &+ h(y(k), \ldots, u(k-1), \ldots, u(k-n_u+1))u(k) \,. \end{aligned}$$

To the affine models, input–output linearization techniques can be applied to obtain a globally linearized process model (Nijmeijer and van der Schaft, 1990). The main idea of this technique is that, if the process is minimum phase, it is possible to linearize it by means of a nonlinear static feedback law. The linearization is exact, i.e., the resulting linear system is valid over the complete operating range. This is a main difference between this technique and the classical linearization through Taylor expansions, where the resulting linear model is valid only in the neighborhood of the nominal operating point. In the absence of constraints, a linear predictive control law can be applied to the linearized process model.[2] The resulting optimization problem is convex, and can be solved by using reliable quadratic programming methods.

Another way to develop a more efficient optimization scheme is to employ the semi-qualitative nature of fuzzy models and to simulate the model on the level of linguistic terms, which can be regarded as qualitative "landmarks" of the process behavior. Since the number of linguistic terms is usually low, it is feasible to consider the optimization problem as a search in a discrete space of the control alternatives. Standard techniques known from operational research and optimal control can be used, for instance, dynamic programming or the branch-and-bound method. In the following section, a method based on the latter technique is outlined.

Branch-and-bound Optimization. The branch-and-bound (B&B) method is a structured search technique that belongs to a general class of combinatorial programming methods (Lawler and Wood, 1966; Mitten, 1970). It is based on the fact that only a small number of the possible solutions need actually be enumerated, while the remaining solutions are eliminated through the application of bounds.

The B&B algorithm involves two basic operations: *branching*, i.e., dividing possible solutions into subsets, and *bounding*, i.e., eliminating those subsets that are known not to contain the solutions. The basic B&B scheme is a recursive application of these two operations. More sophisticated variants employ *search heuristics* to guide the search process and memory storage of the explored states and the corresponding cost functions (Chen and Bushnell, 1996).

Figure 6.9 illustrates the principle of the branch-and-bound method applied to MBPC. The fuzzy model is schematically denoted by

$$\mathbf{x}(k+1) = f\big(\mathbf{x}(k), u(k)\big), \tag{6.32}$$

where $\mathbf{x}(k)$ and $u(k)$ are the state and the input at time step k, respectively. The exact representation of the state is not important, it can be a defuzzified value, or a vector of

[2] In fact any linear control design can be applied. Also, note that the affine model can be exactly inverted, i.e., $u(k)$ can be expressed from eq. (6.31) by: $u(k) = [y(k+1) - g(\cdot)]/h(\cdot)$.

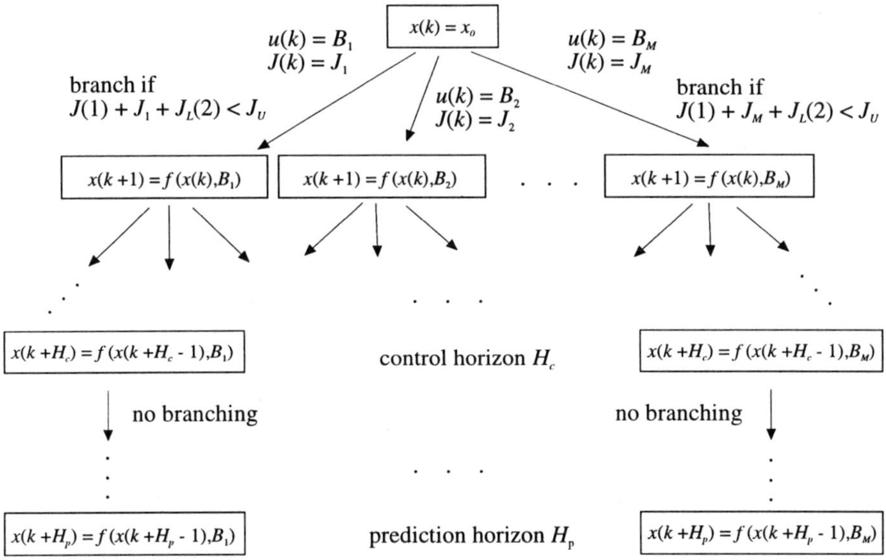

Figure 6.9. Branch-and-bound optimization in predictive control.

membership degrees associated with the linguistic terms defined for the state domain, see Section 2.1.7. The input $u(k)$ takes values from a set of linguistic terms B_j:

$$\mathcal{B} = \{B_1, B_2, \ldots, B_M\} . \tag{6.33}$$

At each time step (level of the tree shown in Figure 6.9), M control alternatives are considered, yielding a maximum of M branches. Let $i = 1, 2, \ldots, H_p$ denote the ith level of the tree ($i = 0$ at the root) and let j denote the branch corresponding to the control alternative B_j. Application of the branching alone would result in the search of the entire tree, i.e., at least M^{H_c} possibilities, which is computationally prohibitive, except for very small problems. A substantial part of the search space can be eliminated by employing lower and upper bounds on the cost function. A particular branch j at level i is followed only if the cumulative cost $J(i)$, given by

$$J(i) = \sum_{m=1}^{i} \alpha_m \left(r(k+m) - \hat{y}(k+m)\right)^2 + \sum_{m=1}^{i} \beta_m \Delta u(k+m-1)^2, \tag{6.34}$$

plus the cost from the level i to the terminal level H_p is lower than an upper bound on the total cost, denoted J_U. The cost from level i to level H_p is unknown, but can be expressed as a sum of two terms. The first one is the cost $J_j(i)$, associated with the transition $\mathbf{x}(k+1) = f\left(\mathbf{x}(k), B_j\right)$, which is computed by evaluating the respective element in the cost function (6.29). The second one is an estimated lower bound on the cost over the remaining steps $i+1, \ldots, H_p$, denoted $J_L(i+1)$. Hence the condition for branching reads:

$$J(i) + J_j(i) + J_L(i+1) < J_U . \tag{6.35}$$

Note that no branching takes place for $i \geq H_c$ (beyond the control horizon), i.e., the last control move $u(k + H_c - 1)$ is applied recursively to the model, until H_p is reached. The efficiency of the bounding mechanism depends on the quality of the bound estimates. The upper bound should be as low as possible and the lower bound as large as possible, in order to decrease the number of control alternatives in the branch-and-bound algorithm. The availability of these estimates depends on the particular problem. If no mechanism for computing the bounds is available, the following strategy can be applied.

The upper bound is initially set to a sufficiently large value. The first path in the tree search follows the "greedy" strategy of choosing the smallest $J_j(i)$ at each level i. This corresponds to the inverse model solution as described in Section 6.1, where the control actions $u(i)$ only take values $B_j, j = 1, 2, \ldots, M$. When following a constant or slowly varying references, the terminal cost $J(H_p)$, in most cases, represents the optimum, or a close upper limit of it. The upper bound is set to this value, i.e., $J_U = J(H_p)$. If at a later stage, $J(H_p)' < J_U$ is found, J_U is replaced by $J(H_p)'$. In the absence of a better estimate, the lower bound is simply set to zero, $J_L(i) = 0$ for all $i = 1, 2, \ldots, H_p - 1$.

Practical experience with this algorithm shows that these worst-case estimates still prevent the algorithm from exploring a large portion of the search space. However, one should not forget that the computational complexity of the algorithm remains exponential, which makes it prohibitively expensive for large H_c and M.

The B&B optimization technique applied to predictive control has three major advantages over other nonlinear optimization methods:

1. The global minimum is always found (intrinsic property of the B&B method). This guarantees optimal performance of the controller, in the discrete space of control alternatives. Some issues connected with this discretization are discussed below.

2. Contrary to iterative methods such as sequential quadratic programming, the maximum number of steps (time) needed to reach a solution is known a priori. This can be considered in the real-time solution of the controller design. By a suitable choice of the number of the control alternatives, a balance can be found between the minimum sampling time and the accuracy of the optimal solution. If required, the search can be aborted at any stage, in which case the current upper bound contains the best solution found so far. The algorithm also does not need any initial guess and hence its performance cannot be negatively influenced by a poor initialization, as in the case of iterative optimization methods.

3. The B&B method implicitly deals with constraints. In fact, the presence of constraints improves the efficiency of bounding, as it restricts the search space by eliminating the control alternatives that result in violation of the constraints. Many other optimization techniques result in worse performance when tight constraints are imposed. The B&B technique also deals with discrete control alternatives in a natural way. In many industrial systems, some of the control variables can only be set at discrete values, which presents problems for numerical techniques based on the computation of gradients.

Apart from the prohibitive computational complexity for large problems, the most serious drawback of the described approach is the restriction of the possible control actions to a set of discrete alternatives (linguistic terms, or other suitably chosen values). For a continuous control space, this discretization may cause oscillations of the outputs around the reference trajectory. To avoid this undesirable feature, a control scheme is given that combines the B&B predictive optimization strategy with the continuous control law by inverting the plant model, described previously. As has already been mentioned, in the absence of constraints, or when no constraints are violated, one-step-ahead prediction is equivalent to inverse model control, and also guarantees optimal performance. However, when constraints are active, the inverse model control results in suboptimal, or even infeasible, performance.

This observation led to the idea of combining both control strategies (Babuška, et al., 1995). When a recursive application of the inverse model control law over the entire prediction horizon results in a violation of a constraint (at any step), the B&B scheme is used, since predictive control will result in better performance. However, if no constraints are violated, the first control action computed by the inverse model is applied to the process. Algorithm 6.2 summarizes this control strategy.

Algorithm 6.2 (Combination of predictive and inverse-model control)

Step 1: **Apply inverse model.** Compute control actions over the prediction horizon using the inverse of the fuzzy model and the fuzzy model itself to predict the process output.

Step 2: **Check constraints.** If some of the constraints are violated at any of the prediction steps, go to Step 3 of this algorithm. Otherwise apply the control action $u(k)$ computed at Step 1 to the process.

Step 3: **Use optimization.** Use the branch-and-bound method to find an optimal solution. Apply $u(k)$ to the process.

This scheme avoids oscillations that would otherwise occur because of the discretization of the control space. With constant or slowly varying references, the constraints are typically not violated, and the inversion can be applied, yielding a continuous (interpolated) control action. For multi-input systems, the inversion cannot be applied, see Section 6.1.2. In such a case, the discrete solution obtained by the B&B method can be used as an initial guess for a numerical optimization method, like sequential quadratic programming, that should converge faster to a refined solution.

Choice of the Discrete Control Input Values. Since the computing time of the branch-and-bound algorithm increases drastically with an increasing number of control alternatives, this number should be as low as possible. However, too coarse a discretization may result in a rough control policy, inferior to those obtained with a finer discretization. Iterative refinement can be applied to reduce the computational load when a fine discretization is desirable (Luus, 1990). Initially, a coarse grid is chosen. A new set $\mathcal{B}'$ with a finer partition is then defined around the obtained control sequence, and a refined solution is computed. This procedure is repeated iteratively. If the control input change is constrained $\mid \Delta u(i) \mid < \Delta u_{MAX}$, discrete control values other than those corresponding to the linguistic terms can be applied, for instance $\mathcal{B} = \{u(i-1) - \Delta u_{MAX}, u(i-1), u(i-1) + \Delta u_{MAX}\}$.

Additional Remarks. The efficiency of the B&B scheme in terms of the required computing time can be improved by combining the B&B method with the concepts of dynamic programming. The explored states and the associated costs can be stored in memory, in order to eliminate the exploration of subsolutions that have already been evaluated. The price one has to pay are the increased memory requirements.

The above idea can also be extended to (on-line) learning of the mapping that corresponds to the input–output function of the optimizer. Clearly, for a given setting of MBPC parameters (horizons, weights in the cost function, etc.), the optimizer establishes a fixed static mapping from the product space of the system states $\mathbf{x}(k)$ and the future reference values $r(k+1), r(k+1), \ldots, r(k+H_p)$ to the space of optimal control actions $u(k)$, see Figure 6.10. This mapping is completely deterministic, time invariant, and, in principle, can be learned off line by simulations or on line, during process operation. This approach is conceptually similar to the idea of learning an optimal control policy on line, by means of reinforcement (Barto, et al., 1983; Anderson, 1987). A similar approach also has been proposed for neural networks (Parisini and Zoppoli, 1995).

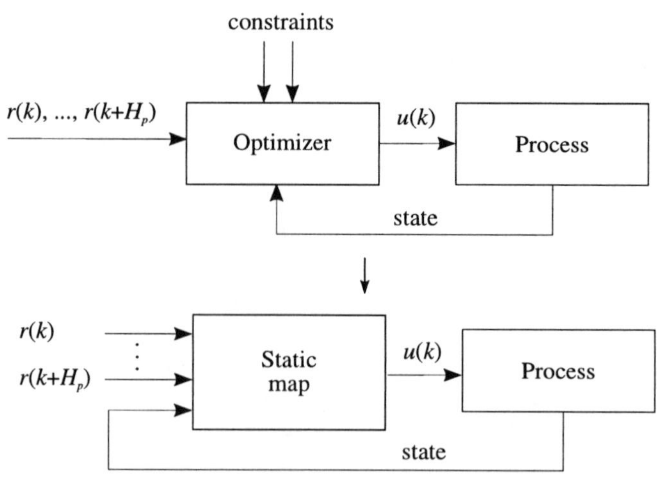

Figure 6.10. Transformation of a predictive controller into a static control law.

Recently, other optimization techniques have been investigated for use in predictive control. Genetic algorithms (GAs) are one of the techniques that can deal with relatively large search spaces in an efficient way. An experimental study by Onnen, et al. (1997) has shown that a GA outperforms the B&B technique only for longer control horizons. For short horizons, the time needed to perform evolutions leading to acceptable solutions is relatively long.

6.2.3 Predictive Control with Fuzzy Objective Function

The quadratic cost objective function (6.29) is only a mathematical approximation of the control objectives. While its quadratic nature is convenient for finding analytical solutions for linear models, it may be less suitable for achieving the real control goals, such as fast rise time, small overshoot, good damping, etc., especially for nonlinear systems. Figure 6.11 illustrates a difference in performance of a predictive controller for two different settings of the prediction horizon. The response shown in Figure 6.11a is smoother, and probably overall more acceptable than the one given in Figure 6.11b, although the latter attains a lower value for the objective function (6.29).

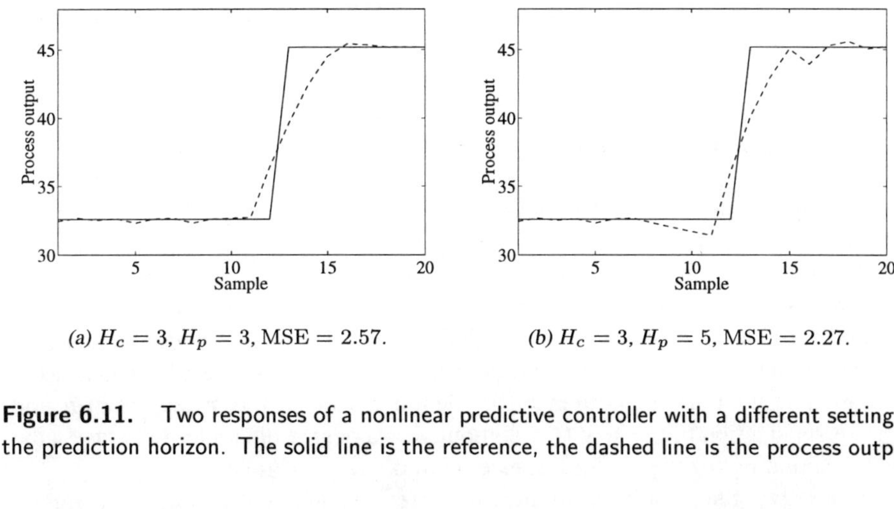

(a) $H_c = 3$, $H_p = 3$, MSE $= 2.57$. (b) $H_c = 3$, $H_p = 5$, MSE $= 2.27$.

Figure 6.11. Two responses of a nonlinear predictive controller with a different setting of the prediction horizon. The solid line is the reference, the dashed line is the process output.

Some authors (Soeterboek, 1992) provide tuning rules that relate the desired performance to the setting of the individual parameters in the cost function (6.29). In practice, however, extra constraints often must be imposed in order to meet the prescribed goals, and to prevent undesirable behavior, such as the nonminimum phase dip illustrated in Figure 6.11b. The quadratic cost (6.29) also minimizes the variance of the process output, which might not be always desirable. In many cases, it is sufficient to keep the controlled variables within certain limits, and more accurate control is not desired, as it may increase the control effort. For these reasons, it is useful to have a direct control over the influence of the individual components of the objective function on the controller performance. It is advantageous if the degree of compensation among the goals and constraints can be specified by the designer. This additional freedom can be achieved by choosing a different representation of the objective function, e.g., as a combination of fuzzy goals and constraints.

Fuzzy Goals and Constraints. Control can be regarded as an activity that aims at the simultaneous satisfaction of control goals and constraints. Rather than making use of mathematical functions, the goals and constraints can be specified as statements in natural language, using fuzzy propositions. An example of a fuzzy goal is "product concentration should be *about 80%*", where concentration is the controlled variable, and the term *about 80%* is represented by a fuzzy set, as shown, for instance, in Figure 6.12a.

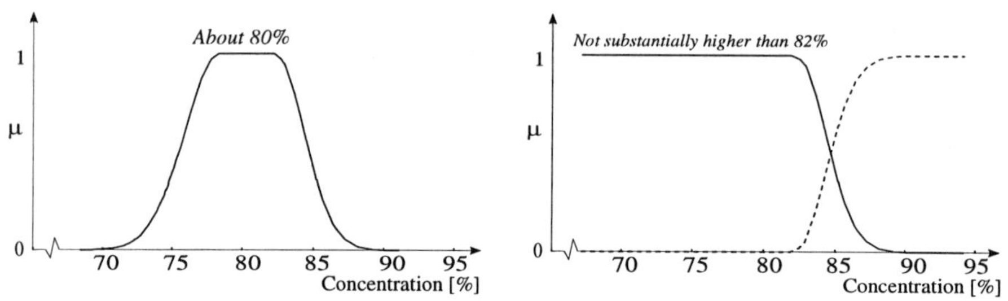

(a) Fuzzy goal. (b) Fuzzy constraint.

Figure 6.12. Example of a fuzzy goal and a fuzzy constraint.

For a crisp measurement x, the degree of satisfaction of a fuzzy goal G is determined by the membership degree of the measurement in the fuzzy set G, $\mu_G(x)$. In some cases, the measurements itself may be represented as a fuzzy set. For example, notions based on human perception can be expressed as rule-based combinations of some measured variables, e.g., comfort may be defined by fuzzy if–then rules that combine different qualitative values of temperature and humidity. Another example is a fuzzy representation of a noisy measurement. By denoting such a fuzzy measuremnt F, the degree of satisfaction of the goal G is computed as a degree of similarity between the two corresponding fuzzy sets, for instance, as $\max(\mu_F(x) \wedge \mu_G(x))$.

Simple fuzzy goals can be combined in more complex goals, using set-theoretic operators, or can be refined by adding other goals and constraints, e.g., "the product concentration should be *about 80%* and *not substantially higher than 82%*". Here the fuzzy constraint *not substantially higher than 82%*, shown in Figure 6.12b, can represent, for instance, a temporary restriction that can be added without removing or modifying the original goal. The logical connective *and* and the negation operator *not* are represented as intersection and complement of the fuzzy sets respectively, $G = G_1 \cap \bar{G}_2$, or in terms of membership degrees, $\mu_G(x) = \mu_{G_1}(x) \wedge (1 - \mu_{G_2}(x))$.

Predictive Control as Fuzzy Decision-making. The idea of decision-making in a fuzzy environment was introduced by Bellman and Zadeh (1970). Fuzzy decision D for n fuzzy goals $G_1, G_2, \ldots, G_n$ and m fuzzy constraints $C_1, C_2, \ldots, C_m$ can be seen as a confluence of these goals and constraints. If simultaneous satisfaction of the goals and constraints is required, the decision D is defined as the intersection of the corresponding fuzzy sets:

$$D = G_1 \cap \ldots \cap G_n \cap C_1 \cap \ldots \cap C_m, \tag{6.36}$$

or in terms of membership degrees

$$\mu_D(x) = \mu_{G_1}(x) \wedge \ldots \wedge \mu_{G_n}(x) \wedge \ldots \wedge \mu_{C_1}(x) \wedge \ldots \wedge \mu_{C_m}(x)\,. \tag{6.37}$$

The *maximizing decision* x_m is an $x \in X$ that maximizes $\mu_D(x)$, i.e.,

$$\mu_D(x_m) = \max_{x \in X} \mu_D(x)\,. \tag{6.38}$$

Other set-theoretic operators can be used to combine the goals and constraints to allow for a certain level of compensation. Optimizing the system's performance over a finite horizon, as in predictive control, corresponds to finding an optimal sequence of decisions in a *multi-stage decision-making process*. Assume that the system under control is described by a state transition equation (6.32). Given the current state $\mathbf{x}(k)$, a sequence of actions $u(k), \ldots, u(k + H_c - 1)$ is sought, which corresponds to the maximizing decision. A control scheme based on decision-making with fuzzy goals and constraints is depicted in Figure 6.13. Detailed treatment of this subject can be found in (Kacprzyk, 1983; Kacprzyk, 1997).

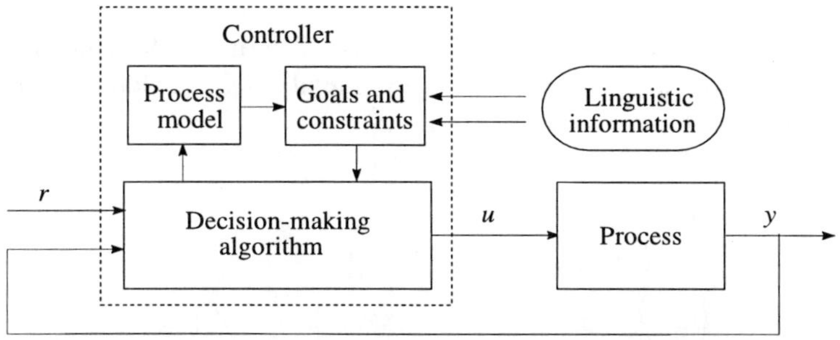

Figure 6.13. Controller based on fuzzy decision making.

The decision-making approach to predictive control was developed and implemented for the Automatic Train Operation (ATO) system of the Sendai city subway in Japan (Yasunobu and Miyamoto, 1985). The fuzzy control criteria include safety, riding comfort, energy consumption and stopping accuracy. Different control modes (with different rules) involve acceleration from zero to a desired speed, constant speed control, braking. The fuzzy predictive controller is in daily use, and is reported to be superior to previously used linear controllers. A recently published comparative study suggests that predictive controllers with fuzzy criteria can be more easily tuned and give better performance than controllers with the standard quadratic objective function (Kaymak, et al., 1997).

6.3 Example: Heat Transfer Process

The model inverse and the predictive control algorithms presented in this chapter were applied to a laboratory heat transfer system. The process consists of a 30 cm tube with a fan and a heating resistor at one end, and a temperature sensor at the other end. The output of the system is the voltage corresponding to the air temperature measurement $y(k) \in [0\text{V}, 10\text{V}]$ and the input is the voltage over a heating resistor $u(k) \in [0\text{V}, 12\text{V}]$. The air flow of the fan can be regulated manually by a valve, considered as an independent measurable variable $v(k) \in [0°, 180°]$. A more detailed description and a first-principle model of this process can be found in (Johansen, 1994).

6.3.1 Fuzzy Modeling

Using the identification method based on fuzzy clustering, the following three TS rules were extracted from process measurements (1300 samples, sample time 1.1 s):

1. **If** $y(k)$ is *Low* and $u(k)$ is *Low*
   then $y(k+1) = 0.9439y(k) + 0.0023u(k-3) + 0.2303$
2. **If** $y(k)$ is *Medium* and $u(k)$ is *Medium*
   then $y(k+1) = 0.9452y(k) + 0.0336u(k-3) + 0.0913$
3. **If** $y(k)$ is *High* and $u(k)$ is *High*
   then $y(k+1) = 0.9280y(k) + 0.0536u(k-3) - 0.0549$

This rule base represents a nonlinear first-order regression model

$$y(k+1) = f\big(y(k), u(k-3)\big)\,. \tag{6.39}$$

The order and the delay of the model have been determined by comparing different model structures. Figure 6.14 shows the membership functions for the output temperature and the input voltage.

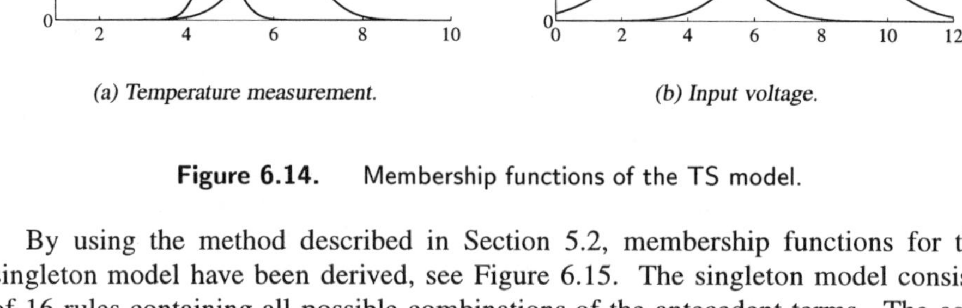

(a) Temperature measurement. (b) Input voltage.

Figure 6.14. Membership functions of the TS model.

By using the method described in Section 5.2, membership functions for the singleton model have been derived, see Figure 6.15. The singleton model consists of 16 rules containing all possible combinations of the antecedent terms. The consequent parameters were estimated using the least-squares method:

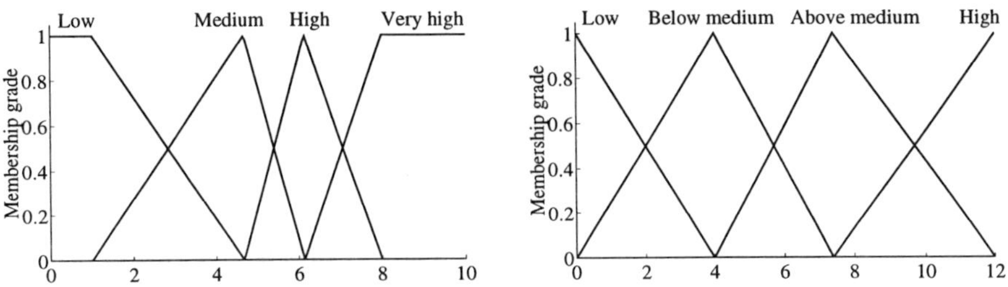

(a) Temperature measurement. (b) Input voltage.

Figure 6.15. Membership functions of the singleton model.

Voltage	Temperature			
	Low	*Medium*	*High*	*Very high*
Low	1.06	3.35	4.76	7.71
Below medium	1.04	3.38	4.76	7.76
Above medium	1.12	3.52	4.87	7.82
High	1.53	3.75	5.10	8.05

The model has been validated by simulation from the input, i.e., $\hat{y}(k+1) = f\big(u(k-3), \hat{y}(k)\big)$. When comparing the fuzzy model performance shown in Figure 6.16a with a linear ARX model shown in Figure 6.16b, one can see that the fuzzy model is much more accurate.

6.3.2 Inverse Model Control

The control algorithm based on the inverse of the singleton fuzzy model, model-based predictive control and the adaptive fuzzy control scheme described earlier in this chapter were tested in simulations of the heat transfer system. The process was simulated using a semi-mechanistic model (Johansen, 1994), with rate limit in the control signal and additive noise at the output.

Figure 6.17 shows the simulation results with the controller based on an inverse model. The open-loop feedback control scheme was applied, i.e., the model's state was updated by the process output at each sample. The reference signal contains several steps of different amplitudes in different operating regions, in order to verify the controller's capability to cope with the process nonlinearity. The controller performance is good, even though the noise at the process output is amplified and causes relatively large variance of the control signal. The influence of noise can be eliminated by the use of a pure open-loop arrangement of the model and controller within an IMC scheme. The purpose of this example is, however, to demonstrate the effect of constraints and disturbance when no feedback is present. The IMC scheme is applied in Section 6.4.

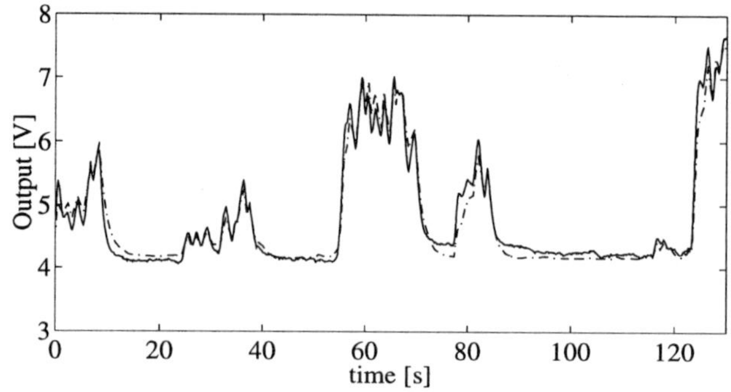

(a) Singleton fuzzy model.

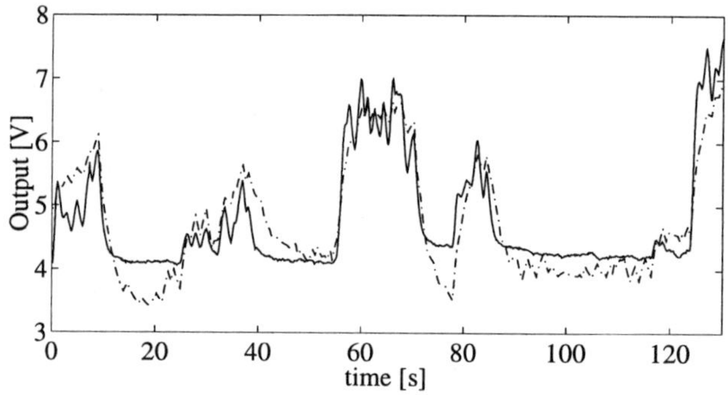

(b) Linear ARX model.

Figure 6.16. Model validation (solid line: process, dashed-dotted line: model).

To demonstrate the effect of constraints, the change of the control action $\Delta u(k)$ was limited to $\Delta u_{MAX} = \pm 0.5\ V$ per sampling period. From the results shown in Figure 6.18, one can see that the inverse model-based controller cannot cope with this rate constraint. The controller generates larger changes in the control action than allowed, and the rate of the control signal must be restricted by a rate limiter which results in overshoots, see Figure 6.18.

6.3.3 *Predictive Control*

Figure 6.19 shows the results obtained with a combined scheme based on predictive control. The prediction and the control horizons of 11 steps (1.21 s) were used. Note that by using the combined control scheme the overshoots are eliminated. Also the summed squared error at the stepwise set-point changes is decreased due to the predictive way of control by 65%. The control input is discretized in three levels, $u(k) \in \{u(k-1) - \Delta u_{MAX}, u(k-1), u(k-1) + \Delta u_{MAX}\}$.

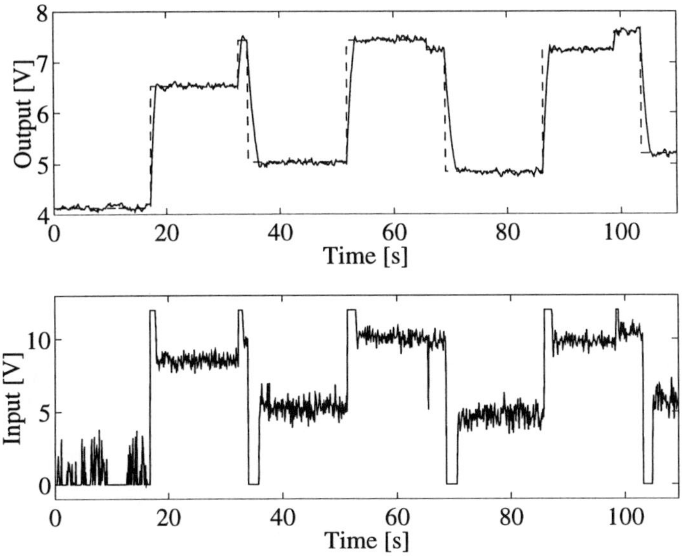

Figure 6.17. Without the rate constraint on the control action, the inverse model-based controller performs well.

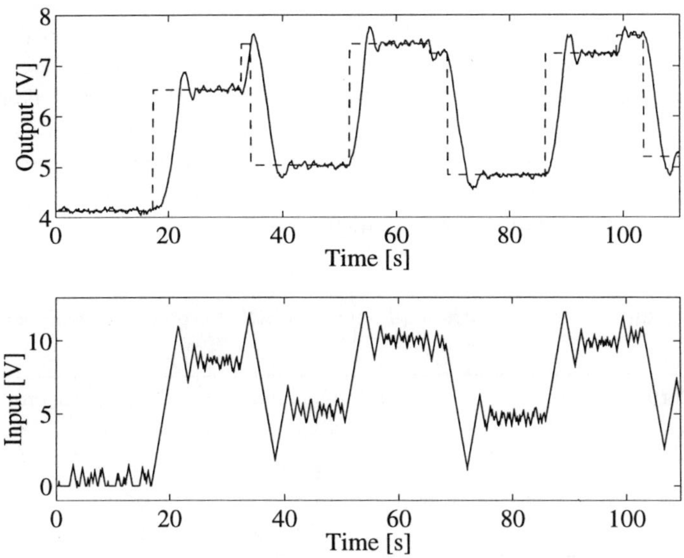

Figure 6.18. In the presence of the rate constraint $|\Delta u(k)| < 0.5\ V/\Delta t$, the performance of the inverse model-based controller decreases and overshoots occur.

The B&B optimization scheme was compared to iterative numerical optimization method known as sequential quadratic programming (SQP). Table 6.1 gives the sum

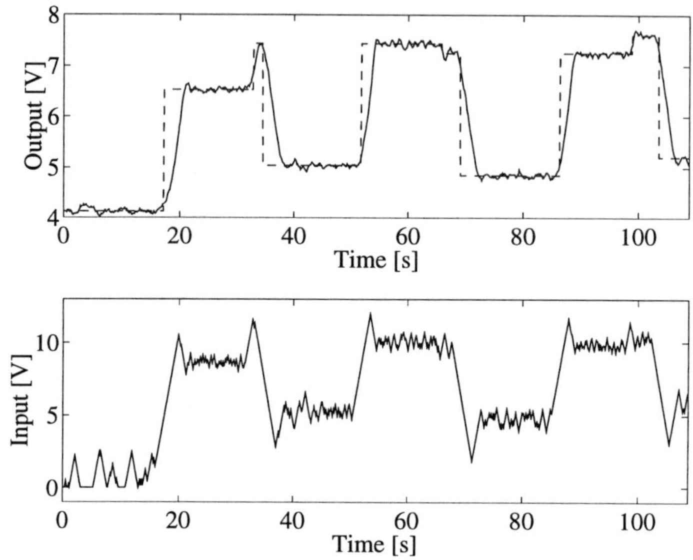

Figure 6.19. In the combined scheme based on predictive control, the rate constraint on the control signal can be easily taken into account and the performance improves.

of squared errors (SSE) as a performance measure, and the of number floating-point operations (FLOPS) as a measure of the computational costs of the two algorithms. The computational requirements of the B&B method at $H_c = 4$ are taken as 100%. The comparison is made for control horizons from 4 to 6 steps. One can see that for the SQP optimization method, the error actually increases with increasing control horizon, due to the fact that local optima are found. Also, the computational costs of SQP are higher than those of the branch-and-bound method.

Table 6.1. Comparison of branch-and-bound and SQP for different prediction horizons.

Control horizon	Branch-and-bound		Seq. Quad. Prog.	
	SSE	FLOPS	SSE	FLOPS
$H_c = 4$	120.57	100%	120.57	100%
$H_c = 5$	92.79	200%	105.97	899%
$H_c = 6$	72.62	359%	111.03	1462%

On the basis of the presented comparison, it can be concluded that the B&B optimization method is superior to SQP with respect to the performance achieved, and also with regard to the computational costs. This conclusion was confirmed by experiments with other systems as well.

6.3.4 Adaptive Predictive Control

The performance of the adaptation mechanism was evaluated by introducing an external disturbance. The outside temperature was decreased by 0.5 $°C$ at time t = 44 s. Figure 6.20 shows the performance of both the fixed controller (dashed line) and the adaptive controller (solid line) for tracking a slowly varying reference. After a short adaptation period of between 44 s and 50 s, the adaptive controller follows again the reference (dashed-dotted line), while the fixed controller exhibits a constant offset (dashed line). Clearly, the negative effects of disturbances can be decreased also by other means, such as the IMC control scheme, shown in Figure 6.8, or an outer-loop controller of an integral type. These solutions are, however, based on a linear feedback which may deteriorate the controller performance due to the process nonlinearity. Moreover, adaptation of the plant model can also cope with changes of the plant parameters.

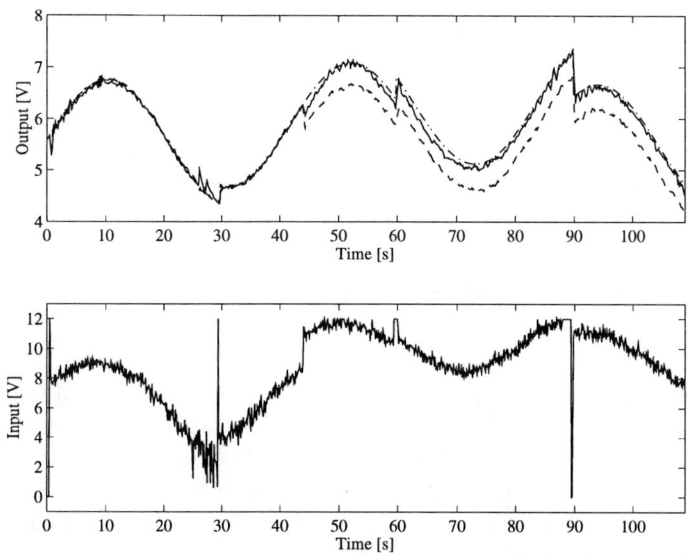

Figure 6.20. Performance of the adaptive controller when a disturbance occurs at t=44 s. The solid line is the adaptive controller, the dashed line is the fixed controller, and the dashed-dotted line is the reference.

Figure 6.21 shows the adaptation of the consequent parameters for the experiment given in Figure 6.20. It illustrates the local nature of the fuzzy model, since some of the rule consequents have been adapted short after the disturbance occurred at time 44 s, and others later, as the system dynamics evolving through the input–state space activated the corresponding rules. Some adaptation also took place before 44 s, because of an initial mismatch between the simulated process and the model identified from the real process data.

The forgetting factor was set to $\lambda = 0.95$, and the covariance matrix was initialized at $\mathbf{P} = 100 \cdot \mathbf{I}$. The covariance matrix was automatically reset each 100 samples, in order

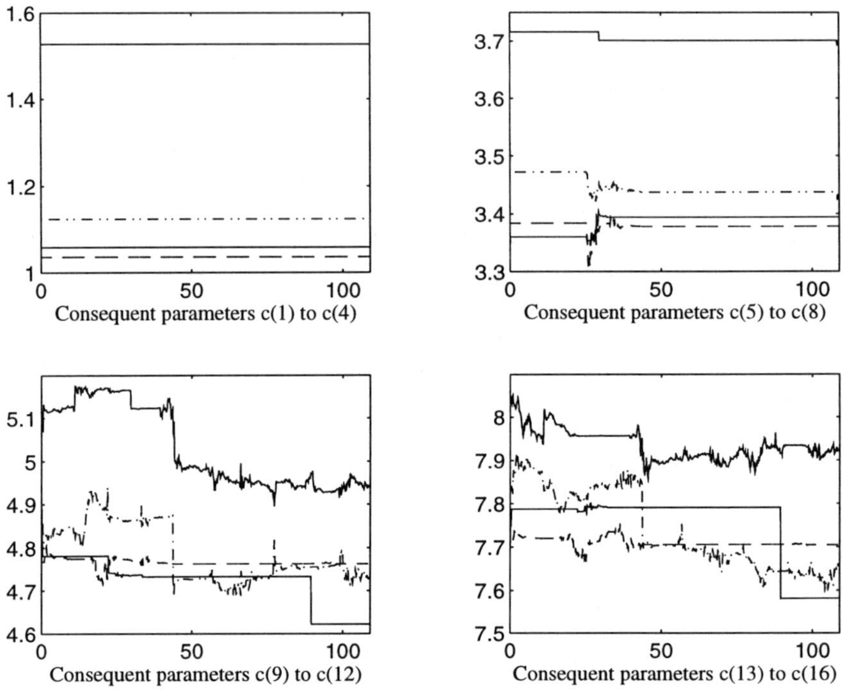

Figure 6.21. Adaptation of the consequent parameters $\mathbf{c}(k)$ for the run shown in Figure 6.20.

to guarantee permanent adaptation of the fuzzy model. The presented scheme allows for *local* adaptation (learning) of the controller, as opposed to *parameter tracking* used in linear adaptive control. A drawback of linear methods is that a balance between the tracking speed and insensitivity to noise is difficult to achieve and the linear controller has no memory, i.e., for a nonlinear system it must continuously re-adapt the parameters as the process state changes.

6.4 Example: pH Control

In this example, the predictive control scheme described in Section 6.2 is compared with an input–output feedback linearization (IOFL) approach (Botto, et al., 1998; Babuška, et al., 1996), using the pH system from Section 5.1.5. The IOFL algorithm uses an affine neural network model with one hidden layer consisting of 3 neurons with sigmoidal activation functions. The purpose is to provide a comparison of two different nonlinear predictive control techniques in terms of modeling, controller performance and computing costs. Both controllers are implemented within the IMC scheme.

The fuzzy predictive controller uses the TS model from Section 5.1.5. The optimization problem is again solved by the branch-and-bound technique. The range of admissible control actions ($\Delta Q \in [-5, 5]$) is discretized in five control actions, and $H_c = H_p$. The cost function does not include any penalty on the control effort.

Table 6.2. Comparison of performance for different control horizons.

H_c	Method	MSE	FLOPS	Control effort
1	B&B	10.38×10^{-1}	9.08×10^{6}	111.16
1	IOFL	10.21×10^{-1}	2.82×10^{5}	89.17
2	B&B	5.54×10^{-1}	1.69×10^{7}	84.37
2	IOFL	5.56×10^{-1}	4.25×10^{5}	82.72
3	B&B	2.43×10^{-1}	2.39×10^{7}	82.86
3	IOFL	2.41×10^{-1}	5.89×10^{5}	85.47
4	B&B	2.57×10^{-1}	3.05×10^{7}	82.06
4	IOFL	3.14×10^{-1}	1.01×10^{6}	95.05
5	B&B	2.69×10^{-1}	3.90×10^{7}	88.04
5	IOFL	2.84×10^{-1}	1.51×10^{6}	90.35

Simulation results achieved with the two control schemes for different control horizons, H_c, are summarized in Table 6.2. The comparison is done in terms of the Mean Squared Error (MSE) over the entire simulation run, the number of floating point operations (FLOPS), and the mean squared control effort, computed as the sum of $\Delta Q(k)^2$ multiplied by the sample time and divided by the number of samples. One can see that in both control schemes the MSE first decreases with increasing H_c until a minimum is reached, and then it starts to increase again, due to a greater prediction error for longer horizons. The computational costs of the branch-and-bound method increase considerably with the control horizon. Figures 6.22 and 6.23 show the control response for the control horizon $H_c = 3$ which gives the best performance for both schemes.

With respect to the MSE performance measure, the two techniques presented give comparable results. Small differences in the MSE values are not significant, and they are often compensated by opposite differences in the control effort. Comparison of the responses reveals that the IOFL scheme results in slightly larger overshoots and steady-state errors, which need to be eliminated by a suitably designed feedback filter. The IOFL method also proves to be very robust with respect to the model-plant mismatch, and has lower computational demands. This makes the technique especially suitable for fast real-time implementations. The branch-and-bound method, however, is more general, as it allows the formulation of the cost function in an arbitrary way (i.e., not only as a quadratic form), in order to meet the various control requirements, such as minimization of absolute error, overshoots, etc.

6.5 Summary and Concluding Remarks

Methods to develop nonlinear controllers that are based on an available fuzzy model of the process under consideration have been presented. The inversion method, described in Section 6.1.2, guarantees exact inversion of singleton and affine TS models.

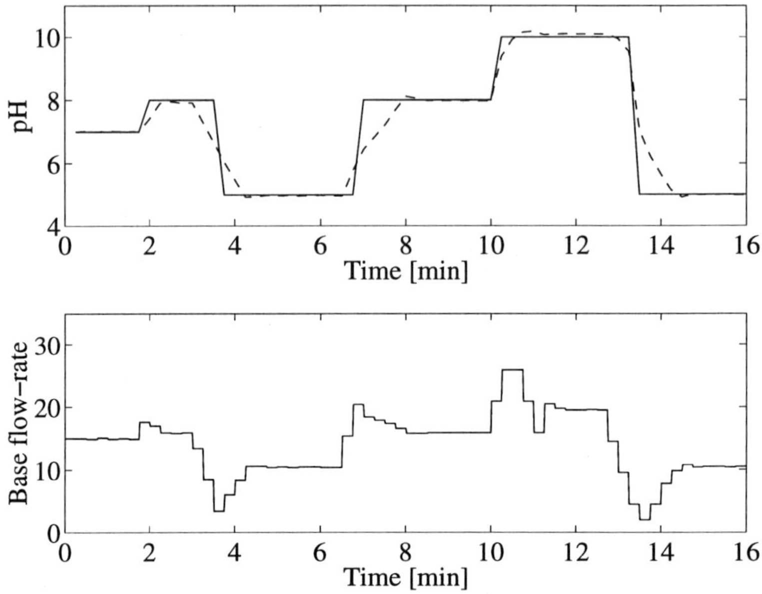

Figure 6.22. Nonlinear predictive controller based on the fuzzy model with $H_c = 3$. The solid line is the reference and the dashed line is the process output.

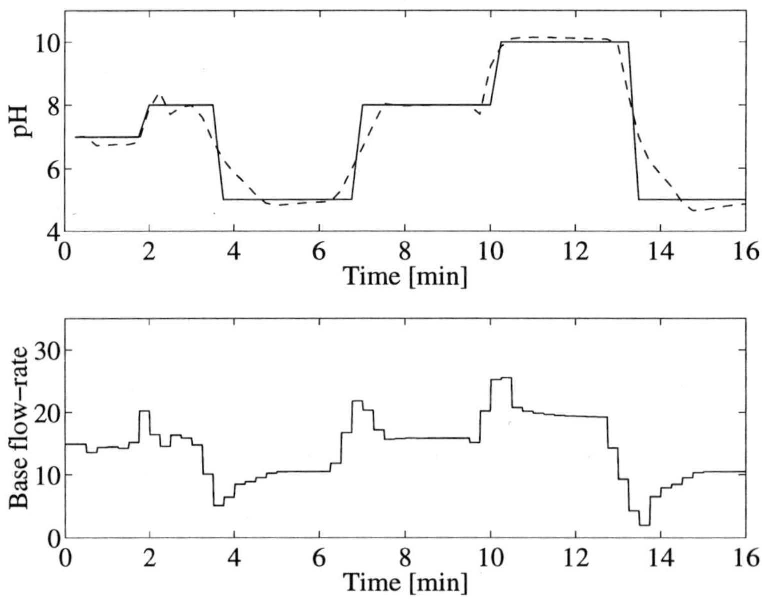

Figure 6.23. Nonlinear predictive controller based on input-output linearization of an affine neural network model with $H_c = 3$. The solid line is the reference and the dashed line is the process output.

The inverse model can be computed by simple analytic manipulations, which is an advantage for real-time implementations. For open-loop stable processes, or processes stabilizable by feedback, whose inverse is stable, a feedforward control scheme is obtained. Such a scheme is inherently stable, and "perfect" control with zero error, both in dynamic transients and in steady states, is achieved. In practice, however, modeling errors and disturbances acting on the process result in a mismatch between the process and model behaviors. This mismatch causes control errors, which must be eliminated by feedback, using, for instance, the IMC control scheme described in Section 6.1.3. The inverse-model approach, however, cannot deal with multiple-input systems in a straightforward way, and also cannot take constraints explicitly into account. In such cases, predictive control provides an effective solution. In Section 6.2, a predictive control scheme based on a fuzzy model and branch-and-bound optimization has been presented.

In contrast to conventional fuzzy control, the controllers developed by using fuzzy models do not require any ad hoc tuning of the rules and membership functions. Some experimentation and iterative tuning may, however, be needed in the modeling and identification phase. Once a fuzzy model of the process is available, it can be simply plugged into the control scheme. The setting of the control and prediction horizons is based on standard rules of thumb. A simple adaptation algorithm based on recursive least-squares estimation can be employed to deal with time-varying systems. Simulation examples have been presented to demonstrate the control performance and the computational aspects of the described algorithms. A comparison with predictive control based on input–output linearization shows that the branch-and-bound technique gives a slightly better performance, but it requires more computations.

7 APPLICATIONS

This chapter describes three selected real-world applications of the fuzzy modeling, identification and control methods presented in this book. Section 7.1 describes the development of a knowledge-based linguistic fuzzy model for predicting the performance and tool wear of a rock-excavation machine. The model has been developed purely on basis of expert knowledge, but it allows for tuning and adjustment by using numerical data. Different facets of the approach, such as translation of the expert knowledge into the linguistic rules and membership functions, the hierarchical organization of knowledge, and validation of the model are discussed. Section 7.2 deals with nonlinear identification based on fuzzy clustering. Takagi-Sugeno, singleton and relational fuzzy models have been developed to model highly nonlinear pressure dynamics. Experimental results of real-time predictive control based on the fuzzy model are presented. The last application, given in Section 7.3, demonstrates a modeling approach based on the combination of a fuzzy model with a first-principles model. A linguistic fuzzy model which represents the kinetic term in enzymatic Penicillin–G conversion is developed from numerical data. This fuzzy model is then incorporated in macroscopic balance equations describing the overall conversion process. It is shown that this approach gives good predictions, and at the same time allows for qualitative interpretation of the unknown relationships learnt from data.

7.1 Performance Prediction of a Rock-cutting Trencher

A linguistic fuzzy model for performance prediction of a rock-cutting trencher is described.[1] The rule bases and the membership functions were constructed entirely from experts' knowledge. Hierarchical organization of knowledge into several rule bases helps to keep the complexity of the model at a reasonable level. In order to improve the model's prediction accuracy, the consequent membership functions can be adjusted with measured data, using least-squares estimation. The model has been validated on a limited number of available field measurements.

7.1.1 The Trencher and Its Performance

A trencher is a rock-excavation machine, which uses a rotating chain equipped with bits to excavate large slots (trenches) in rock and soil for underground cables and pipelines. Figure 7.1a shows a photograph of the T-850 trencher (Vermeer, USA), considered in this study. The trencher's chain is covered by baseplates on which steel bits are located in V-shaped patterns as shown in Figure 7.1b. The chain is pushed into the ground, and as it rotates, the bits cut through the rock.

In order to estimate the expected costs of a given excavation project, the performance of the trencher has to be predicted with reasonable accuracy. This performance, and consequently the costs of a specific excavation project, are determined by the trencher's production (excavation) rate and by the bit consumption rate (due to the bit wear and breakage). Both these factors depend on the properties of the excavated rock material, and on the trencher characteristics. Derivation of rigorous mathematical models based on the underlying physical mechanism is difficult and inefficient, because the interactions between the machine tool and the environment are complex and uncertain. Moreover, the limited amount of available data also does not allow to use conventional statistical methods to establish equations to predict the trencher's performance. However, available expert knowledge and experience, acquired during field and laboratory experiments (Deketh, 1995), enable to use a knowledge-based approach to developing a prediction model. Fuzzy logic provides a means to smoothly interface the qualitative expert knowledge contained in the rule base to numerical input data. Moreover, it is sufficient to express the predictions in the form of qualitative linguistic values, rather than by precise numbers whose accuracy may be artificial and deceiving, given the uncertainty of the problem.

7.1.2 Knowledge-based Fuzzy Model

This section describes the selection of the input variables and illustrates how the domain-specific expert knowledge is converted into the linguistic if–then rules and membership functions.

[1] This section is a shortened version of the article (den Hartog, et al., 1997).

(a) The T-850 trencher.

(b) Chain equipped with bits.

Figure 7.1. The T-850 trencher and a detail of the cutting chain.

Selection of Model Inputs. Out of many different factors that may possibly affect the trencher's performance, the experts selected the following six variables as inputs to the model:

- spacing of the fractures in the rock, called *joint spacing*, (three variables, which determine the size and shape of blocks which can be loosened from the rock mass),
- dimensions of the hard rock body in the trench,
- strength of the rock material,

- orientation of the joints (fractures).

Other rock material parameters such as the ductility, the volume percentage of abrasive minerals, or the grain size and shape are not included, because the experts consider them less important, and the aim is to keep the model at a reasonable complexity.

The outputs of the model are the prediction of the trencher's production and bit consumption rates in terms of qualitative linguistic values and also defuzzified numerical values. A separate prediction model is established for each of these variables.

The number of selected inputs to the model (6) is quite large, if all of the variables are considered as inputs to one rule base. Since this would lead to a large number of complicated rules, the model has been decomposed into several smaller submodels with a manageable number of variables. This decomposition follows naturally from the structure of the expert knowledge, as shown below. The result is a two-level hierarchical knowledge base which is described in the following sections.

Rule Bases for Bit Consumption Rate. The bit consumption rate is the number of broken bits per m^3 of rock material excavated. Breakage occurs when the impact forces exerted on the bits exceed the strength of the bit material. The main factors for the assessment of the impact forces are: the size (mass) of the rock blocks and the strength of these blocks. When the blocks are small, they do not break the bits regardless of the rock strength. An example of a rule which represents this situation is:

If Block Size is Small **then** Bit Consumption is Very low.

The larger the blocks are, the more the strength of the material affects the bit consumption rate, which leads to rules like:

If Block Size is Medium **and** Strength is Medium
then Bit Consumption is Medium.

When excavating a massive rock (block diameter larger than about 60 cm), the rock blocks cannot be loosened from the rock mass and the bits cut the rock. In such a case, the bit consumption rate depends mainly on the severity of the chain vibrations, which in turn depend on the strength of the rock material. Severe vibrations of the chain result in impact motion of the bits and can cause breakage. An example of a rule corresponding to this situation is:

If Block Size is Large **and** Strength is High
then Bit Consumption is High.

The block size is not directly measured, but it is determined from the spacing of three orthogonal joint sets, J_1, J_2 and J_3 using a separate rule base ('Rule base 1'), as shown in Figure 7.2.

The spacing of the joint sets is divided into five linguistic terms: Very small (VS), Small (SM), Medium (ME), Large (LA) and Very large (VL). The boundaries of the membership functions, shown in Figure 7.3a, are determined by the experts, based on the observations of the different T-850 trencher projects. It is expected that when more data become available, methods for automated extraction or adjustment of the membership functions can be applied.

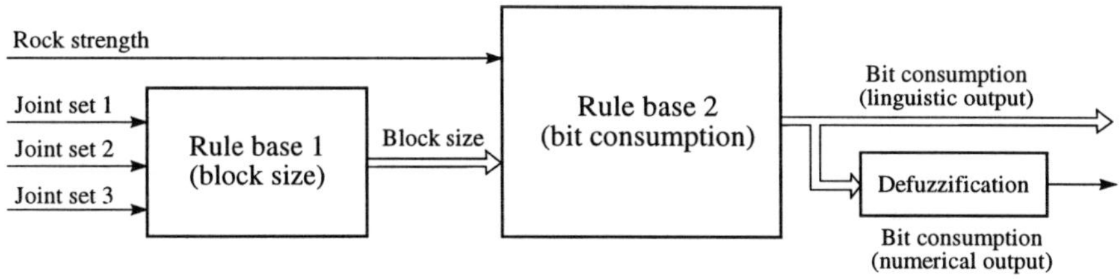

Figure 7.2. Rule bases for predicting the bit consumption. Single lines represent numerical variables, while double lines represent fuzzy variables.

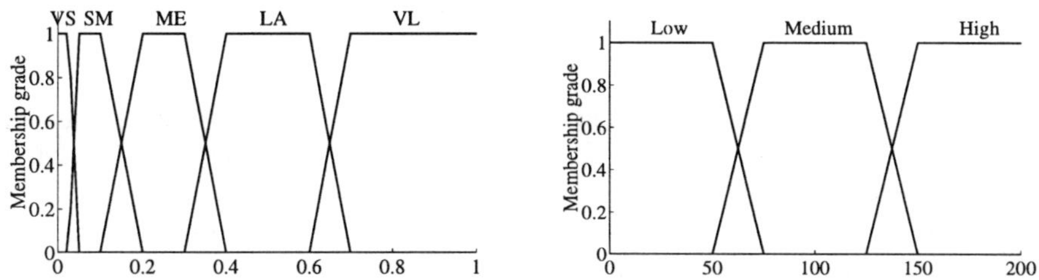

(a) Joint spacing. (b) Strength.

Figure 7.3. Membership functions for joint spacing and strength.

Eleven linguistic terms are used for the block size. No membership functions are associated with these terms, since the block size is an internal variable, directly connected to the bit consumption rule base as depicted Figure 7.2. The number of linguistic classes is rather large, because they describe not only the size of the blocks but also their shape. Since this information is also used for determining the production rate, 'Rule base 1' is common for both prediction models.

The rock strength is a direct numerical input to 'Rule base 2'. Three linguistic terms, Low, Medium and High are defined for this variable, as shown in Figure 7.3b. The predicted bit consumption is given in linguistic terms and the corresponding membership degrees. The following five terms are used: Very low (VL), Low (LO), Medium (ME), High (HI) and Very high (VH). Membership functions for these terms are shown in Figure 7.4. Fluctuations of the bit consumption rate for different excavation projects are larger for projects with high bit consumption rates. This is reflected in the width of the corresponding consequent membership functions.

The number of rules defined in 'Rule base 1' (block size) is 25 and in 'Rule base 2' (bit consumption) is 21. To obtain a numerical prediction, the output fuzzy set is defuzzified, using the weighted fuzzy-mean method (2.35).

Rule Bases for Production Rate. The production rate of the trencher mainly depends on the mode of excavation, which is ripping (loosening and fracturing of the rock), or cutting. Small rock blocks which can be loosened from the rock mass can

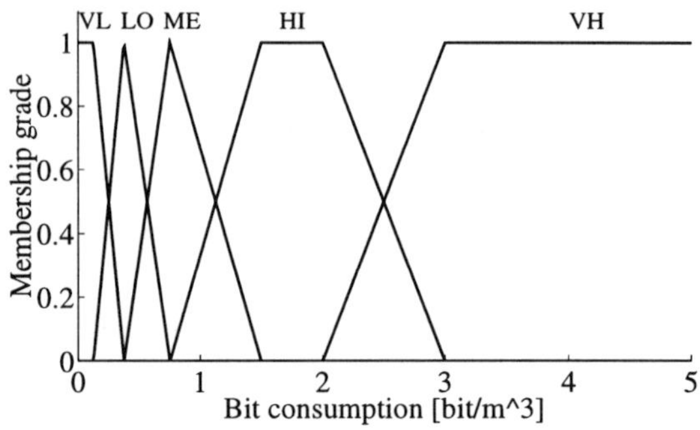

Figure 7.4. Membership functions for bit consumption.

easily be transported out of the trench, which results in a high production rate. The cutting efficiency of the bits increases with the feed (the displacement of the bits in the rock material per chain revolution).

For massive rock where the rock blocks cannot be loosened from the rock mass (rock block diameter larger than about 60 cm), both the strength of the rock material and the dimensions of the rock body significantly affect the feed. An increase in the strength results in a decrease of the feed. The dimensions of the rock body (i.e., the width of the trench and the thickness of the layer to be excavated) affect the feed, and hence also the production rate of the trencher. For rock blocks which can be loosened from the rock mass the influence of the dimensions of the rock body on the feed is of minor importance, compared to the influence of the strength of the material and the size of the rock blocks. As an example, consider the excavation of a massive rock of medium strength where the dimensions of the rock body are large; this involves a cutting process at a low feed level. The resulting production is low:

If Rock Dimension is Large **and** Strength is Medium **then** Feed is Low.
If Block Size is Large **and** Feed is Low **then** Production is Low.

As one can see from this example, the reasoning proceeds in two stages, first the feed is determined from the strength and the rock dimensions and then, based on the feed and the block size, the production rate is predicted. This leads to a two-layered knowledge base structure shown in Figure 7.5.

The membership functions for the rock strength are the same as in the bit-consumption rule base (see Figure 7.3b). The rock dimension variable also uses three linguistic terms with equally sized membership functions. The output of 'Rule base 3', which is the feed, is given in five linguistic terms. Like the block size, also the feed is an internal linguistic variable, and thus no membership functions need to be specified. The last input variable, used directly in 'Rule base 4', is the orientation of the joints determined by the orientation of the joint set with the smallest spacing (J_1). It is specified in three qualitative terms: Favorable, Unfavorable and Very unfavorable.

The output of 'Rule base 4' is the prediction of the production rate which is given in six linguistic terms. The width of the consequent membership functions increases with the production rate to reflect the decreasing accuracy of the predictions. The number

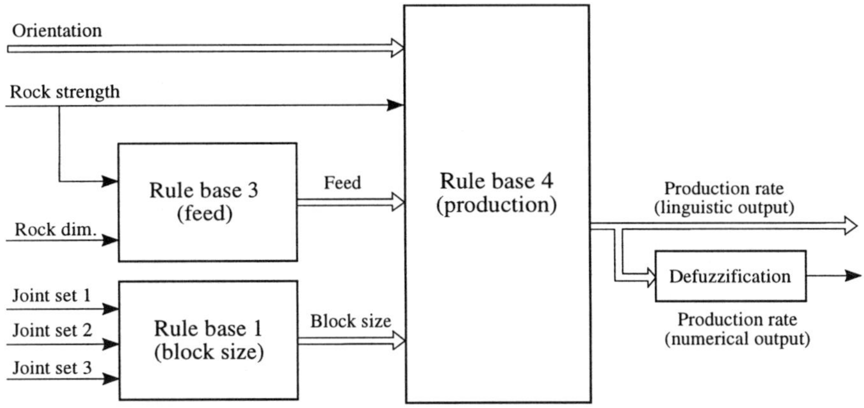

Figure 7.5. Rule bases for predicting production rate. Single lines represent numerical variables, while double lines represent fuzzy variables.

of rules defined in 'Rule base 3' is nine, and in 'Rule base 4' is 41. 'Rule base 1' contains 25 rules and is the same as for the bit consumption model.

7.1.3 Applied Methods and Algorithms

This section describes the applied algorithms for fuzzy inference, rule chaining, defuzzification, and adaptation of the consequents.

The Structure of the Fuzzy Model. Each rule base is a collection of linguistic if–then rules with multiple antecedents and a single consequent:

$$R_i: \ \textbf{If } x_1 \text{ is } A_{i1} \textbf{ and }, \ldots, \textbf{ and } x_p \text{ is } A_{ip} \textbf{ then } y \text{ is } B_i, \quad i = 1, \ldots, K, \tag{7.1}$$

where $x_1, x_2, \ldots, x_p$ are the input (antecedent) variables, p is the number of inputs and y is the output (consequent) variable. A_{ij} are instances of the *reference linguistic terms*, defined for a given antecedent variable x_j: $\mathcal{A}_j = \{A_{jk} | k = 1, 2, \ldots, N_j\}$. Similarly, $\mathcal{B} = \{B_k | k = 1, 2, \ldots, M\}$ is the set of reference terms defined for the output variable y. Two types of variables are distinguished in the considered fuzzy model; they depend on the universe of discourse (domain) in which they are defined.

Crisp (Numerical) Variables. The domain of these variables is a subset of the real line, $X \subset \mathbb{R}$. The linguistic terms A_{jk} for these variables are defined by trapezoidal membership functions (A.4) whose parameters are specified by the expert. Numerical variables include measurements of the rock properties, such as the strength, the joint spacing, the rock body dimensions in the trench, etc.

Fuzzy Variables. These variables are defined as fuzzy sets on the set of reference linguistic terms. They are represented as vectors in $[0, 1]^{N_j}$, where N_j denotes the number of linguistic terms defined for the given variable. Fuzzy variables need not be associated with any numerical universe. This can be because their values are qualitative in nature, or are considered as qualitative for convenience. An example of a variable that is an input to the fuzzy model in linguistic terms is the orientation of joints with the linguistic term set:

$$\text{orientation} \in \{\text{Favorable, Unfavorable, Very unfavorable}\}\,.$$

The values of orientation are fuzzy singletons $[1,0,0]$ for Favorable, $[0,1,0]$ for Unfavorable and $[0,0,1]$ for Very unfavorable. Generally, a fuzzy variable x_j can attain any value $[\mu_1, \mu_2, \ldots, \mu_{N_j}]$, $\mu_k \in [0,1]$, expressing various shades among the reference terms. In the above example, fuzzy set $[1, 0.4, 0]$ may, for instance, represent More or less favorable orientation. Other examples of fuzzy variables are the internal variables passed among the rule bases, such as the feed or the block size.

Inference Mechanism. The inference mechanism is based on the max-min inference (Algorithm 2.1 in Section 2.1.4). The reason for choosing the max-min inference is its simple implementation in the hierarchy of the rule bases. The use of logical inference would require the use of the relational calculus (see Section 2.1.4), and hence a discretization of the domains. Membership functions would have to be defined for the intermediate variables, which would require additional information from the expert, which he or she might not be able to supply.

The hierarchical structure of the rule bases requires that information is passed from the output of one rule base to the input of another rule base. This is accomplished by feeding the *fuzzy set* at the output of the first rule base directly (without defuzzification) to the second rule base (see Section 2.1.7). As an example assume that the inference in 'Rule base 3' results in the feed given by the following fuzzy set:

$$\text{feed} = \{0/\text{Very low},\ 0.7/\text{Low},\ 0.1/\text{Medium},\ 0/\text{High},\ 0/\text{Very high}\}\,.$$

The membership degree of the propositions "If feed is Low" in 'Rule base 4' is thus 0.7, the membership degree of the propositions "If feed is Medium" is 0.1, and the propositions with the remaining linguistic terms such as "If feed is High" have the membership degree equal to zero. This approach does not require any additional information from the user, since no membership functions are needed for the intermediate variable. Also the computational load is lower than when using defuzzification and subsequent *fuzzification* at the input of the second rule base.

Adaptation of Consequent Membership Functions. To obtain numerical values at the output of the fuzzy model, a defuzzification method must be applied. The consequent fuzzy sets B_i are first individually defuzzified by the mean-of-maxima method, yielding the crisp representative values $b_i = \text{mom}(B_i)$. The numerical y_0 output of the fuzzy model is then calculated as the weighted fuzzy mean of the b_i's (2.35).

The consequent membership functions can be adjusted to improve the numerical accuracy of the fuzzy model. Off-line or recursive least-squares methods are first applied to the b_i parameters. Then, the consequent membership functions B_i are modified such that the partition is preserved. Figure 7.6 gives an example of initial and adapted membership functions of the bit consumption rate, using the off-line least-squares method (5.58).

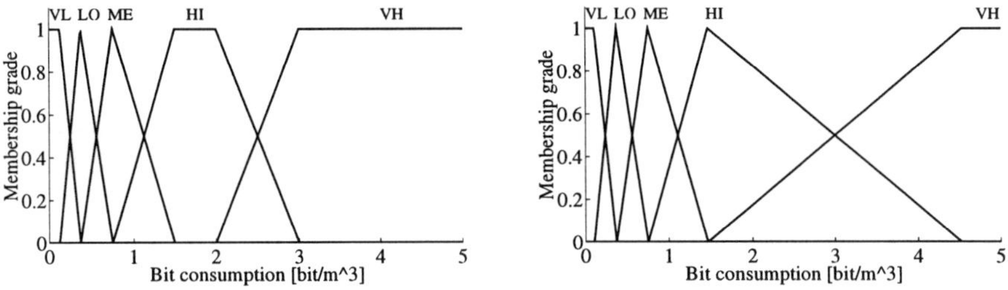

(a) Initial membership functions. *(b) Adapted membership functions.*

Figure 7.6. Initial and adapted membership functions of the consequent linguistic terms.

Care must be exercised with this method if few data samples are available. In such a case, some rules may be activated only for a single data point in the data set, which results in a perfect fit for the given point, but possibly a decreased performance on new data. The numerical adaptation of the consequent membership functions should always be complemented by expert validation of the meaning of the adapted linguistic terms.

7.1.4 Model Validation and Results

The aim of validation is to verify whether the developed knowledge-based model performs as required. Two types of validation have been applied: qualitative validation and quantitative validation.

Qualitative Validation. The purpose of qualitative validation is to check the overall correctness of the rules with respect to the background expert knowledge and to discover possible errors in the rule bases. Since the performance of the model depends mainly on the validity of the knowledge, qualitative validation plays, in our case, a more important role. At the moment, very few measurements are available, and this hinders us from performing a thorough quantitative validation. Dependency analysis was performed by changing the values of the input variables and observing their qualitative effects on the system output, and checking the reasoning paths of the inference mechanism.

Quantitative Validation. Available field data items are used as inputs to the system and the predicted numerical outputs (defuzzified values) are compared with the measured values from the field tests. The data set consists of two parts. The first 11 measurements were available during the development stage of the model, and although the data was not directly used to built the model, it cannot be trusted as a validation data set. The last five measurements were acquired after the model was developed and can serve as true validation data. Note that the number of available measurements is extremely small, due to practical difficulties and costs associated with gathering the data.

Large errors occur for large values of the bit consumption and production rates, which was expected. In the future, additional factors may need to be included in the model, to improve its performance for large outputs. The root mean square error is 0.67 (14% of the maximal value) for the bit consumption rate and 21.69 (16% of the maximal value) for the production rate.

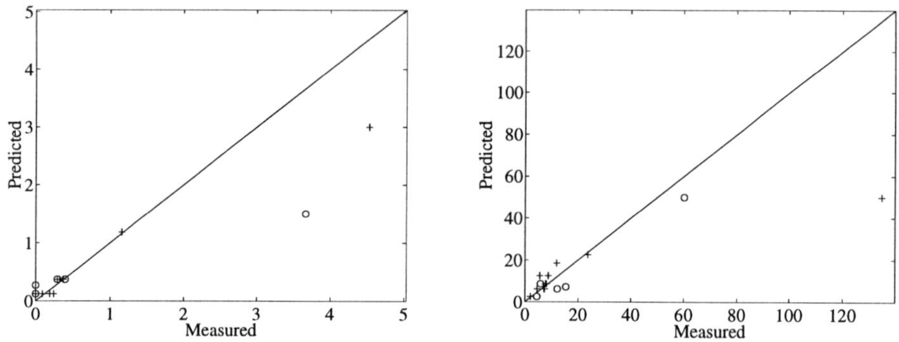

(a) *Bit consumption* [bit/m^3].　　(b) *Production* [m^3/hour].

Figure 7.7. Numerical performance of the prediction model. Solid line is the ideal prediction, '+' denotes prediction on the data available during model development, 'o' denotes prediction on new data.

Table 7.1 gives the linguistic output and compares the defuzzified output of the model with the measured bit consumption rate for the different trencher projects. The defuzzified prediction values correlate reasonably well with the actual bit consumption rate and also the production rate.

The results obtained with the fuzzy model for the production rate were compared with the output of the computer program developed by the Vermeer Manufacturing Company. Only six project can be compared, since the program can predict the production rate for certain types of rock only. These projects and the corresponding predicted values are given in Table 7.2. It can be concluded that with respect to all three performance indices (root mean square error, standard deviation of error, and maximum absolute error), the accuracy of the fuzzy model is significantly better than the accuracy of the Vermeer program.

7.1.5 Discussion

A fuzzy model has been developed for prediction of the T-850 trencher performance and tool wear prediction. The model is based on expert knowledge and experience from 11 investigated trenching projects and laboratory experiments. The model uses six input variables: rock strength, joint spacing for three different sets, joint orienta-

Table 7.1. Results for the bit consumption rate. The last five projects represent data which were not available during the development phase.

project name	linguistic prediction	numerical prediction [bits/m^3]	measured value [bits/m^3]
Avoriaz (F)	Very high (1.00)	3.00	4.53
Billingsbach (D)	Low (1.00)	0.38	0.31
Fabrègues (F)	Very low (0.50)	0.13	0.24
Langmeil (D)	Very low (0.60)	0.13	0.00
La Lauze (F)	Low (1.00)	0.38	0.43
Monaco (F)	Low (0.28), High (0.72)	1.19	1.16
Oberndorf (D)	Very low (1.00)	0.13	0.00
Schnarrtanne 1 (D)	Very low (1.00)	0.13	0.19
Schnarrtanne 2 (D)	Low (0.50)	0.38	0.37
Sybillenbad (D)	Very low (1.00)	0.13	0.09
Wölland (D)	Very low (1.00)	0.13	0.00
Sybillenbad 2 (D)	Very low (1.00)	0.13	0.00
Castelneau (F)	Low (0.50)	0.13	0.29
La Couronne (F)	High (1.00)	1.50	3.66
Millau (F)	Low (1.00)	0.38	0.39
Buschdorf 2 (D)	Very low (0.4), Low (0.60)	0.28	0.00

Table 7.2. Comparison of the measured production rates and production rates predicted by the fuzzy model and by the Vermeer program.

project name	measured value [bits/m^3]	fuzzy model [bits/m^3]	Vermeer program [bits/m^3]
Fabrègues (F)	23.60	22.50	6.20
Langmeil (D)	11.90	18.50	16.76
Monaco (F)	7.17	6.25	8.04
Castelneau (F)	12.15	6.25	21.80
Millau (F)	5.80	8.75	17.40
Buschdorf 2 (D)	15.33	7.25	17.44
root mean square error	—	5.07	9.65
standard deviation of error	—	5.43	10.36
maximum absolute error	—	8.08	17.40

tion and trench dimensions. The knowledge is hierarchically organized in four rule bases. The prediction is obtained both in linguistic terms and numerical values (after defuzzification). The predicted bit consumption and production rates for the different trencher projects correlate reasonably well with the actual performance. The developed fuzzy model outperforms a computer program which is currently used to advise the dealers on trencher selection, bit consumption and production rates, to enable them to predict the expected costs of excavation projects (Vermeer International, Goes, the Netherlands). However, more field data are required in the future to fully validate the model.

7.2 Pressure Modeling and Control

This section describes the development of different types of fuzzy models for highly nonlinear pressure dynamics in a laboratory fermenter. The models were constructed from input–output process data, using the clustering-based identification methods described in Sections 5.1 and 5.2. Experimental results of a real-time predictive controller based on the fuzzy model are presented.

7.2.1 Process Description

The process under consideration is a 40 l laboratory fermenter which contains 25 l of water. At the bottom of the fermenter, air is fed into the water at a specified flow rate which is kept at a desired value by a local mass-flow controller. The air pressure in the head space can be controlled by the position of an outlet valve at the top of the fermenter, see Figure 7.8a.

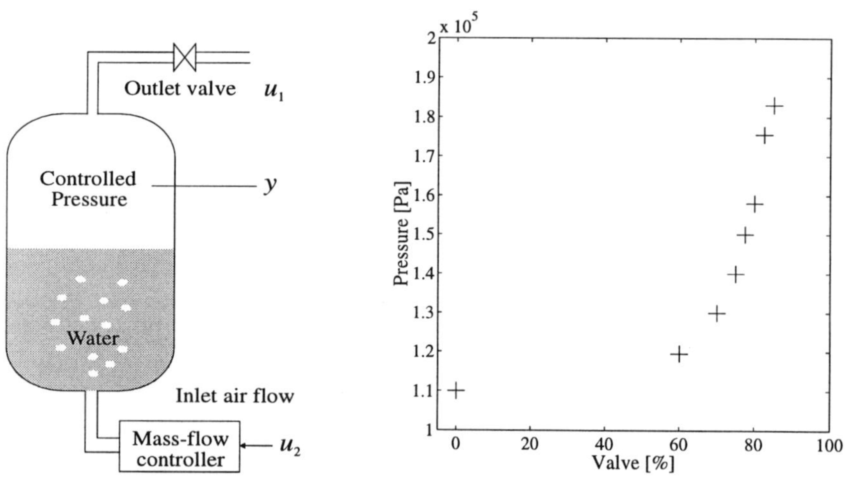

(a) Fermentation tank. (b) Steady-state characteristic.

Figure 7.8. Experimental setup (a) and the nonlinear steady-state relation between the outlet valve position and pressure (b).

This process has two inputs: the position of the outlet valve, denoted by u_1, and the inlet air flow rate, denoted by u_2, and one output: the pressure in the head space, denoted by y. The inlet flow rate can be kept constant, in which case the process is a single-input, single-output system. Because of the underlying physical mechanisms, and because of the nonlinear characteristic of the outlet valve, the process has a strongly nonlinear behavior, both static and dynamic. The steady-state characteristic is shown in Figure 7.8b.

7.2.2 Data Collection

Experiments were designed to collect the identification and validation data. The inputs of the system were excited with multi-sinusoidal signals, such that the range of amplitudes and frequencies of interest was covered (van Can, et al., 1995). The upper plots in Figure 7.9 show the input signals and the bottom plot shows the corresponding output. For the SISO variant of the process, a similar signal was used for the valve position and the air-flow rate was kept constants at 22.5 l/min. The time constant of the process is about 45 s, which allows for a sample time of $T = 5$ s. This choice is also suitable with regard to the measurement noise. The identification data sets consist of 482 samples.

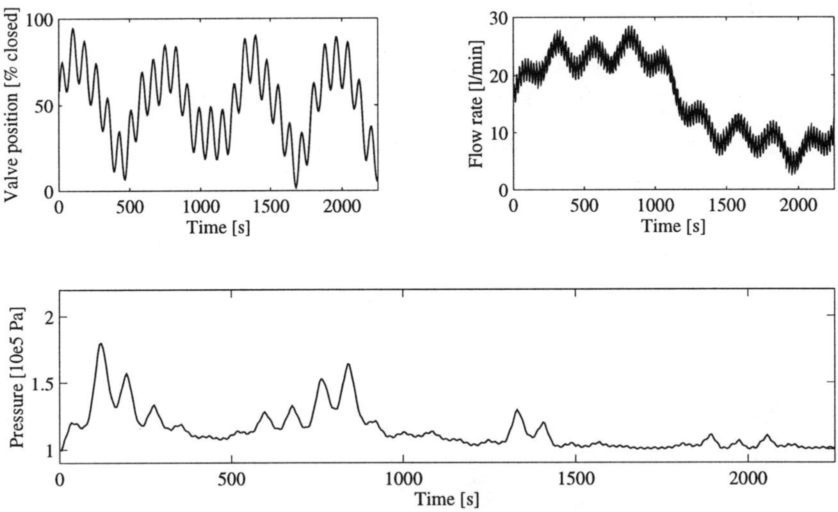

Figure 7.9. Identification data set.

7.2.3 SISO Fuzzy Model

Takagi–Sugeno Model. Based on the prior knowledge about the process, a first-order NARX structure was selected, resulting in TS rules of the following form:

$$\textbf{If } y(k) \text{ is } A_i \textbf{ and } u(k) \text{ is } B_i \textbf{ then } y(k+1) = a_i y(k) + b_i u(k) + c_i, \tag{7.2}$$

where $y(k)$ and $u(k)$ are the pressure and the valve position at time k, respectively. The membership functions of the antecedent linguistic terms A_i and B_i, as well as the consequent parameters a_i, b_i and c_i were estimated from the data by fuzzy clustering.

First the data matrix $\mathbf{Z}$ is formed, which contains the regressors $y(k)$, $u(k)$ and the regressand $y(k+1)$:

$$\mathbf{Z}^T = \begin{bmatrix} y(1) & u(1) & y(2) \\ y(2) & u(2) & y(3) \\ \vdots & \vdots & \vdots \\ y(N-1) & u(N-1) & y(N) \end{bmatrix}.$$

The GK clustering algorithm (Section 3.4.1) with the compatible cluster merging procedure (Section 4.5.2) was applied to the data, initially with $c_{\max} = 9$ clusters. After merging four and two clusters in sequence, the procedure arrived at the final partition of three clusters. The CCM threshold was $\gamma = 0.55$. Membership functions for the antecedent variables were obtained by fitting the cluster projections by parametric exponential functions, as described in Section 5.1.1. The consequent parameters are estimated by the global least-squares technique given in Section 5.5. The following rules are obtained:

1. **If** $y(k)$ is Low **and** $u(k)$ is Open
   then $y(k+1) = 0.67y(k) + 0.0007u(k) + 0.35.$
2. **If** $y(k)$ is Medium **and** $u(k)$ is Half closed
   then $y(k+1) = 0.80y(k) + 0.0028u(k) + 0.07.$
3. **If** $y(k)$ is High **and** $u(k)$ is Closed
   then $y(k+1) = 0.90y(k) + 0.0071u(k) - 0.39.$

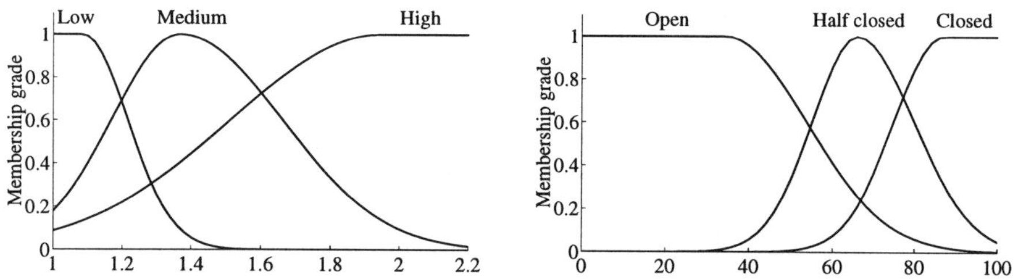

(a) Pressure [10^5 Pa].   (b) Valve position [% closed].

Figure 7.10. Membership functions for the pressure and valve position.

The membership functions for the antecedent linguistic terms are given in Figure 7.10. The consequents represent stable linear models which are in good agreement with local linear models obtained from local step responses measured on the process. One can see that the gain of the consequent models increases with the pressure, which corresponds with the prior knowledge that at higher pressure values the sensitivity of the process is in the orders of magnitude higher. This can also be seen from Figure 7.9.

Validation was performed on a different data set than the one used for identification. The model is simulated from the input: $\hat{y}(k+1) = f(\hat{y}(k), u(k))$, which imposes stronger requirements on its accuracy than one-step-ahead prediction which uses the process output $y(k)$ instead of the predicted value $\hat{y}(k)$. From Figure 7.11 one can see that the TS model follows the measured process output with a reasonable accuracy,

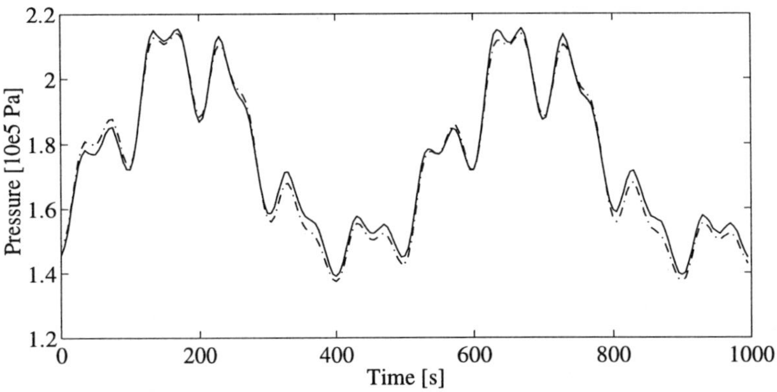

Figure 7.11. Validation of the TS fuzzy model. Solid line: process, dashed-dotted line: model.

Singleton and Relational Fuzzy Model. The TS fuzzy model for the pressure dynamics presented in the previous section was converted into a Mamdani model with singleton consequents, and into a fuzzy relational model by using the methods presented in Section 5.2. From the membership functions shown in Figure 7.10, triangular antecedent membership functions for the singleton model were derived, see Figure 7.12.

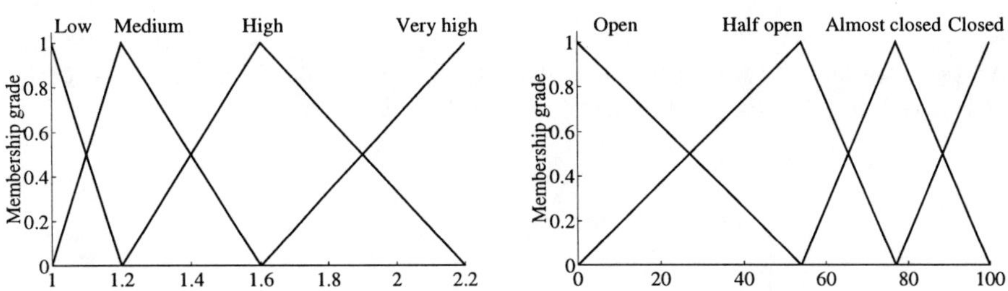

(a) *Pressure [10^5 Pa].*

(b) *Valve position [% closed].*

Figure 7.12. Antecedent membership functions of the singleton model.

The rule base consists of 16 rules, containing all possible combinations of the antecedent terms. The consequent singletons were estimated by the least-squares method. The rule base is given in Table 7.3. This model is, in fact, a look-up table with irregularly spaced index vectors, determined by the cores of fuzzy sets in the antecedent partitions. In our example, the index vector for the pressure is $y_i = [1.0, 1.2, 1.6, 2.2]$ and for the valve position $u_i = [0, 54, 77, 100]$. For values between the index vector elements, the output is linearly interpolated. In simulation, the singleton model is slightly more accurate than the TS model (the root mean-squared errors are $1.72 \cdot 10^3$ Pa and $2.26 \cdot 10^3$ Pa respectively). The reason for this improvement is the approximation error of the TS model, as discussed in Section 2.3.2, and probably also the fact that the singleton model has more free parameters.

Table 7.3. Rule base of the singleton model.

	Pressure			
Valve	*Low*	*Medium*	*High*	*Very high*
Open	1.06	1.13	1.46	1.70
Half open	1.05	1.19	1.51	2.03
Almost closed	1.07	1.23	1.63	2.12
Closed	1.11	1.33	1.79	2.34

Instead of using real-valued consequents, the model can be represented in a relational form (see Section 5.2.3). The consequent singletons from the above table are simply expressed in terms of their degree of membership in the fuzzy sets for pressure, shown in Figure 7.12a. The obtained three-dimensional fuzzy relation is given in Table 7.4. An approximate linguistic model can be obtained by considering only the labels with the largest membership degrees.

7.2.4 *MISO Fuzzy Model*

The MISO model is a straightforward extension of the SISO model presented in Section 7.2.3. The process dynamics are represented by: $y(k+1) = f(y(k), u_1(k), u_2(k))$, where $y(k)$ is the pressure at time k, $u_1(k)$ and $u_2(k)$ are the model inputs, the outlet valve position and the inlet flow rate respectively. From the process measurements $\{y(k), u_1(k), u_2(k)\}$, the data matrix for clustering was formed:

$$\mathbf{Z}^T = \begin{bmatrix} y(1) & u_1(1) & u_2(1) & y(2) \\ y(2) & u_1(2) & u_2(2) & y(3) \\ \cdots & \cdots & \cdots & \cdots \\ y(N-1) & u_1(N-1) & u_2(N-1) & y(N) \end{bmatrix}.$$

The data were clustered by the GK algorithm with compatible cluster merging, using the threshold $\gamma = 0.65$. The algorithm was initialized with nine clusters and arrived to the final partition containing three clusters.

Figures 7.13a to 7.13c shows the exponential membership functions fitted to the projection of the fuzzy partition onto the antecedent variables. After estimating the

Table 7.4. Fuzzy relation obtained from the singleton model. The entries in the table are the membership degrees for the different combinations of the linguistic values of $u(k)$, $y(k)$ and $y(k+1)$.

	Pressure $y(k)$ is Low				Pressure $y(k)$ is Medium			
	Pressure $y(k+1)$				Pressure $y(k+1)$			
Valve	*Low*	*Medium*	*High*	*V. high*	*Low*	*Medium*	*High*	*V. high*
Open	0.75	0.25	0	0	0.33	0.67	0	0
Half open	0.68	0.32	0	0	0.05	0.95	0	0
Almost cl.	0.60	0.40	0	0	0	0.92	0.08	0
Closed	0.43	0.57	0	0	0	0.69	0.31	0
	Pressure $y(k)$ is High				Pressure $y(k)$ is Very high			
	Pressure $y(k+1)$				Pressure $y(k+1)$			
Valve	*Low*	*Medium*	*High*	*V. high*	*Low*	*Medium*	*High*	*V. high*
Open	0	0.41	0.59	0	0	0	0.91	0.09
Half open	0	0.32	0.68	0	0	0	0.29	0.71
Almost cl.	0	0.04	0.96	0	0	0	0.14	0.86
Closed	0	0	0.74	0.26	0	0	0	1.00

consequent parameters with the least-squares method, the following rule base was obtained:

1. **If** $y(k)$ is Low **and** $u_1(k)$ is Open **and** $u_2(k)$ is Low
   then $y(k+1) = 0.52y(k) + 5 \cdot 10^{-5}u_1(k) + 3.2 \cdot 10^{-3}u_2(k) + 0.44$

2. **If** $y(k)$ is Medium **and** $u(k)$ is Half closed **and** $u_2(k)$ is High
   then $y(k+1) = 0.76y(k) + 2.4 \cdot 10^{-4}u_1(k) + 4.9 \cdot 10^{-3}u_2(k) + 0.04$

3. **If** $y(k)$ is High **and** $u(k)$ is Closed **and** $u_2(k)$ is Very high
   then $y(k+1) = 0.85y(k) + 7.4 \cdot 10^{-4}u_1(k) + 5.3 \cdot 10^{-3}u_2(k) - 0.45$

Figure 7.13d shows validation of the model by simulation from the inputs only. Visual comparison of the process and model outputs as well as the root mean-squared error of $1.42 \cdot 10^3$ Pa indicate that even this simple fuzzy model can serve as a reliable long-range predictor for this process.

7.2.5 Predictive Control Based on the Fuzzy Model

The SISO TS fuzzy model developed in Section 7.2.3 was incorporated into a nonlinear predictive control scheme with the IMC structure, see Section 6.1.3. It was verified experimentally that choosing $H_p = H_c = 1$ is sufficient for the accurate control of the process for different reference signals. The rate of change of the valve position is constrained to 10% per sample time. Figures 7.14 and 7.15 show real-time control

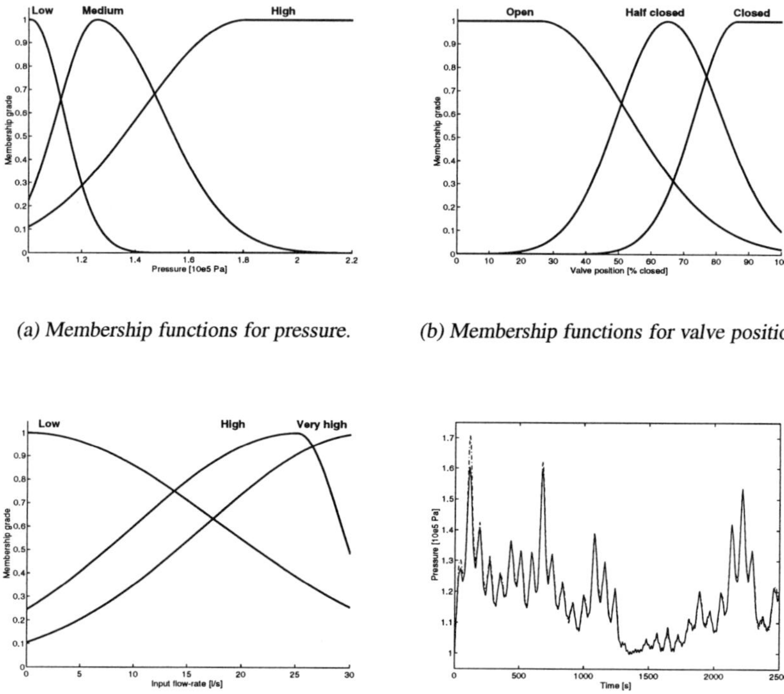

(a) *Membership functions for pressure.*

(b) *Membership functions for valve position.*

(c) *Membership functions for flow rate.*

(d) *Model validation.*

Figure 7.13. Membership functions (a, b, c) and model validation (d). Solid line is the process output, dash-dotted line is the model.

results obtained with the predictive controller for step-like and sinusoidal references, respectively.

Additional experiments were carried out in order to check the controller's robustness. The input flow rate was changed from its nominal value of 22.5 l/min to 15 l/min and to 30 l/min. Such a disturbance results in a considerable mismatch between the output of the process and of the model. However, the feedback from the internal model corrects the steady-state error that would otherwise occur, and the control results are satisfactory, as shown in Figure 7.16.

7.2.6 Discussion

Takagi–Sugeno, singleton and relational fuzzy models were developed for a highly nonlinear dynamic system. The obtained rules and membership functions provide insight into the process dynamics, which can be locally analyzed. In the case of the affine TS model, tools from linear system theory can be used. The relational fuzzy model can be validated in qualitative terms. The obtained fuzzy models are

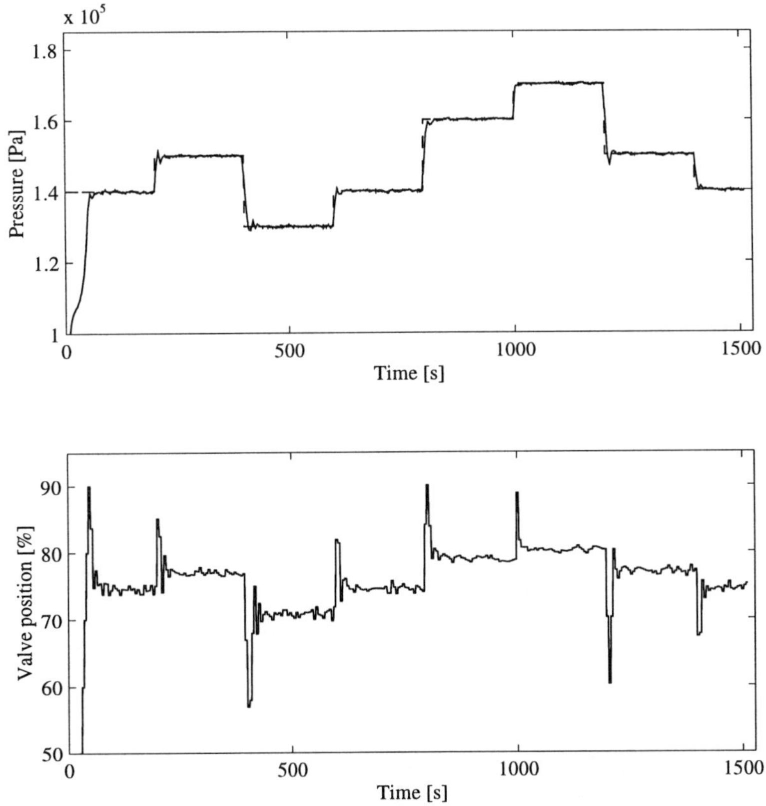

Figure 7.14. Real-time control response for a step-wise reference. Dashed line is the reference and solid line the process output.

accurate numerical predictors which can be incorporated into a model-based predictive controller.

In (Babuška, et al., 1996), the predictive controller based on a fuzzy model is compared to a direct fuzzy controller of the Mamdani type, fuzzy supervised PI controller[2] and a predictive controller based on a neural network model. The comparison is based on several practical aspects, including the development time, the type and amount of prior information needed for the controller design, the tuning requirements, and the closed-loop performance under different conditions (reference signals, disturbances). The conclusions are that the predictive controller outperforms the model-free nonlinear control schemes and also a linear PI controller, which must be detuned in order to be stabilize the process across the entire operating range. Another advantage of the predictive control approach is that it can be directly used for the multivariable version of the process, using the MISO fuzzy model.

[2] See for instance (Driankov, et al., 1993) for details on direct and supervisory fuzzy control.

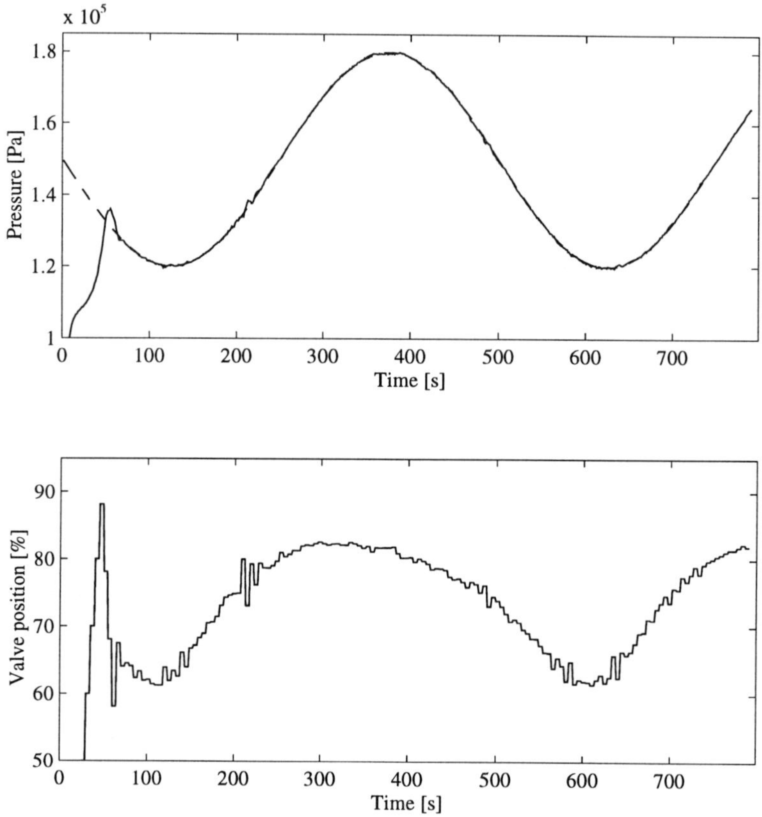

Figure 7.15. Real-time control response for a sinusoidal reference. Dashed line is the reference and solid line the process output.

7.3 Fuzzy Modeling of Enzymatic Penicillin–G Conversion

A linguistic fuzzy model of enzymatic Penicillin–G conversion kinetics has been developed from data.[3] The model describes qualitatively how the enzyme kinetics depend on the concentrations of the components involved in the conversion. The rules and the membership functions were extracted from experimental data, using the identification method described in Section 5.2. The fuzzy model has been incorporated into a white-box model based on macroscopic balances.

7.3.1 Introduction

In biotechnology, new processes are often based on modified microorganisms or new enzymes. In order to speed up the development of the new processes and their control

[3] This section is an extended version of the article (Babuška, et al., 1996).

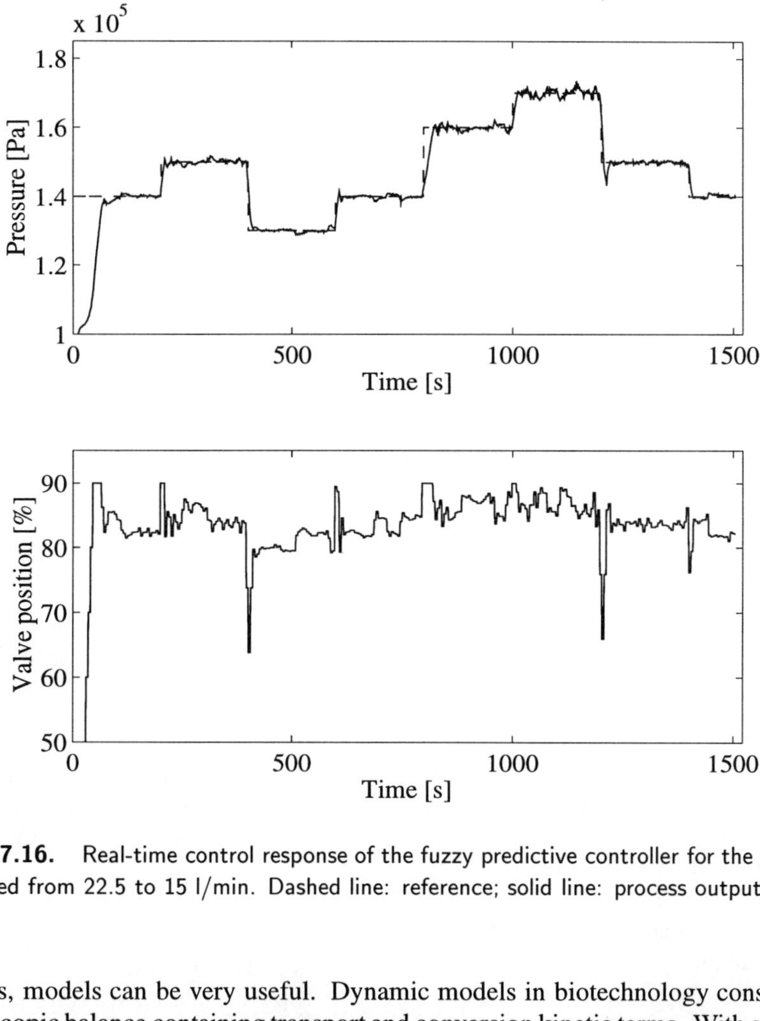

Figure 7.16. Real-time control response of the fuzzy predictive controller for the gas-flow decreased from 22.5 to 15 l/min. Dashed line: reference; solid line: process output.

systems, models can be very useful. Dynamic models in biotechnology consist of a macroscopic balance containing transport and conversion kinetic terms. With standard process equipment, the transport kinetics can be represented empirically using prior experience, but accurate kinetic models are not directly available. Experiments under relevant conditions are needed to obtain information about the kinetics. Data from experiments can be used to estimate parameters of a given kinetic model, but it is not always clear how this model should be chosen, especially for new, or not very well-understood, conversion processes.

Semi-mechanistic (hybrid) modeling approaches have been proposed that combine first-principles with nonlinear black-box models, using neural networks (Thompson and Kramer, 1994; Psichogios and Ungar, 1992; Su, et al., 1992). A neural network is typically used as a general nonlinear function approximator that "learns" the unknown relationships from data, and serves as a predictor of unmeasured process quantities that are difficult to model from first principles. Due to their approximation capabilities, neural networks are suitable for representation of these generally nonlinear relationships. A major drawback of this approach is, however, that a posteriori analysis of the neural network model is difficult, and validation is restricted to numerical simulation.

This section describes a different approach, which is based on fuzzy modeling. A fuzzy model developed from data serves as a good numerical predictor, and at the same time provides a posteriori qualitative knowledge about the process. The usual quantitative validation performed on data is complemented by a qualitative method, which enhances the reliability of the model. For an unknown process, the rules provide a qualitative description of the unknown relationships, which may have a more general scope of validity, e.g., when changing the scale of the process. The membership functions, on the other hand, serve as a numeric-to-symbolic interface, and may depend on the particular process parameters. The expert can also modify the rules that are not in agreement with his/her experience and can supply additional rules for those domains of the process variables where insufficient data are available.

7.3.2 Process Description

The process under study is the enzymatic conversion (hydrolysis) of Penicillin–G (PenG) to 6–aminopenicillanic acid (APA) and phenyl acetic acid (PhAH) at pH of 8.0 and temperature of 310 K by the enzyme penicillin acylase:

$$\text{Benzyl Penicillin (PenG)} + H_2O \xrightarrow{\text{Penicillin Acylase}} \begin{array}{l}\text{6 Amino Penicillinic Acid (APA)}\\ + \text{Phenyl Acetate (PhAH).}\end{array}$$

This process has been chosen for the study here presented, because it can be performed in a (repeated) batch mode with well-reproducible experiments, it has complex nonlinear kinetics, which is typical for many (bio)chemical processes, and finally, it is a well-understood process which allows for the validation of the results both qualitatively and numerically by comparison with some known kinetic models (Carleysmith, et al., 1973). A similar modeling approach based on neural networks was applied to the same process (van Can, et al., 1997), giving an additional opportunity to evaluate the advantages and drawbacks of the individual methods.

A white-box description of the known macroscopic balances is based on the basic process knowledge. At pH=8 and $T = 310$ K, the following holds (van Can, et al., 1997):

- The stoichiometry of the reaction (PenG:APA:PhAH) equals 1:1:1.
- One mol of base is needed per mol converted PenG (or produced APA or PhAH) to keep the pH at 8.
- Enzyme deactivation and chemical side reactions for PenG and APA are insignificant in the time frame of one experiment (two to three hours).
- The conversion rate is proportional to the enzyme concentration.

This knowledge can be formalized in the following macroscopic balance equations:

$$B_{k+1} = B_k + \Delta T \frac{[E]_k V_k r_e}{M_B}, \tag{7.3a}$$

$$
\begin{aligned}
V_{k+1} &= V_k + \Delta T \frac{[E]_k V_k r_e}{M_B}, && (7.3b)\\
[\text{PenG}]_{k+1} &= \frac{V_k}{V_{k+1}} \left([\text{PenG}]_k - \Delta T R_{\text{G}} [E]_k r_e\right), && (7.3c)\\
[\text{APA}]_{k+1} &= \frac{V_k}{V_{k+1}} \left([\text{APA}]_k + \Delta T R_{\text{A}} [E]_k r_e\right), && (7.3d)\\
[\text{PhAH}]_{k+1} &= \frac{V_k}{V_{k+1}} \left([\text{PhAH}]_k + \Delta T R_{\text{P}} [E]_k r_e\right), && (7.3e)\\
[E]_{k+1} &= \frac{V_k}{V_{k+1}} [E]_k, && (7.3f)
\end{aligned}
$$

where the symbols have the following meaning:

[PenG]	:	concentration of penicillin–G (mol/l),
[APA]	:	concentration of 6–aminopenicillanic acid (mol/l),
[PhAH]	:	concentration of phenyl acetic acid (mol/l),
$[E]$	:	concentration of enzyme (U/l),
V	:	volume (l),
B	:	consumed base (l),
ΔT	:	sample time (s),
r_e	:	conversion rate (mol/U/s),
M_B	:	concentration of added base (mol/l),
R_x	:	stoichiometric coefficient between H^+ and component x (–),
k	:	discrete time.

The kinetic term r_e depends on the concentrations of PenG, APA and PhAH, but the structure of this relationship is generally not known. For the process under study, however, the Michaelis–Menten type equation can be used:

$$
r_e = \frac{r_{\max}[\text{PenG}]}{K_M + [\text{PenG}] + \frac{K_M[\text{APA}]}{K_{\text{A}}} + \frac{K_M[\text{PhAH}]}{K_{\text{P}}} + \frac{K_M[\text{PhAH}][\text{APA}]}{K_{\text{P}}K_{\text{A}}} + \frac{[\text{PenG}][\text{APA}]}{K_{\text{A}}}}, \tag{7.4}
$$

with:

r_e	:	conversion rate (mol/U/s),
$r_{\max}$	:	maximum conversion rate (mol/U/s),
K_M	:	Michaelis–Menten coefficient (mol/l),
K_{A}	:	inhibition coefficient for APA (mol/l),
K_{P}	:	inhibition coefficient for PhAH (mol/l).

It should be noted, however, that the estimation of these parameters from data is a nontrivial optimization problem which requires good initial values to converge (van Can, et al., 1997). It will be shown later that, with fuzzy modeling, no assumptions about the structure of the kinetics term are necessary, and that initialization does not play an important role either. This allows the approach to be used also for poorly understood processes.

7.3.3 Experimental Set-up

The data were obtained from laboratory experiments performed in a stirred, thermostated laboratory bioreactor with a maximum volume of 1500 cm^3. The laboratory set-up is depicted in Figure 7.17. Due to the acid–base equilibrium of PenG, APA and PhAH, one mol of H^+ is released per one mol of converted PenG. In order to keep the pH constant, base must be added to the reactor. The base added is the only measured quantity. The sampling period is $\Delta T = 60$ s.

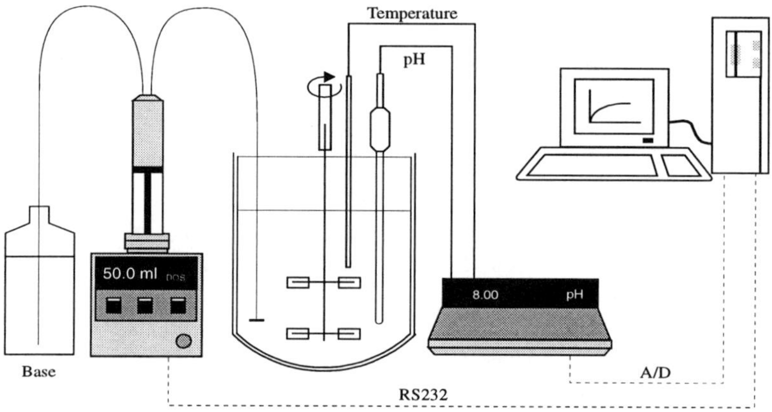

Figure 7.17. Experimental set-up.

The concentrations and the conversion rate are computed from the amount of added base, using the following equations, derived from the balance equations (7.3):

$$
\begin{aligned}
V_k &= V_0 + B_k && \text{(7.5a)}\\
[\mathrm{PenG}]_k &= \frac{[\mathrm{PenG}]_0 V_0 - M_B B_k}{V_k}, && \text{(7.5b)}\\
[\mathrm{APA}]_k &= \frac{[\mathrm{APA}]_0 V_0 + M_B B_k}{V_k}, && \text{(7.5c)}\\
[\mathrm{PhAH}]_k &= \frac{[\mathrm{PhAH}]_0 V_0 + M_B B_k}{V_k}, && \text{(7.5d)}\\
[\mathrm{E}]_k &= \frac{[\mathrm{E}]_0 V_0}{V_k}, && \text{(7.5e)}\\
r_{e,k} &= \frac{M_B (B_{k+1} - B_k)}{[\mathrm{E}]_k V_k \Delta T}. && \text{(7.5f)}
\end{aligned}
$$

7.3.4 Fuzzy Modeling

It is expected that the specific conversion rate r_e depends on the concentrations [PenG], [APA] and [PhAH] in a nonlinear way, but the exact form of this dependence in considered unknown:

$$r_e = f\big([\mathrm{PenG}], [\mathrm{APA}], [\mathrm{PhAH}]\big).$$

Data from 10 batch experiments started at different initial conditions were recorded, yielding a total of $N = 816$ data points for identification. Additional experiments were carried out to obtain data for model validation. Some typical experiments are shown in Figure 7.18. The concentrations and the conversion rate are calculated from the amount of added base, using equations (7.5).

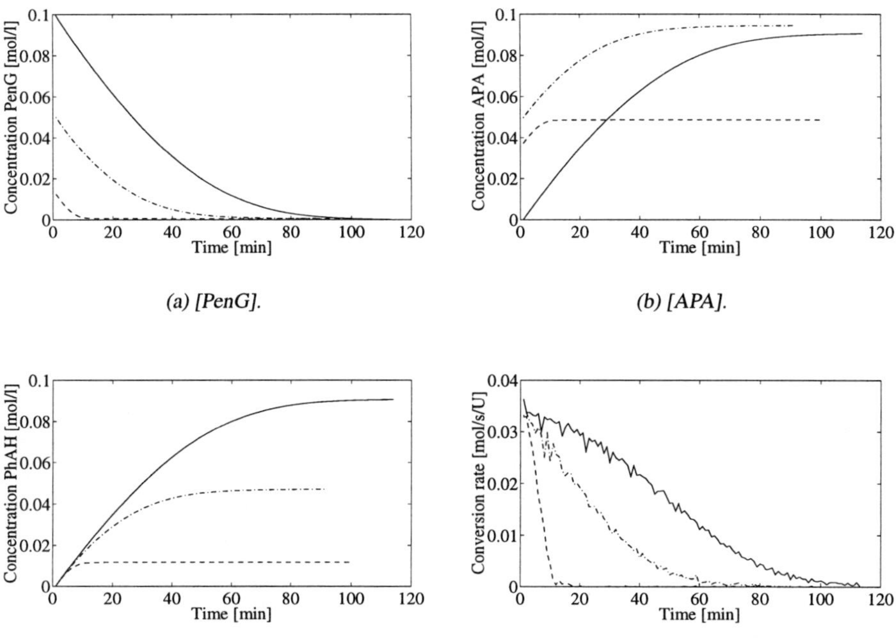

(a) [PenG]. (b) [APA].

(c) [PhAH]. (d) Conversion rate r_e.

Figure 7.18. Experimental data from some typical batch experiments. The three curves in each graph represent experiments started from different initial conditions.

The data sequences are arranged in the following matrix

$$\mathbf{Z}^T = \begin{bmatrix} [\text{PenG}]_1 & [\text{APA}]_1 & [\text{PhAH}]_1 & r_{e,1} \\ [\text{PenG}]_2 & [\text{APA}]_2 & [\text{PhAH}]_2 & r_{e,2} \\ \vdots & \vdots & \vdots & \vdots \\ [\text{PenG}]_N & [\text{APA}]_N & [\text{PhAH}]_N & r_{e,N} \end{bmatrix},$$

and clustered by the GK algorithm (see Section 3.4.1). The number of clusters is set to 3, since the number and information content of the data do not allow for more clusters (numerical problems appear in inversion of the cluster covariance matrix). As shown later, the model derived from the three clusters is sufficiently accurate. If this were not the case, more measurements would be needed, based on experiments at different initial conditions. The total number of fuzzy sets was reduced, since some of the projected fuzzy sets cover the entire domain (and therefore can be neglected), and others project into similar membership functions, see Section 5.1.3. As a result, the domain of each input variable was partitioned into two regions in which the conversion rate can be approximated as a linear function of the component concentrations. By applying the method to derive a linguistic model from Section 5.2, three triangular membership functions per variable are obtained, see Figure 7.19.

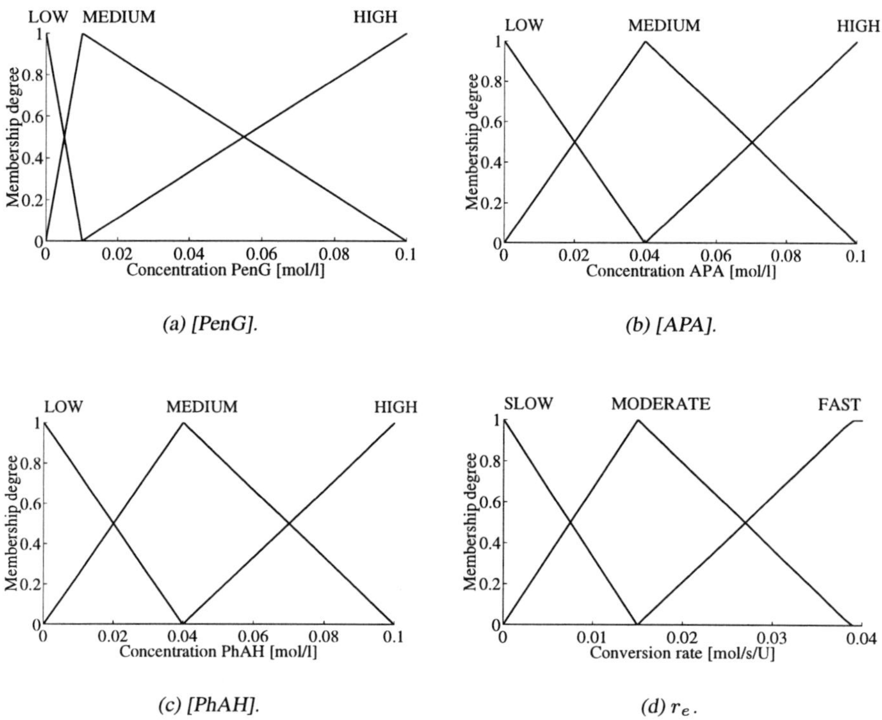

Figure 7.19. Membership functions.

The relationship between the concentrations and the specific conversion rate is described by *linguistic rules* of the following form:

If [PenG] is High **and** [APA] is Low **and** [PhAH] is Low **then** r_e is Fast. (7.6)

The rule base contains rules for all possible combinations of the antecedent linguistic terms, and is given in Tab. 7.5. The rule base was first established as a fuzzy relation, using the φ-composition method described in Section 5.2. The consequent of the ith rule is then chosen as the linguistic term corresponding to the largest element $r_{i,j}$ in the ith row of the fuzzy relation. If two terms have equal $r_{i,j}$, both are given. The numbers in brackets preceding the rules are certainty factors (CF) obtained as the difference between the maximum and minimum r_{ij} in each row of the fuzzy relational matrix.

The certainty factors indicate to which degree each particular rule is able to describe the relationship between the antecedent and the consequent variables. Rules with low certainty factors are less reliable than rules with high ones. Low certainty factors indicate insufficient or contradictory data in the particular operating region defined by the respective membership functions. Certainty factors of 0.5 and above (printed in bold) indicate reliable rules.

Table 7.5. Rule base of the fuzzy model.

#	CF	[PenG]	[APA]	[PhAH]	conversion rate
1.	(0.48)	LOW	LOW	LOW	FAST
2.	**(0.55)**	LOW	LOW	MEDIUM	SLOW/MOD
3.	(0.11)	LOW	LOW	HIGH	SLOW
4.	**(0.54)**	LOW	MEDIUM	LOW	MODERATE
5.	**(0.74)**	LOW	MEDIUM	MEDIUM	SLOW
6.	(0.25)	LOW	MEDIUM	HIGH	SLOW
7.	(0.10)	LOW	HIGH	LOW	SLOW
8.	**(0.76)**	LOW	HIGH	MEDIUM	SLOW
9.	**(0.71)**	LOW	HIGH	HIGH	SLOW
10.	**(0.93)**	MEDIUM	LOW	LOW	FAST
11.	**(0.63)**	MEDIUM	LOW	MEDIUM	FAST
12.	(0.04)	MEDIUM	LOW	HIGH	MOD/FAST
13.	**(0.75)**	MEDIUM	MEDIUM	LOW	FAST
14.	**(0.82)**	MEDIUM	MEDIUM	MEDIUM	MODERATE
15.	(0.22)	MEDIUM	MEDIUM	HIGH	MODERATE
16.	(0.19)	MEDIUM	HIGH	LOW	MOD/FAST
17.	**(0.72)**	MEDIUM	HIGH	MEDIUM	MODERATE
18.	(0.47)	MEDIUM	HIGH	HIGH	MODERATE
19.	**(0.90)**	HIGH	LOW	LOW	FAST
20.	(0.19)	HIGH	LOW	MEDIUM	FAST
21.	(0.00)	HIGH	LOW	HIGH	—
22.	(0.37)	HIGH	MEDIUM	LOW	FAST
23.	**(0.50)**	HIGH	MEDIUM	MEDIUM	FAST
24.	(0.06)	HIGH	MEDIUM	HIGH	MOD/FAST
25.	(0.08)	HIGH	HIGH	LOW	MOD/FAST
26.	(0.06)	HIGH	HIGH	MEDIUM	MOD/FAST
27.	(0.04)	HIGH	HIGH	HIGH	MOD/FAST

The rules (without the certainty factors) and the membership functions were presented to an expert, who confirmed the overall correctness of most of the rules and also the relevance of the membership functions. Major disagreement concerned the rules with low certainty factors, which were not considered reliable. This is caused by the fact that insufficient data were available for high PenG concentrations. Additional experiments can be designed in order to obtain more data in this region. The expert was not able to directly assess the numerical values of the membership function parameters, but he confirmed that the distribution of the membership functions over the domains with respect to the relative influence of APA and PhAH on the conversion kinetics is realistic.

7.3.5 Semi-mechanistic Model

The obtained fuzzy model is used as a predictor of the conversion rate and is incorporated into the macroscopic balance for a batch process, taking into account the dilution due to the added base, see equations (7.3). This semi-mechanistic model is schematically depicted in Figure 7.20. Figure 7.21a compares the conversion rate estimated by the fuzzy model with the rate computed from the experimental data for two different validation data sets which were not used for the identification. Figure 7.21b compares the base added, as simulated by the model and measured for the same two experiments. It should be stressed that the values in Figure 7.21 are calculated recursively with only the initial concentrations given.

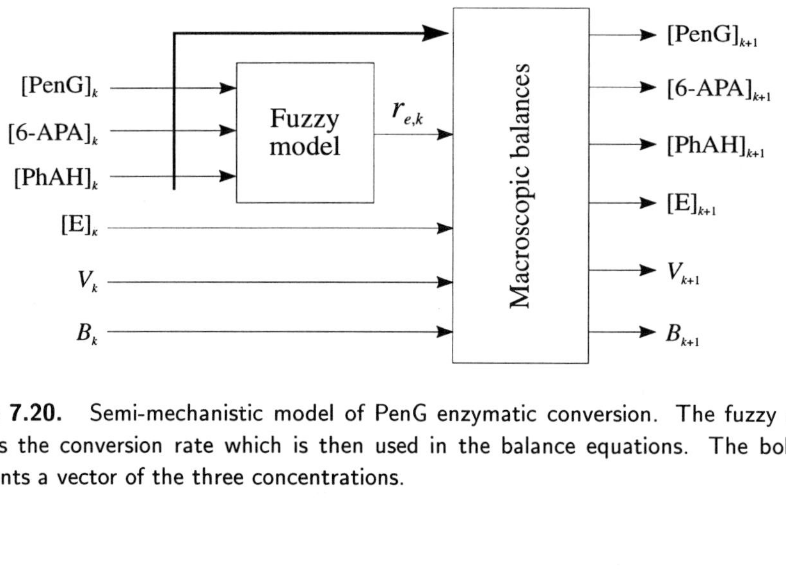

Figure 7.20. Semi-mechanistic model of PenG enzymatic conversion. The fuzzy model predicts the conversion rate which is then used in the balance equations. The bold line represents a vector of the three concentrations.

7.3.6 Discussion

A linguistic fuzzy model to predict the enzyme kinetics in the Penicillin–G conversion was developed from experimental data. A posteriori analysis of the model indicates that most of the obtained rules are in good agreement with the expert's knowledge. The fuzzy model also provides good predictions of the conversion kinetics. When combined with the white-box model of the process (macroscopic balances) accurate prediction of the overall process behavior is obtained. The model can be used for controller design and process optimization, without the need to reveal the exact underlying kinetic mechanisms that would be represented in the Michaelis–Menten type model. Hence, the fuzzy modeling approach has a great potential for new, or poorly understood,

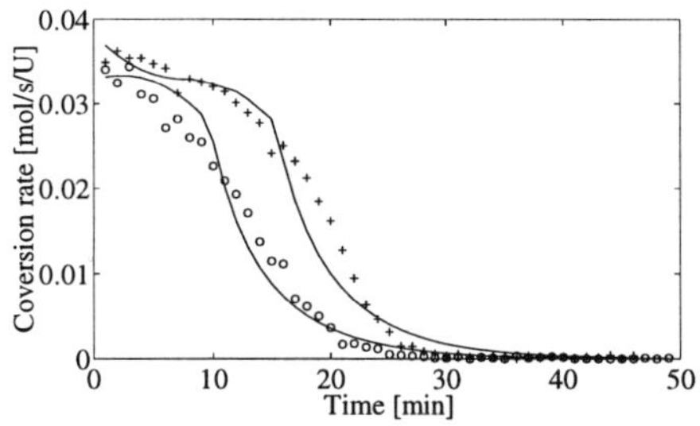

(a) Conversion rate.

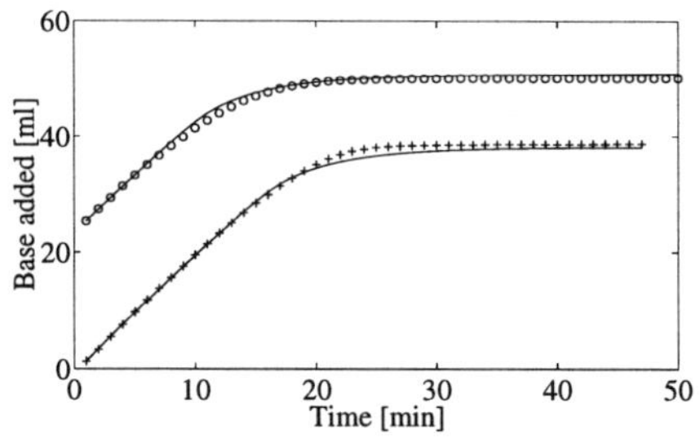

(b) Base added.

Figure 7.21. Graph (a): conversion rate computed by the fuzzy model (solid line) and calculated from the process data (markers). Graph (b): base added, simulated using the fuzzy model (solid line) and measured (markers).

processes, as it can be used both to develop prediction models and to improve the understanding of the process characteristics.

The numerical accuracy of the model is comparable with that obtained with a similar hybrid approach using a neural network (van Can, et al., 1997). A drawback of the latter approach is that the neural network remains a black-box model and does not provide any additional information about the process. The validity of the neural network must be assessed on the basis of numerical simulations only. Van Can, et al. 1997 also compared their approach with numerical optimization of the parameters in the Michaelis–Menten equation (7.4). They concluded that given a priori the correct relation for the kinetics, comparable accuracy of the numerical results was obtained, however, in general, the correct relation for the kinetics is not known a priori and proper identification of the related parameters is difficult.

7.4 Summary and Concluding Remarks

Three applications of fuzzy modeling and model-based control were described in order to demonstrate different aspects of the methodologies presented in the previous chapters. Also other applications that those presented here were realized. For example, Kaymak (1994) applied the identification method based on fuzzy clustering to the modeling of an enzymatic washing process. The modeling of ecological systems is reported by (Setnes, et al., 1997). Fuzzy clustering has also been applied in the biomedical field to the estimation of respiratory parameters from pressure–flow records of mechanically ventilated patients (Knijnenburg, 1997). Sousa, et al. (1997) describe an application of predictive control based on a fuzzy model to an air-conditioning system.

Most of the applications are relatively complex nonlinear systems, not easily amenable to conventional mechanistic modeling or linear identification. The possibility to interpret and qualitatively validate the obtained fuzzy models proved to be an important advantage of the approach, compared to other techniques, such as neural networks.

For an efficient use of the presented algorithms, a MATLAB toolbox has been developed (Babuška, 1995), see also Appendix B.

Appendix A
Basic Concepts of Fuzzy Set Theory

This appendix gives the definitions of the concepts of fuzzy set theory, which are used in this book. For a comprehensive treatment of fuzzy set theory, see, for instance, (Zimmermann, 1996; Klir and Yuan, 1995).

A.1 Fuzzy Sets

Definition A.1 (fuzzy set) A fuzzy set A on universe (domain) X is defined by the membership function $\mu_A(x)$ which is a mapping from the universe X into the unit interval:

$$\mu_A(x)\colon X \to [0,1]\,. \tag{A.1}$$

$\mathcal{F}(X)$ denotes the set of all fuzzy sets on X.

Fuzzy set theory allows for a partial membership of an element in a set. If the value of the membership function, called the membership degree (grade), equals one, x belongs completely to the fuzzy set. If it equals zero, x does not belong to the set. If the membership degree is between 0 and 1, x is a partial member of the fuzzy set. In the fuzzy set literature, the term *crisp* is often used to denote nonfuzzy quantities, e.g., a crisp number, a crisp set, etc.

A.2 Membership Functions

In a discrete set $X = \{x_i \mid i = 1, 2, \ldots, n\}$, a fuzzy set A may be defined by a list of ordered pairs: membership degree/set element:

$$A = \{\mu_A(x_1)/x_1,\ \mu_A(x_2)/x_2,\ \ldots, \mu_A(x_n)/x_n\}, \tag{A.2}$$

or in the form of two related vectors:

$$\mathbf{x} = [x_1, x_2, \ldots, x_n]^T, \quad \boldsymbol{\mu} = [\mu_A(x_1), \mu_A(x_2), \ldots, \mu_A(x_n)]^T\,. \tag{A.3}$$

In continuous domains, fuzzy sets are defined analytically by their membership functions. In this book, the following forms of membership functions are used:

Trapezoidal membership function:

$$\mu(x;a,b,c,d) = \max\left(0, \min(\frac{x-a}{b-a}, 1, \frac{d-x}{d-c})\right), \tag{A.4}$$

where a, b, c and d are coordinates of the trapezoid apexes. When $b = c$, a *triangular* membership function is obtained.

Piece-wise exponential membership function:

$$\mu(x;c_l,c_r,w_l,w_r) = \begin{cases} \exp(-(\frac{x-c_l}{2w_l})^2), & \text{if } x < c_l, \\ \exp(-(\frac{x-c_r}{2w_r})^2), & \text{if } x > c_r, \\ 1, & \text{otherwise,} \end{cases} \tag{A.5}$$

where c_l and c_r are the left and right shoulder, respectively, and w_l, w_r are the left and right width, respectively. For $c_l = c_r$ and $w_l = w_r$, the Gaussian membership function is obtained.

A.3 Basic Definitions

Definition A.2 (α-cut) The α-cut A_α of a fuzzy set A is the crisp subset of the universe of discourse X whose elements all have membership degrees greater than or equal to α:

$$A_\alpha = \{x \mid \mu_A(x) \geq \alpha\}, \quad \alpha \in [0,1]. \tag{A.6}$$

The α-cut operator is also denoted by $\alpha\text{-cut}(A, \alpha)$.

Definition A.3 (strict α-cut) An α-cut A_α is strict if $\mu_A(x) \neq \alpha$ for each $x \in A_\alpha$.

Definition A.4 (convex fuzzy set) A fuzzy set defined in $\mathbb{R}^n$ is convex if each of its α-cuts is a convex set.

Definition A.5 (support) The support of a fuzzy set A is the crisp subset of X whose elements all have nonzero membership grades:

$$\text{supp}(A) = \{x \mid \mu_A(x) > 0\}. \tag{A.7}$$

Definition A.6 (core) The core of a fuzzy set A is a crisp subset of X consisting of all elements with membership grades equal to one:

$$\text{core}(A) = \{x \mid \mu_A(x) = 1\}. \tag{A.8}$$

In the literature, the core is sometimes also denoted as the kernel, ker(A).

Definition A.7 (cardinality) Cardinality of a fuzzy set $A = \{\mu_A(x_i)/x_i \mid i = 1, 2, \dots, n\}$ is defined as the sum of the membership degrees:

$$|A| = \sum_{i=1}^{n} \mu_A(x_i). \tag{A.9}$$

Definition A.8 (height) The height of a fuzzy set A is the supremum of the membership grades of elements in A:

$$\text{hgt}(A) = \sup_{x \in X} \mu_A(x) \,. \tag{A.10}$$

Definition A.9 (normal fuzzy set) A fuzzy set A is normal if $\exists x \in X$ such that $\mu_A(x) = 1$. The operator $\text{norm}(A)$ denotes normalization of a fuzzy set, i.e., $A' = \text{norm}(A) \Leftrightarrow \mu_{A'}(x) = \mu_A(x)/\,\text{hgt}(A),\ \forall x$.

A.4 Operations on Fuzzy Sets

Definitions of operations on sets extend from ordinary set theory to fuzzy sets. In most cases, there are various ways to extend these operations. This section presents the basic definitions of fuzzy intersection, union and complement, and definitions of some other operations used in this book.

Definition A.10 (intersection of fuzzy sets) Let A and B be two fuzzy sets in X. The intersection of A and B is a fuzzy set C, denoted $C = A \cap B$, such that for each $x \in X$:

$$\mu_C(x) = \min\big(\mu_A(x), \mu_B(x)\big) \,. \tag{A.11}$$

The minimum operator is also denoted by '$\wedge$', i.e., $\mu_C(x) = \mu_A(x) \wedge \mu_B(x)$.

Definition A.11 (union of fuzzy sets) Let A and B be two fuzzy sets in X. The union of A and B is a fuzzy set C, denoted $C = A \cup B$, such that for each $x \in X$:

$$\mu_C(x) = \max\big(\mu_A(x), \mu_B(x)\big) \,. \tag{A.12}$$

The maximum operator is also denoted by '$\vee$', i.e., $\mu_C(x) = \mu_A(x) \vee \mu_B(x)$.

Fuzzy intersection of two fuzzy sets can be specified in a more general way by a binary operation on the unit interval, i.e., a function of the form:

$$i\colon [0,1] \times [0,1] \to [0,1] \,. \tag{A.13}$$

In order for a function i to qualify as a fuzzy intersection, it must have appropriate properties. Functions known as t-norms posses the properties required for the intersection. Similarly, functions called t-conorms can be used for the fuzzy union.

Definition A.12 (t-norm/fuzzy intersection) A t-norm i is a binary operation on the unit interval that satisfies at least the following axioms for all $a, b, c \in [0,1]$ (Klir and Yuan, 1995):

$$i(a,1) = a \quad \text{(boundary condition)}, \tag{A.14a}$$
$$b \le c \ \text{implies}\ \ i(a,b) \le i(a,c) \quad \text{(monotonicity)}, \tag{A.14b}$$
$$i(a,b) = i(b,a) \quad \text{(commutativity)}, \tag{A.14c}$$
$$i(a, i(b,c)) = i(i(a,b),c) \quad \text{(associativity)}\,. \tag{A.14d}$$

Some frequently used t-norms are:

standard intersection: $i(a, b) = \min(a, b)$
(algebraic) product: $i(a, b) = ab$
bold (Łukasiewicz) intersection: $i(a, b) = \max(0, a + b - 1)$

Definition A.13 (t-conorm/fuzzy union) A t-conorm u is a binary operation on the unit interval that satisfies at least the following axioms for all $a, b, c \in [0, 1]$ (Klir and Yuan, 1995):

$$u(a, 0) = a \quad \text{(boundary condition)}, \tag{A.15a}$$

$$b \leq c \ \text{ implies } \ u(a, b) \leq u(a, c) \quad \text{(monotonicity)}, \tag{A.15b}$$

$$u(a, b) = u(b, a) \quad \text{(commutativity)}, \tag{A.15c}$$

$$u(a, u(b, c)) = u(u(a, b), c) \quad \text{(associativity)}. \tag{A.15d}$$

Some frequently used t-conorms are:
standard union: $u(a, b) = \max(a, b)$
(algebraic) sum: $u(a, b) = a + b - ab$
bold (Łukasiewicz) union: $u(a, b) = \min(1, a + b)$

Definition A.14 (complement of a fuzzy set) Let A be a fuzzy sets in X. The complement of A is a fuzzy set, denoted $\bar{A}$, such that for each $x \in X$:

$$\mu_{\bar{A}}(x) = 1 - \mu_A(x)\,. \tag{A.16}$$

A.5 Fuzzy Relations

Definition A.15 (fuzzy relation) An n-ary fuzzy relation is a mapping

$$R: X_1 \times X_2 \times \cdots \times X_n \rightarrow [0, 1], \tag{A.17}$$

which assigns membership grades to all n-tuples $(x_1, x_2, \ldots, x_n)$ from the Cartesian product $X_1 \times X_2 \times \cdots \times X_n$.

A fuzzy relation is, in fact, a fuzzy set in the Cartesian product $X_1 \times X_2 \times \cdots \times X_n$. The membership grades represent the degree of association (correlation) among the elements of the different domains X_i. For computer implementations, R is conveniently represented as an n-dimensional array: $R = [r_{i_1, i_2, \ldots, i_n}]$.

Example: Consider a fuzzy relation R describing the relationship $x \approx y$ ("x is approximately equal to y") by means of the following membership function $\mu_R(x, y) = e^{-(x-y)^2}$. Figure A.1 shows a mesh plot of this relation.

A.6 Projections and Cylindrical Extensions

Definitions in this section are adopted from (Kruse, et al., 1994).

Definition A.16 (n-dimensional universe) A family $\mathcal{U} = (X^{(i)})_{i \in \mathbb{N}_n}$ of non-empty domains $X^{(i)}$, $i = 1, 2, \ldots, n$, $n \in \mathbb{N}$, is called a universe of dimension n. $\mathbb{N}_n = \{1, 2, \ldots, n\}$ is the index set related to this universe. For any non-empty index subset $I \subseteq \mathbb{N}_n$ the product space is defined by:

$$X^I = \times_{i \in I} X^{(i)}$$

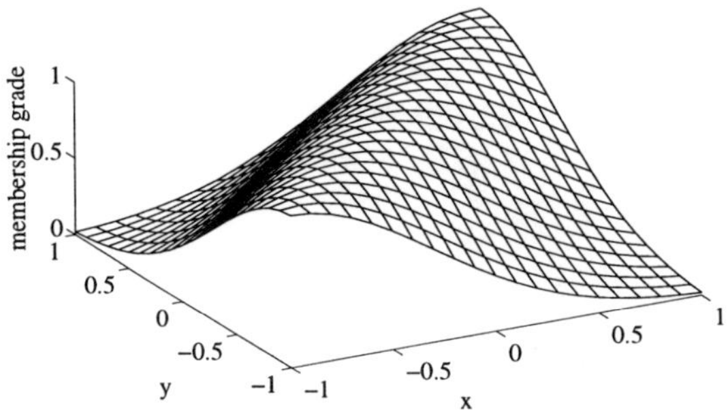

Figure A.1. Fuzzy relation $\mu_R(x, y) = e^{-(x-y)^2}$.

The product space $X^{\mathbb{N}_n}$ is denoted by X for short. A fuzzy set defined on a multidimensional universe is called a *multidimensional fuzzy set.*

Definition A.17 (point-wise projection) Let $\mathcal{U} = (X^{(i)})_{i \in \mathbb{N}_n}$ be a universe of dimension n and let C, S and T be index subsets of $\mathbb{N}_n$ which satisfy the conditions $T = S \cup C$, $S \cap C = \emptyset$ and $S \neq \emptyset$. Point-wise projection of X^T onto X^S is the mapping $\text{red}_S^T \colon X^T \to X^S$ defined by:

$$\text{red}_S^T(x^T) = x^S \text{ with } \left(\forall i \in S \colon (x^S)^{(i)} = (x^T)^{(i)}\right) . \tag{A.18}$$

Definition A.18 (projection of a fuzzy set) Let $\mathcal{U} = (X^{(i)})_{i \in \mathbb{N}_n}$ be a universe of dimension n, M an index set with $\emptyset \neq M \subseteq \mathbb{N}_n$. The projection of A onto X^M is the mapping $\text{proj}_M \colon \mathcal{F}(X) \to \mathcal{F}(X^M)$ defined by

$$\text{proj}_M\left(\mu(x)\right) = \sup\left\{\mu(x') \mid x' \in X \ \wedge \ x = \text{red}_M^{\mathbb{N}_n}(x')\right\} . \tag{A.19}$$

An example of projection from $\mathbb{R}^2$ to $\mathbb{R}$ is given in Figure A.2.

Definition A.19 (cylindrical extension of a fuzzy set) Let $\mathcal{U} = (X^{(i)})_{i \in \mathbb{N}_n}$ be a universe of dimension n. Let M and N be index subsets with $\emptyset \neq M \subseteq N \subseteq \mathbb{N}_n$. The cylindrical extension of A to X^N is the mapping $\text{ext}_M^N \colon \mathcal{F}(X^M) \to \mathcal{F}(X^N)$ defined by

$$\text{ext}_M^N\left(\mu(x)\right) = \mu\left(\text{red}_M^N(x)\right) . \tag{A.20}$$

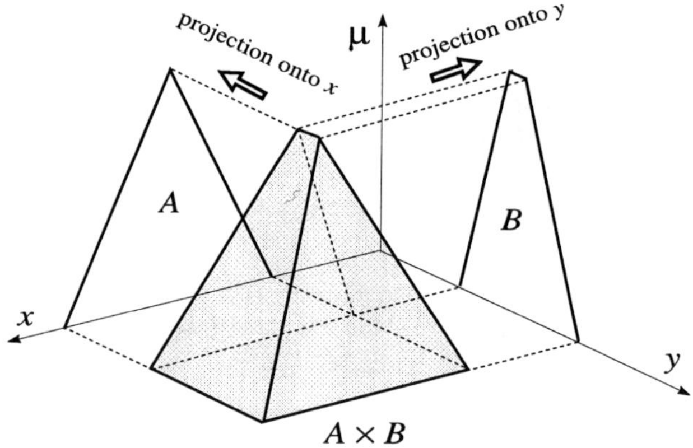

Figure A.2. Example of projection from $\mathbb{R}^2$ to $\mathbb{R}$.

Appendix B
Fuzzy Modeling and Identification Toolbox for MATLAB

This appendix briefly describes a toolbox for the use with MATLAB[1], which was developed to support the fuzzy modeling and identification techniques described in this book.

B.1 Toolbox Structure

The functions in the toolbox can be divided into the following categories:

- Fuzzy models of the Mamdani and Takagi–Sugeno types.
- Fuzzy clustering tools: fuzzy c-means, possibilistic c-means, Gustafson-Kessel algorithm, validity measures, etc.
- Identification based on product-space fuzzy clustering, simulation of dynamic fuzzy models.
- User interface: rule base editor and parser, membership functions editor.

Most of the functions are described in a reference manual and demonstrations are provided. In the following sections, a brief description of the identification tools is given.

B.2 Identification of MIMO Dynamic Systems

A MIMO model is represented as a set of coupled input–output MISO models of the Takagi-Sugeno type. Consider a system with n_i inputs: $\mathbf{u} \in U \subset \mathbb{R}^{n_i}$, and n_o outputs: $\mathbf{y} \in Y \subset \mathbb{R}^{n_o}$. Denote q^{-1} the backward shift operator: $q^{-1}y(k) \stackrel{\text{def}}{=} y(k-1)$. Denote by A and B polynomials in q^{-1}, e.g., $A = a_0 + a_1 q^{-1} + a_2 q^{-2} + \ldots$. Given two integers, $m < n$, define an ordered sequence of delayed samples of the signal y as:

$$\{y(k)\}_m^n \stackrel{\text{def}}{=} [y(k-m), y(k-m-1), \ldots, y(k-m-n+1)]\,. \tag{B.1}$$

[1] MATLAB is a registered trademark of The Mathworks, Inc.

The MISO models are of the input–output NARX type:

$$y_l(k+1) = F_l\left(\mathbf{x}_l(k)\right), \quad l = 1, 2, \ldots, n_o, \tag{B.2}$$

where the regression vector $\mathbf{x}_l(k)$ is given by:

$$\begin{aligned} \mathbf{x}_l(k) \;=\; & \left[\{y_1(k)\}_0^{n_{yl1}}, \{y_2(k)\}_0^{n_{yl2}}, \ldots, \{y_{n_o}(k)\}_0^{n_{yln_o}}, \right. \\ & \left. \{u_1(k+1)\}_{n_{dl1}}^{n_{ul1}}, \{u_2(k+1)\}_{n_{dl2}}^{n_{ul2}}, \ldots, \{u_{n_i}(k+1)\}_{n_{dln_i}}^{n_{uln_i}}\right] . \end{aligned} \tag{B.3}$$

Recall that n_y and n_u are the number of delayed outputs and inputs, respectively, and n_d is the number of pure (transport) delays from the input to the output. n_y is an $n_o \times n_o$ matrix, and n_u, n_d are $n_o \times n_u$ matrices. F_l are fuzzy models of the TS type (5.1). The individual MISO models are estimated independently of each other, using the techniques from Section 5.1.

B.3 Matlab implementation

The core of the identification routine is the GK clustering algorithm. Its basic implementation is given below:

```
function [U,V,F] = gk(Z,U0,m,tol)
% Clustering with fuzzy covariance matrix (Gustafson-Kessel algorithm)
%
% [U,V,F] = GK(Z,U0,m,tol)
%--------------------------------------------------------------------
% Input:  Z     ... N by n data matrix
%         U0    ... initial fuzzy partition matrix or number of clusters
%         m     ... fuzziness exponent (m > 1)
%         tol   ... termination tolerance
%--------------------------------------------------------------------
% Output: U     ... fuzzy partition matrix
%         V     ... cluster means (centers)
%         F     ... concatenated cluster covariance matrices:
%                   F = [F1;F2;...Fc], where c is the number of clusters

%----------------- prepare matrices ----------------------------------

[mz,nz] = size(Z);                                % data matrix size
c = size(U0,2);
if c == 1, c = U0; end;                           % number of clusters
mZ1 = ones(mz,1);                                 % auxiliary variable
nZ1 = ones(nz,1);                                 % auxiliary variable
V1c = ones(1,c);                                  % auxiliary variable
U = zeros(mz,c);                                  % partition matrix
d = U;                                            % distance matrix
F = zeros(c*nz,nz);                               % covariance matrix

%----------------- initialize U --------------------------------------

if size(U0,2) == 1,
  meanZ = mean(Z);
  aa = max(abs(Z - ones(mz,1)*meanZ));
  V = 2*(ones(c,1)*aa).*(rand(c,nz)-0.5) + ones(c,1)*meanZ;
  for j = 1 : c,
    ZV = Z - mZ1*V(j,:);
```

```
    d(:,j) = sum((ZV.^2)')';
  end;
  d = (d+1e-10).^(-1/(m-1));
  U0 = (d ./ (sum(d')'*V1c));
end;

%----------------- iterate -----------------------------------------

while  max(max(U0-U)) > tol                        % auxiliary variable
  U = U0; Um = U.^m; sumU = sum(Um);               % aux vars
  V = (Um'*Z)./(nZ1*sumU)';                        % calculate centers
  for j = 1 : c,                                   % for all clusters
    ZV = Z - mZ1*V(j,:);                           % aux var
    f = nZ1*Um(:,j)'.*ZV'*ZV/sumU(j);              % covariance matrix
    d(:,j)=sum((ZV*(det(f)^(1/nz)*inv(f)).*ZV)')';  % distances
  end;
  d = (d+1e-10).^(-1/(m-1));
  U0 = (d ./ (sum(d')'*V1c));                       % partition matrix
end

%----------------- update final F and U ----------------------------

U = U0; Um = U.^m; sumU = nZ1*sum(Um);
for j = 1 : c,
    ZV = Z - mZ1*V(j,:);
    F((j-1)*nz+(1:nz),:) = nZ1*Um(:,j)'.*ZV'*ZV/sumU(1,j);
end;

%----------------- end of function ---------------------------------
```

In the toolbox, this basic GK iteration scheme is enhanced with an on-line graphical output, and some additional output arguments are returned.

The GK algorithm is a part of a higher-level routine `fmclust`[2] which first transforms the time sequences of data into n_o regression problems. For each of them, a TS model is then constructed. The function has the following synopsis:

```
FM = fmclust(U,Y,c,FMtype,m,tol,seed,ny,nu,nd).
```

The data sequences are given in columns of the matrices `U` and `Y`, respectively. The number of required clusters, `c`, is a scalar for MISO systems and a vector for MIMO systems (each MISO model may have a different number of clusters). The remaining parameters are optional. `FMtype` specifies whether the antecedent membership functions of the resulting model are computed analytically in the antecedent product space by using eq. (5.13), or are derived by projection (5.3). Product-space membership functions give faster but often less accurate models. `m` is the fuzziness exponent with the default value `m` = 2. Larger values imply fuzzier (more overlapping) clusters. The termination tolerance for the clustering algorithm can be given in `tol` (default `tol` = 0.01). The fuzzy partition matrix is initialized at random. In order to obtain reproducible results, the random generator may be seeded by supplying the `seed` parameter.

[2]The development of this tool was supported by the Esprit LTR project FAMIMO (Fuzzy Algorithms for the Control of Multi-Input, Multi-Output Processes), No. 21911.

Default value is `sum(100*clock)`. The `ny`, `nu` and `nd` parameters are the delay matrices defined in Section B.2.

The output of `fmclust` is the `FM` matrix which contains the parameters of the obtained fuzzy model (under MATLAB 5.1, `FM` is a structure). The obtained fuzzy model can be simulated by using function `fmsim` with the following synopsis:

```
[Ym,VAF] = fmsim(U,Y,FM)\,.
```

This function simulates the fuzzy model `FM` from the input data `U` and plots the simulated output `Ym` along with the true output `Y`. The first n_y values of `Y` are used to initialize `Ym`. The output argument `VAF` is the performance index of the model, computed as the *variance accounted for* (VAF) given by:

$$\mathrm{VAF} = 100\% \cdot \left[1 - \left(\mathrm{var}(\mathtt{Y} - \mathtt{Ym})\right)/\mathrm{var}(\mathtt{Y})\right] . \tag{B.4}$$

As additional output arguments, the degrees of fulfillment and the local outputs of the individual rules can be obtained. A simple example of a sinusoidal function approximated by the TS fuzzy model with five rules is:

```
u = (0:0.02:1)';
y = sin(7*u);
FM = fmclust(u,y,5);
[ym,VAF,Yloc] = fmsim(u,y,FM);
```

As another example, consider a two-input, two-output dynamic model with the following structure:

$$\begin{aligned} y_1(k+1) &= f_1(y_1(k), y_1(k-1), u_1(k), u_2(k)), \\ y_2(k+1) &= f_2(y_1(k), y_2(k), u_2(k-2))\,. \end{aligned}$$

The corresponding structural parameters are defined as follows:

```
ny = [2 0;                 % output lags
      1 1];
nu = [1 1;                 % input lags
      0 1];
nd = [1 1;                 % transport delays
      1 3];
```

Assume that the identification data are available in matrices `U1` and `Y1`. The system can be identified by:

```
c = [3 3];                 % number of clusters
m = 2.2;                   % fuzziness parameter
tol = 0.001;               % termination criterion
FMtype = 2;                % projected membership f.
FM = fmclust(U1,Y1,c,FMtype,m,tol,0,ny,nu,nd);
```

Here, values different from the defaults were supplied for the `m`, `tol` and `FMtype` parameters. The number of clusters is three in each MISO model. The obtained model can be validated by using data sets `U2` and `Y2`:

```
[ym,VAF] = fmsim(U2,Y2,FM);
```

In addition to the above functions, utilities are provided, e.g., for automatically generating a LaTEX documentation for the obtained model (`fm2tex`). The toolbox can be obtained from the author at the following address:

Dr. Robert Babuška
Control Engineering Laboratory
Faculty of Information Technology and Systems
Delft University of Technology
Mekelweg 4, P.O. Box 5031, 2600 GA Delft
the Netherlands

tel: +31 15 2785117
fax: +31 15 2626738
e-mail: R.Babuska@ET.TUDelft.NL
http:// lcewww.et.tudelft.nl/ ~babuska

Appendix C
Symbols and Abbreviations

Printing Conventions. Lower case characters in bold print denote column vectors. For example, $\mathbf{x}$ and $\mathbf{a}$ are column vectors. A row vector is denoted by using the transpose operator, for example $\mathbf{x}^T$ and $\mathbf{a}^T$. Lower case characters in italics denote elements of vectors and scalars. Upper case bold characters denote matrices, for instance, $\mathbf{X}$ is a matrix. Upper case italic characters such as A denote crisp and fuzzy sets. Upper case calligraphic characters denote families (sets) of sets.

No distinction is made between variables and their values, hence x may denote a variable or its value, depending on the context. No distinction is made either between a function and its value, e.g., μ may denote both a membership function and its value (a membership degree). Superscripts are sometimes used to index variables rather than to denote a power or a derivative. Where confusion could arise, the upper index is enclosed in parentheses. For instance, in fuzzy clustering $\mu_{ik}^{(l)}$ denotes the ikth element of a fuzzy partition matrix, computed at the lth iteration. $(\mu_{ik}^{(l)})^m$ denotes the mth power of this element.

General mathematical symbols

$A, B, \ldots$	fuzzy sets
$\mathcal{A}, \mathcal{B}, \ldots$	families (sets) of fuzzy sets
$\mathbf{A}, \mathbf{B}, \mathbf{C}, \mathbf{D}$	system matrices
$\mathcal{F}(X)$	set of all fuzzy sets on X
$\mathbf{I}$	identity matrix of appropriate dimensions
K	number of rules in a rule base
$P(A)$	power set of A
R	fuzzy relation
$S(\cdot, \cdot)$	similarity measure
$\mathbf{X}$	matrix containing regressor data
X, Y	domains (universes) of variables x and y
$\mathbf{a}, b$	consequent parameters in a TS model
n	dimension of the vector $[\mathbf{x}^T, y]$
p	dimension of $\mathbf{x}$
$u(k), y(k)$	input and output of a dynamic system at time k

$x(k)$	state of a dynamic system
$\mathbf{x}$	regression vector
y	regressand
$\mathbf{y}$	vector containing regressand data
β	degree of fulfillment of a rule
γ	normalized degree of fulfillment
μ, $\mu(\cdot)$	membership degree, membership function
τ	time constant
$\mathbb{N}_k$	set of natural numbers $1, 2, \ldots, k$
$\mathbb{R}$	set of real numbers
$\mathbf{0}$	matrix of appropriate dimensions with all entries equal to zero
$\mathbf{1}$	matrix of appropriate dimensions with all entries equal to one

Symbols related to fuzzy clustering

$\mathbf{F}$	cluster covariance matrix
$\mathbf{H}$	projection matrix
M_{fc}	fuzzy partitioning space
M_{hc}	hard partitioning space
M_{pc}	possibilistic partitioning space
$\mathbf{U} = [\mu_{ik}]$	fuzzy partition matrix
$\mathbf{V}$	matrix containing cluster prototypes (means)
$\mathbf{Z}$	data (feature) matrix
c	number of clusters
$d(\cdot, \cdot)$	distance measure
m	weighting exponent (determines fuzziness of the partition)
$\mathbf{v}$	cluster prototype (center)
$\mathbf{z}$	data vector
ϕ	eigenvector of $\mathbf{F}$
λ	eigenvalue of $\mathbf{F}$
$\mu_{i,k}$	membership of data vector $\mathbf{z}_k$ into cluster i

Operators:

$\cap$	(fuzzy) set intersection (conjunction)
$\cup$	(fuzzy) set union (disjunction)
$\wedge$	intersection, logical AND, minimum
$\vee$	union, logical OR, maximum
$\lvert A \rvert$	cardinality of (fuzzy) set A
$\mathbf{X}^T$	transpose of matrix $\mathbf{X}$
$\bar{A}$	complement (negation) of A
∂	partial derivative
$\circ$	sup-t (max-min) composition
$\langle \mathbf{x}, \mathbf{y} \rangle$	inner product of $\mathbf{x}$ and $\mathbf{y}$
$\mathrm{cog}(A)$	center of gravity defuzzification of fuzzy set A
$\mathrm{core}(A)$	core of fuzzy set A
det	determinant of a matrix
diag	diagonal matrix
$\mathrm{ext}_M^N(A)$	cylindrical extension of A from M to N
$\mathrm{hgt}(A)$	height of fuzzy set A

$\text{mom}(A)$	mean of maxima defuzzification of fuzzy set A
$\text{norm}(A)$	normalization of fuzzy set A
$\text{proj}_M(A)$	point-wise projection of A onto M
$\text{rank}(\mathbf{X})$	rank of matrix $\mathbf{X}$
$\text{supp}(A)$	support of fuzzy set A

Abbreviations

AI	artificial intelligence
B&B	branch-and-bound technique
CMAC	cerebellar model articulation controller
COG	center of gravity
FLOP	floating point operations
MBPC	model-based predictive control
MIMO	multiple–input, multiple–output
MISO	multiple–input, single–output
MOM	mean of maxima
(N)ARMAX	(nonlinear) autoregressive moving average with exogenous inputs
(N)ARX	(nonlinear) autoregressive with exogenous inputs
(N)FIR	(nonlinear) finite impulse response model
(N)OE	(nonlinear) output error model
OLS	ordinary least-squares method
PID	proportional integral derivative (controller)
SISO	single–input, single–output
SQP	sequential quadratic programming
TLS	total least-squares method

Abbreviations of clustering algorithms and validity measures

ACF	average cluster flatness
APD	average partition density
AWCD	average within-cluster distance
CCM	compatible cluster merging
FCE	fuzzy c-elliptotypes
FCM	fuzzy c-means
FCR	fuzzy c-regression
FCV	fuzzy c-varieties
FHV	fuzzy hypervolume
FMLE	fuzzy maximum likelihood estimates clustering
GK	Gustafson–Kessel algorithm
PCM	possibilistic c-means
PGK	possibilistic GK algorithm

References

Akaike, H. (1974). A new look at the statistical model identification. *IEEE Trans. Automatic Control 19*(6), 716–723.

Albus, J. (1975). A new approach to manipulator control: the cerebellar model articulation controller (CMAC). *Trans. of the ASME*, 220–227.

Ali, S. and J. Heuer (1995). Knowledge-based multimodels as a basis for the design of man-machine systems in process control. In *Proceedings 14-th European Annual Conference on Human Decision Making and Manual Control*, Delft, the Netherlands, pp. Session 5–2.

Anderson, C. (1987). Strategy learning with multilayer connectionist representations. In *Proceedings Fourth International Workshop on Machine Learning*, Irvine, USA, pp. 103–114.

Åström, K. and B. Wittenmark (1989). *Adaptive Control.* Addison–Wesley.

Babu, G. and M. Murty (1994). Clustering with evolution strategies. *Pattern Recognition 27*(2), 321–329.

Babuška, R. (1995). Fuzzy modeling and cluster analysis toolbox for MATLAB. In *Proceedings Third European Congress on Intelligent Techniques and Soft Computing EUFIT'95*, Aachen, Germany, pp. 1479–1483.

Babuška, R., M. Botto, J. S. da Costa and H. Verbruggen (1996). Neural and fuzzy modeling in nonlinear predictive control: *A Comparative Study.* In *Proceedings CESA, Lille*, pp. 1049–1054.

Babuška, R., C. Fantuzzi, U. Kaymak and H. Verbruggen (1996). Improved inference for Takagi–Sugeno models. In *Proceedings Fifth IEEE International Conference on Fuzzy Systems*, New Orleans, USA, pp. 701–706.

Babuška, R., M. Setnes, U. Kaymak and H. van Nauta Lemke (1996). Rule base simplification with similarity measures. In *Proceedings Fifth IEEE International Conference on Fuzzy Systems*, New Orleans, USA, pp. 1642–1647.

Babuška, R., J. Sousa and H. Verbruggen (1995). Model-based design of fuzzy control systems. In *Proceedings Third European Congress on Intelligent Techniques and Soft Computing EUFIT'95*, Aachen, Germany, pp. 837–841.

Babuška, R., H. te Braake, A. Krijgsman and H. Verbruggen (1996). Comparison of intelligent control schemes for real-time pressure control. *Control Engineering Practice 4*(11), 1585–1592.

Babuška, R., H. van Can and H. Verbruggen (1996). Fuzzy modeling of enzymatic Penicillin–G conversion. In *Preprints 13th IFAC World Congress*, Volume N, San Francisco, USA, pp. 479–484.

Babuška, R. and H. Verbruggen (1994a). Applied fuzzy modeling. In *Proceedings IFAC Symposium on Artificial Intelligence in Real Time Control*, Valencia, Spain, pp. 61–66.

Babuška, R. and H. Verbruggen (1995c). Identification of composite linear models via fuzzy clustering. In *Proceedings European Control Conference*, Rome, Italy, pp. 1207–1212.

Babuška, R. and H. Verbruggen (1995d). New approach to constructing fuzzy relational models from data. In *Proceedings Third European Congress on Intelligent Techniques and Soft Computing EUFIT'95*, Aachen, Germany, pp. 583–587.

Babuška, R. and H. Verbruggen (1995b). A new identification method for linguistic fuzzy models. In *Proceedings FUZZ-IEEE/IFES'95*, Yokohama, Japan, pp. 905–912.

Babuška, R. and H. Verbruggen (1996e). An overview of fuzzy modeling for control. *Control Engineering Practice 4*(11), 1593–1606.

Babuška, R. and H. Verbruggen (1997f). Fuzzy set methods for local modeling and identification. In R. Murray-Smith and T. A. Johansen (Eds.), *Multiple Model Approaches to Nonlinear Modeling and Control*, pp. 75–100. London, UK: Taylor & Francis.

Backer, E. (1995). *Computer-assisted Reasoning in Cluster Analysis*. New York: Prentice Hall.

Baldwin, J., T. Martin and B. Pilsworth (1995). *FRIL – Fuzzy and Evidential Reasoning in Artificial Intelligence*. Taunton: Research Studies Press.

Barto, A., R. Sutton and C. Anderson (1983). Neuron like adaptive elements that can solve difficult learning control problems. *IEEE Trans. Systems, Man and Cybernetics 13*(5), 834–846.

Bellman, R. and L. Zadeh (1970). Decision making in a fuzzy environment. *Management Science 17*(4), 141–164.

Berenji, H. and P. Khedar (1993). Clustering in product space for fuzzy inference. In *Proceedings of Second International Conference on Fuzzy Systems*, San Francisco, California, U.S.A., pp. 1402–1407.

Bezdek, J. (1980). A convergence theorem for the fuzzy isodata clustering algorithms. *IEEE Trans. Pattern Anal. Machine Intell. PAMI-2*(1), 1–8.

Bezdek, J. (1981). *Pattern Recognition with Fuzzy Objective Function*. Plenum Press, New York.

Bezdek, J., C. Coray, R. Gunderson and J. Watson (1981a). Detection and characterization of cluster substructure, I. Linear structure: Fuzzy c-lines. *SIAM J. Appl. Math. 40*(2), 339–357.

Bezdek, J., C. Coray, R. Gunderson and J. Watson (1981b). Detection and characterization of cluster substructure, II. Fuzzy c-varieties and convex combinations thereof. *SIAM J. Appl. Math. 40*(2), 358–372.

Bezdek, J. and J. Dunn (1975). Optimal fuzzy partition: a heuristic for estimating the parameters in a mixture of normal distributions. *IEEE Trans. Computers C-24*, 835–838.

Bezdek, J., R. Hathaway, R. Howard, C. Wilson and M. Windham (1987). Local convergence analysis of a grouped variable version of coordinate descent. *Journal of Optimization Theory and Applications 54*(3), 471–477.

Bezdek, J. and S. Pal (Eds.) (1992). *Fuzzy Models for Pattern Recognition*. New York: IEEE Press.

Botto, M., T. van den Boom, A. Krijgsman and J. S. da Costa (1998). Predictive control based on neural network models with I/O feedback linearization. To appear in International Journal of Control.

Boyd, S., L. Ghaoui and E. Feron (1994). *Linear Matrix Inequalities in Systems and Control Theory*. Philadelphia: SIAM.

Brown, M. and C. Harris (1994). *Neurofuzzy Adaptive Modelling and Control*. New York: Prentice Hall.

Buckley, J. and Y. Hayashi (1994). Fuzzy neural networks. In R. R. Yager and L. A. Zadeh (Eds.), *Fuzzy Sets, Neural Networks and Soft Computing*, pp. 233–249. New York: van Nostrand Reinhold.

Camacho, E. and C. Bordons (1995). *Model Predictive Control in the Process Industry*. Berlin, Germany: Springer Verlag.

Carleysmith, S., P. Dunhill and D. Lilly (1973). Kinetics behaviour of immobilized penicillin amidase. *Biotechnology and Bioengineering 15*, 13–25.

Chen, S. and S. Billings (1989). Representation of nonlinear systems: The NARMAX model. *International Journal of Control 49*, 1013–1032.

Chen, S., C. Cowan and P. Grant (1991). Orthogonal least squares learning algorithm for radial basis function networks. *IEEE Trans. Neural Networks* 2(2), 302–309.

Chen, X. and M. Bushnell (1996). *Efficient Branch and Bound Search with Applications to Computer-Aided Design*. Boston, Dordrecht, London: Kluwer Academic Publishers.

Clarke, D., C. Mohtadi and P. Tuffs (1987). Generalised predictive control. part 1: The basic algorithm. part 2: Extensions and interpretations. *Automatica 23*(2), 137–160.

Dave, R. (1992). Boundary detection through fuzzy clustering. In *IEEE International Conference on Fuzzy Systems*, San Diego, USA, pp. 127–134.

de Boor, C. (1978). *A Practical Guide to Splines*. Springer-Verlag, New York.

de Oliveira, J. V. (1993). Neuron inspired rules for fuzzy relational structures. *Fuzzy Sets and Systems 57*(1), 41–55.

de Vlieger, J. (1983). *Minimum-Time Control of Systems with Saturation Bounds*. PhD dissertation, Delft University of Technology, Delft, the Netherlands.

Deketh, H. (1995). *Wear of rock cutting tools. Laboratory experiments on the abrasivity of rock*. Rotterdam, the Netherlands: Balkema Publishers.

den Hartog, M., R. Babuška, H. Deketh, M. A. Grima, P. Verhoef and H. Verbruggen (1997). Knowledge-based fuzzy model for performance prediction of a rock-cutting trencher. *International Journal of Approximate Reasoning 16*(1), 43–66.

DeSarbo, W. (1982). Gennclus: New models for general nonhierarchical clustering analysis. *Psychometrika 47*(4), 449–76.

Driankov, D., H. Hellendoorn and M. Reinfrank (1993). *An Introduction to Fuzzy Control.* Springer, Berlin.

Duda, R. and P. Hart (1973). *Pattern Classification and Scene Analysis.* New York: John Wiley & Sons.

Dunn, J. (1974a). A fuzzy relative of the ISODATA process and its use in detecting compact well-separated clusters. *J. Cybern. 3*(3), 32–57.

Dunn, J. (1974b). A graph theoretic analysis for pattern classification via Tamura's fuzzy relation. *IEEE Trans. Systems, Man, and Cybernetics 4*(3), 310–313.

Economou, C., M. Morari and B. Palsson (1986). Internal model control. 5. Extension to nonlinear systems. *Ind. Eng. Chem. Process Des. Dev. 25*, 403–411.

Esragh, F. and E. Mamdani (1979). A general approach to linguistic approximation. *Int. J. Man-Machine Studies 11*, 501–519.

Fantuzzi, C. and R. Rovatti (1996). On the approximation capabilities of the homogeneous takagi–sugeno model. In *Proceedings Fifth IEEE International Conference on Fuzzy Systems*, New Orleans, USA, pp. 1067–1072.

Filev, D. (1996). Model based fuzzy control. In *Proceedings Fourth European Congress on Intelligent Techniques and Soft Computing EUFIT'96*, Aachen, Germany.

Friedman, J. (1991). Multivariate adaptive regression splines. *The Annals of Statistics 19*(1), 1–141.

Gaines, B. and J. Boose (1988). *Knowledge Acquisition for Knowledge-Based Systems.* London: Academic Press.

Gath, I. and A. Geva (1989). Unsupervised optimal fuzzy clustering. *IEEE Trans. Pattern Analysis and Machine Intelligence 7*, 773–781.

Gill, P., W. Murray and M. Wright (1981). *Practical Optimization.* New York and London: Academic Press.

Godfrey, K. (Ed.) (1993). *Perturbation Signals for System Identification.* New York: Prentice Hall.

Gunderson, R. (1983). An adaptive FCV clustering algorithm. *Int. J. Man-Machine Studies 19*, 97–104.

Gustafson, D. and W. Kessel (1979). Fuzzy clustering with a fuzzy covariance matrix. In *Proc. IEEE CDC*, San Diego, CA, USA, pp. 761–766.

Hall, R. and D. Seborg (1989). Modelling and selft-tuning control of a multivariable pH neutralization process. part I: Modelling and multiloop control. In *Proceedings American Control Conference*, Volume 2, Pittsburgh, U.S.A., pp. 1822–1827.

Hathaway, R. and J. Bezdek (1991a). Grouped coordinate minimization using Newton's method for inexact minimization in one vector coordinate. *Journal of Optimization Theory and Applications 71*(3), 503–516.

Hathaway, R. and J. Bezdek (1993b). Switching regression models and fuzzy clustering. *IEEE Trans. Fuzzy Systems 1*(3), 195–204.

Haykin, S. (1994). *Neural Networks.* New York: Macmillan Maxwell International.

Henson, M. and M. Kurtz (1994). Input–output linearization of constrained nonlinear processes. In *Paper 230d of the 1994 Anual AIChE Meeting*, San Francisco, U.S.A.

Huber, P. (1985). Projection pursuit (with discussion). *The Annals of Statistics 13*, 435–475.

Hunt, K., D. Sbarbaro, R. Zbikowki and P. Gawthrop (1992). Neural networks for control systems – a survey. *IEEE Trans. Neural Networks 28*, 1083–1112.

Ikoma, N. and K. Hirota (1993). Nonlinear autoregressive model based on fuzzy relation. *Information Sciences 71*, 131–144.

Isermann, R. (1977). *Digital Control Systems*. Berlin: Springer–Verlag.

Jager, R. (1995). *Fuzzy Logic in Control*. PhD dissertation, Delft University of Technology, Delft, The Netherlands.

Jager, R., H. Verbruggen and P. Bruijn (1992). The role of defuzzification methods in the application of fuzzy control. In *Proceedings IFAC Symposium on Intelligent Components and Instruments for Control Applications 1992*, Málaga, Spain, pp. 111–116.

Jain, A. and R. Dubes (1988). *Algorithms for Clustering Data*. Englewood Cliffs: Prentice Hall.

Jang, J.-S. (1992). Self-learning fuzzy controllers based on temporal back propagation. *IEEE Trans. Neural Networks 3*(5), 714–723.

Jang, J.-S. (1993). ANFIS: Adaptive-network-based fuzzy inference systems. *IEEE Transactions on Systems, Man & Cybernetics 23*(3), 665–685.

Jang, J.-S. and C.-T. Sun (1993). Functional equivalence between radial basis function networks and fuzzy inference systems. *IEEE Transactions on Neural Networks 4*(1), 156–159.

Jang, J.-S., C.-T. Sun and E. Mizutani (1997). *Neuro-Fuzzy and Soft Computing; a Computational Approach to Learning and Machine Intelligence*. Upper Sadle River: Prentice-Hall.

Johansen, T. (1994). *Operating Regime Based Process Modelling and Identification*. PhD dissertation, The Norwegian Institute of Technology – University of Trondheim, Trondheim, Norway.

Kacprzyk, J. (1983). *Multistage Decision-Making under Fuzziness*. Verlag TUV Rheinland.

Kacprzyk, J. (1997). *Multistage Fuzzy Control; a Model-Based Approach to Fuzzy Control and Decision Making*. Chichester: Wiley.

Kavli, T. (1993). ASMOD – an algorithm for adaptive spline modeling of observation data. *International Journal of Control 58*(4), 947–967.

Kaymak, U. (1994). Application of fuzzy methodologies to a washing process. Chartered designer thesis, Delft University of Technology, Control Lab., Faculty of El. Eng., Delft.

Kaymak, U. and R. Babuška (1995). Compatible cluster merging for fuzzy modeling. In *Proceedings FUZZ-IEEE/IFES'95*, Yokohama, Japan, pp. 897–904.

Kaymak, U., J. Sousa and H. Verbruggen (1997). A comparative study of fuzzy and conventional criteria in model-based predictive control. In *Proceedings FUZZ-IEEE'97*, Barcelona, Spain, pp. 907–914.

Kaymak, U. and H. van Nauta Lemke (1993). A parametric generalized goal function for fuzzy decision making with unequally weighted objectives. In *Proceedings of the Second IEEE International Conference on Fuzzy Systems*, Volume 2, San Francisco, USA, pp. 1156–1160.

Klawonn, F., J. Kinzel and R. Kruse (1994). Modification of genetic algorithms for designing and optimizing fuzzy controllers. In *Proceedings of the First IEEE Conference on Evolutionary Computation*, Orlando, U.S.A., pp. 28–33.

Klir, G. and B. Yuan (1995). *Fuzzy sets and fuzzy logic; theory and applications*. Prentice Hall.

Knijnenburg, O. (1997). Estimation of respiratory mechanics with multiple linear regression and fuzzy clustering. MSc thesis, Delft University of Technology, Delft, the Netherlands.

Kosko, B. (1994). Fuzzy systems as universal approximators. *IEEE Trans. Computers 43*, 1329–1333.

Krishnapuram, R. (1994). Generation of membership functions via possibilistic clustering. In *IEEE World Congress on Computational Intelligence*, Volume 2, Orlando, U.S.A., pp. 902–908.

Krishnapuram, R. and C.-P. Freg (1992). Fitting an unknown number of lines and planes to image data through compatible cluster merging. *Pattern Recognition 25*(4), 385–400.

Krishnapuram, R. and J. Keller (1993). A possibilistic approach to clustering. *IEEE Trans. Fuzzy Systems 1*(2), 98–110.

Krone, A. and H. Kiendl (1994). Automatic generation of positive and negative rules for two-way fuzzy controllers. In *Proceedings Second European Congress on Intelligent Techniques and Soft Computing EUFIT'94*, Aachen, Germany, pp. 1206–1210.

Kruse, R., J. Gebhardt and F. Klawonn (1994). *Foundations of Fuzzy Systems*. Chichester: John Wiley and Sons.

Ku, R. and M. Athans (1973). On the adaptive control of linear systems using the open-loop feedback optimal approach. In *Proceedings 3rd IFAC Symposium*.

Lakoff, G. (1973). Hedges: a study in meaning criteria and the logic of fuzzy concepts. *Journal of Philosofical Logic 2*, 458–508.

Lawler, E. and E. Wood (1966). Branch-and-bound methods: A survey. *Journal of Operations Research 14*, 699–719.

Lee, C. (1990a). Fuzzy logic in control systems: fuzzy logic controller - part I. *IEEE Trans. Systems, Man and Cybernetics 20*(2), 404–418.

Lee, C. (1990b). Fuzzy logic in control systems: fuzzy logic controller - part II. *IEEE Trans. Systems, Man and Cybernetics 20*(2), 419–435.

Leonaritis, I. and S. Billings (1985). Input-output parametric models for non-linear systems. *International Journal of Control 41*, 303–344.

Lin, C. (1994). *Neural Fuzzy Control Systems with Structure and Parameter Learning*. Singapore: World Scientific.

Lindskog, P. and L. Ljung (1994). Tools for semiphysical modeling. In *Proceedings SYSID'94*, Volume 3, pp. 237–242.

Ljung, L. (1987). *System Identification, Theory for the User*. New Jersey: Prentice-Hall.

Luus, R. (1990). Optimal control by dynamic programming using systematic reduction in grid size. *Int. J. Control 51*, 995–1013.

Mamdani, E. (1977). Application of fuzzy logic to approximate reasoning using linguistic systems. *Fuzzy Sets and Systems 26*, 1182–1191.

McGraw, K. and K. Harbisson-Briggs (1989). *Knowledge Acquisition: Principles and Guidelines*. Englewood Cliffs: Prentice-Hall.

Mesarovic, M. and Y. Takahara (1989). *Abstract Systems Theory*. Berlin, Germany: Springer-Verlag.

Mitten, L. (1970). Branch-and-bound methods: General formulation and properties. *Journal of Operations Research 18*, 24–34.

Murray-Smith, R. and T. A. Johansen (Eds.) (1997). *Multiple Model Approaches to Nonlinear Modeling and Control*. London, UK: Taylor & Francis. (in press).

Nakamori, Y. and M. Ryoke (1994). Identification of fuzzy prediction models through hyperellipsoidal clustering. *IEEE Trans. Systems, Man and Cybernetics 24*(8), 1153–73.

Narendra, K., J. Balakrishnan and M. Ciliz (1995). Adaptation and learning using multiple models, switching, and tuning. *IEEE Control Systems 15*(3), 37–51.

Nevistić, V. and M. Morari (1995). Constraint control of feedback-linearizable systems. In *Proceedings Third European Control Conference*, Rome, Italy, pp. 1726–1731.

Nijmeijer, H. and A. van der Schaft (1990). *Nonlinear Dynamical Control Systems*. New York, USA: Springer-Verlag.

Novák, V. (1989). *Fuzzy Sets and their Applications*. Bristol: Adam Hilger.

Novák, V. (1996). A horizon shifting model of linguistic hedges for approximate reasoning. In *Proceedings Fifth IEEE International Conference on Fuzzy Systems*, New Orleans, USA, pp. 423–427.

Onnen, C., R. Babuška, U. Kaymak, J. Sousa, H. Verbruggen and R. Isermann (1997). Genetic algorithms for optimization in predictive control. *Control Engineering Practice 5*(10), 1363–1372.

Pal, N. and J. Bezdek (1995). On cluster validity for the fuzzy c-means model. *IEEE Trans. Fuzzy Systems 3*(3), 370–379.

Palm, R., D. Driankov and H. Hellendoorn (1997). *Model Based Fuzzy Control*. Springer, Berlin.

Parisini, T. and R. Zoppoli (1995). A receding-horizon regulator for nonlinear systems and a neural approximation. *Automatica 31*(10), 1443–1451.

Pedrycz, W. (1984). An identification algorithm in fuzzy relational systems. *Fuzzy Sets and Systems 13*, 153–167.

Pedrycz, W. (1985). Applications of fuzzy relational equations for methods of reasoning in presence of fuzzy data. *Fuzzy Sets and Systems 16*, 163–175.

Pedrycz, W. (1993). *Fuzzy Control and Fuzzy Systems (second, extended, edition)*. John Willey and Sons, New York.

Pedrycz, W. (1995). *Fuzzy Sets Engineering*. Boca Raton, Fl.: CRC Press.

Psichogios, D. and L. Ungar (1992). A hybrid neural network – first principles approach to process modeling. *AIChE J. 38*, 1499–1511.

Richalet, J. (1993). Industrial applications of model based predictive control. *Automatica 29*, 1251–1274.

Rissanen, J. (1978). Modeling by shortest data description. *Automatica 14*, 465–471.

Rovatti, R. (1996). Takagi–sugeno models as approximators in Sobolev norms: the SISO case. In *Proceedings Fifth IEEE International Conference on Fuzzy Systems*, New Orleans, USA, pp. 1060–1066.

Ruspini, E. (1970). Numerical methods for fuzzy clustering. *Inf. Sci. 2*, 319–350.

Schram, G., A. Krijgsman and M. Verhaegen (1996). System identification with orthogonal basis functions and neural networks. In *Preprints 13th IFAC World Congress*, San Francisco, USA, pp. 221–226.

Seber, G. and C. Wild (1989). *Nonlinear Regression*. New York: John Wiley & Sons.

Setnes, M. (1995). *Fuzzy Rule Base Simplification Using Similarity Measures*. M.Sc. thesis, Delft University of Technology, Delft, the Netherlands. (A.95.023).

Setnes, M., R. Babuška, H. Verbruggen, M. Sánchez and H. van den Boogaard (1997). Fuzzy modeling and similarity analysis applied to ecological data. In *Proceedings FUZZ-IEEE'97*, Barcelona, Spain, pp. 415–420.

Sjöberg, J., H. Hjalmarsson and L. Ljung (1994). Neural networks in system identification. In *Proceedings SYSID'94*, Volume 2, pp. 49–72.

Sjöberg, J., Q. Zhang, L. Ljung, A. Benveniste, B. Delyon, P.-Y. Glorennec, H. Hjalmarsson and A. Juditsky (1995). Nonlinear black-box modeling in system identification: a unified overview. *Automatica 31*(12), 1691–11724.

Smith, S., B. Nokleby and D. Comer (1994). A computational approach to fuzzy logic controller design and analysis using cell state space methods. In Kandel and Langholz (Eds.), *Fuzzy Control Systems*, pp. 398–427. Boca-Raton, Fl.: CRC Press.

Soeterboek, R. (1992). *Predictive Control: A Unified Approach*. New York, USA: Prentice Hall.

Sousa, J., R. Babuška and H. Verbruggen (1997). Fuzzy predictive control applied to an air-conditioning system. *Control Engineering Practice 5*(10), 1395–1406.

Strang, G. (1976). *Linear Algebra and Its Applications*. New York, U.S.A.: Academic Press.

Su, H.-T., N. Bhat and T. McAvoy (1992). Integrated neural networks with first principles models for dynamic modeling. In *Preprints IFAC DYCORD+'92*, College Park, Maryland.

Sugeno, M. and G. Kang (1988). Structure identification of fuzzy model. *Fuzzy Sets and Systems 28*, 15–33.

Sugeno, M. and K. Tanaka (1991). Successive identification of a fuzzy model and its application to prediction of a complex system. *Fuzzy Sets and Systems 42*, 315–334.

Sugeno, M. and T. Yasukawa (1993). A fuzzy-logic-based approach to qualitative modeling. *IEEE Trans. Fuzzy Systems 1*, 7–31.

Takagi, T. and M. Sugeno (1985). Fuzzy identification of systems and its application to modeling and control. *IEEE Trans. Systems, Man and Cybernetics 15*(1), 116–132.

Tanaka, H., S. Uejima and K. Asai (1982). Linear regression analysis with fuzzy model. *IEEE Trans. Systems, Man & Cybernetics 12*(6), 903–907.

Tanaka, K., T. Ikeda and H. Wang (1996). Robust stabilization of a class of uncertain nonlinear systems via fuzzy control: Quadratic stability, H_∞ control theory and linear matrix inequalities. *IEEE Transactions on Fuzzy Systems 4*(1), 1–13.

Tanaka, K. and M. Sugeno (1992). Stability analysis and design of fuzzy control systems. *Fuzzy Sets and Systems 45*(2), 135–156.

te Braake, H., H. van Can, G. van Straten and H. Verbruggen (1997). Two step approach in training of regulated activation weights neural networks (rawn). *Engineering Applications of AI*, 157–170.

Thompson, M. and M. Kramer (1994). Modeling chemical processes using prior knowledge and neural networks. *AIChE J. 40*, 1328–1340.

van Can, H., H. te Braake, C. Hellinga, A. Krijgsman, H. Verbruggen, K. Luyben and J. Heijnen (1995). Design and real-time testing of a neural model predictive controller for a nonlinear system. *Chemical Engineering Science 50*(15), 2419–2430.

van Can, H., H. te Braake, C. Hellinga, K. Luyben and J. Heijnen (1997). An efficient model development strategy for bioprocesses based on neural networks in macroscopic balances. *Biotechnology and Bioengineering 54*, 549–566.

van den Hof, P., P. Heuberger and J. Bokor (1995). System identification with generalized orthonormal basis functions. *Automatica 31*(12), 1821–1834.

van Huffel, S. and J. Vandewalle (1991). *The Total Least Squares Problem; Computational Aspects and Analysis*. Philadelphia, U.S.A.: Frontiers in Applied Mathematics, SIAM.

Verhaegen, M. and P. Dewilde (1992). Subspace model identification. Part I: the output-error state space model identification class of algorithms. *International Journal of Control 56*, 1187–1210.

Voisin, A., L. Rondeau, R. Ruelas, G. Dubois and M. Lamotte (1995). Conditions to establish and equivalence between a fuzzy relational model and a linear model. In *Proceedings Third European Congress on Intelligent Techniques and Soft Computing EUFIT'95*, Aachen, Germany, pp. 523–528.

Walsh, G. (1975). *Methods of optimization*. New York, USA: John Wiley & Sons.

Wang, H., K. Tanaka and M. Griffin (1995). Parallel distributed compensation of nonlinear systems by Takagi–Sugeno fuzzy model. In *Proceedings FUZZ-IEEE/IFES'95*, Yokohama, Japan, pp. 531–538.

Wang, L.-X. (1992). Fuzzy systems are universal approximators. In *Proc. IEEE Int. Conf. on Fuzzy Systems 1992*, San Diego, USA, pp. 1163–1170.

Wang, L.-X. (1994). *Adaptive Fuzzy Systems and Control, Design and Stability Analysis*. New Jersey: Prentice Hall.

Weiss, A. and C. Kulikowski (Eds.) (1991). *Computer Systems that Learn*. San Mateo, CA: Morgan Kaufman.

Westwick, D. and M. Verhaegen (1996). Identifying MIMO Wiener systems using subspace model identification methods. *Signal Processing 52*(2), 235–258.

Yager, R. (1994). Alternative structures for knowledge representation in fuzzy logic controllers. In Kandel and Langholz (Eds.), *Fuzzy Control Systems*, pp. 99–137. Boca-Raton, Fl.: CRC Press.

Yager, R. and D. Filev (1994). *Essentials of Fuzzy Modeling and Control*. New York: John Wiley.

Yang, M.-S. (1993). A survey of fuzzy clustering. *Mathl. Comput. Modelling 18*(11), 1–16.

Yasunobu, S. and S. Miyamoto (1985). Automatic train operation system by predictive fuzzy control. In M. Sugeno (Ed.), *Industrial Applications of Fuzzy Control*, pp. 1–18. North-Holland.

Yi, S. and M. Chung (1993). Identification of fuzzy relational model and its application to control. *Fuzzy Sets and Systems 59*, 25–33.

Yoshinari, Y., W. Pedrycz and K. Hirota (1993). Construction of fuzzy models through clustering techniques. *Fuzzy Sets and Systems 54*, 157–165.

Zadeh, L. (1973). Outline of a new approach to the analysis of complex systems and decision processes. *IEEE Trans. Systems, Man, and Cybernetics 1*, 28–44.

Zeng, X. and M. Singh (1994a). Approximation theory of fuzzy systems – SISO case. *IEEE Trans. Fuzzy Systems 2*, 162–176.

Zeng, X. and M. Singh (1995b). Approximation theory of fuzzy systems – MIMO case. *IEEE Trans. Fuzzy Systems 3*(2), 219–235.

Zhang, I. (1980). A smoothing-out technique for min-max optimization. *Math. Prog. 19*, 61–77.

Zhao, J. (1995). *Fuzzy logic in modeling and control.* PhD dissertation, CESAME, Louvain la Neuve, Belgium.

Zhao, J., V. Wertz and R. Gorez (1994). A fuzzy clustering method for the identification of fuzzy models for dynamical systems. In *9th IEEE International Symposium on Intelligent Control*, Columbus, Ohio, USA.

Zimmermann, H.-J. (1996). *Fuzzy Set Theory and its Application* (Third ed.). Boston: Kluwer.

Author Index

W

Y

Z

Subject Index

U

V

W